# STUDENT'S SOLUTIONS MANUAL

## PAUL LORCZAK

*with contributions from*

### BEVERLY FUSFIELD            CARRIE GREEN

# APPLIED BASIC MATHEMATICS

## William Clark

*Harper College*

## Robert Brechner

*Miami-Dade College*

PEARSON

Addison
Wesley

Boston   San Francisco   New York
London  Toronto  Sydney  Tokyo  Singapore  Madrid
Mexico City  Munich  Paris  Cape Town  Hong Kong  Montreal

Reproduced by Pearson Addison-Wesley from electronic files supplied by the author.

Copyright © 2008 Pearson Education, Inc.
Publishing as Pearson Addison-Wesley, 75 Arlington Street, Boston, MA 02116.

ISBN-13: 978-0-321-48651-6
ISBN-10: 0-321-48651-X

2 3 4 5 6 OPM 11 10 09 08

PEARSON

Addison
Wesley

# CONTENTS

# Chapter 1  Whole Numbers

## Section 1.1  Understanding the Basics of Whole Numbers

### Concept Check

1.   The numbers 0, 1, 2, 3, 4, 5, 6, 7,8, and 9 are called  _digits_ .

3.   Any of the numbers 1, 2, 3, 4, 5, 6, 7, 8, 9, 10, 11, 12, 13, 14, 15, ... are called the  _natural_  or  _counting_ numbers.

5.   A representation for a number in which each period is separated by a comma is known as _standard_ notation.

7.   A  _rounded_ number is an approximation of an exact number.

9.   In rounding, each digit to the right of the place value to which a number is to be rounded is changed to a _zero_ .

### Guide Problems

11.  For the number 128, identify the digit in each place.

Recall the rightmost digit in a whole number is in the ones place, the next digit to the left is in the tens place, the next in the hundreds place, then thousands, and so on.

a. ones place _8_ .

b. tens place _2_ .

c. hundreds place _1_ .

*In problems 13-27, the place value of the underlined digit is given as the answer. Recall that whole numbers are specified using periods of three place values each. The periods up to the billions period are listed below from leftmost to rightmost. The three place values within each period are also listed from left to right.*

| Period | Place values | | |
|---|---|---|---|
| billions | hundred billions | ten billions | billions |
| millions | hundred millions | ten millions | millions |
| thousands | hundred thousands | ten thousands | thousands |
| ones | hundreds | tens | ones |

13.  In 67<u>5</u> the 5 is in the *ones* place.

15.  In <u>4</u>889 the 4 is in the *thousands* place.

17.  In 56,<u>2</u>37 the 2 is in the *hundreds* place.

19. In 1<u>5</u>1,436 the 5 is in the *ten thousands* place.

21. In 78<u>0</u>,984 the 0 is in the *thousands* place.

23. In <u>4</u>01,804 the 4 is in the *hundred thousands* place. Note the underlined 4 is in the hundred thousands place. The rightmost 4 is in the ones place.

25. In <u>4</u>,938,286 the 4 is in the *millions* place.

27. In 8,4<u>7</u>2,711,337 the 7 is in the *ten millions* place. Note there is also a 7 in the hundred thousands place and a 7 in the ones place.

## Guide Problems

29. In word form, 729 is "  seven    hundred   twenty-nine." The 7 is in the hundreds place.

31. In standard form, 32809 is written   32,809 . Commas are used to separate the three digit periods when a number has five or more digits. In word form, this number is "thirty-two thousand, eight hundred  nine ." In the word form, the last digit written is the ones digit which is "nine" here. Note the word form does not use the word *and*.

*In Exercises 33 through 43, the standard notation for the given number is found by separating each period of three digits with a comma whenever the number has more than four digits. If a number has exactly four digits a comma need not be used but both forms are acceptable. The word form of the number is found by writing the name of the number occupying each period (using your knowledge of writing one-, two-, and three digit numbers) followed by the name of the period (in all but the case of the ones period.) In the word form the periods are separated by commas. A period consisting entirely of zeros does not contribute to the word form.*

33. 26  standard notation: 26
    word form: twenty-six

35. 812  standard notation: 812
    word form: eight hundred twelve

37. 9533  standard notation: 9533 or 9,533
    word form: nine thousand, five hundred thirty-three

39. 81184  standard notation: 81,184
    word form: eighty-one thousand, one hundred eighty-four

41. 58245  standard notation: 58,245. For the word form, consider the digits in each period.
         digits: 58      word form: fifty-eight      period: thousands
         digits: 245      word form: two hundred forty-five      period: ones
    Word form for 58245 : fifty-eight thousand, two hundred forty-five

43. 498545  standard notation: 498,545. For the word form, consider the digits in each period.
         digits: 498      word form: four hundred ninety-eight      period: thousands
         digits: 545      word form: five hundred forty-five      period: ones
    Word form for 498545 : four hundred ninety-eight thousand, five hundred forty-five

## Guide Problems

45. Write 271 in expanded form.
    271 has 2 hundreds, 7 tens, and 1 one.
    Expanded form:   $200 + 70 + 1$

*In Exercises 47-53, an expanded form is first found as follows. For each digit that is not zero, write the digit followed by the name of the place it occupies separating each digit-word pair with a plus sign. The other expanded form is found by writing each digit-word pair in the sum as a number. For example, "8 tens" is 80 and "5 ten-thousands" is 50,000.*

47. In expanded form 73 is
    7 tens + 3 ones or  70 + 3

49. In expanded form 2746 is
    2 thousands + 7 hundreds + 4 tens + 6 ones
    or   2000 + 700 + 40 + 6

51. In expanded form 25,370 is
    2 ten thousands + 5 thousands + 3 hundreds
        + 7 tens
    or   20,000 + 5000 + 300 + 70

53. In expanded form 896,905 is
    8 hundred thousands + 9 ten thousands
        + 6 thousands + 9 hundreds + 5 ones
    or   800,000 + 90,000 + 6000 + 900 + 5

## Guide Problems

55. Round 853 to the nearest hundred.

    a. What digit is the hundreds place?
       That is, what digit is the place to which
       you want to round. Here, it is 8.

    b. Which place determines what me must
       do in the hundreds place?
       The place that determines how we round
       is the place immediately to the right of the
       place we are rounding to. Here the place
       to the right of hundreds is the <u>tens</u> place.

    c. What digit is in that place?  5

    d. Explain what to do next.
       Increase the specified digit, 8, by one,
       because the digit in the tens place is 5
       or more. Next, change each digit to the
       right of the specified place value to zero.

    e. Write the rounded number. <u>900</u>
       As described in part d., the 8 in the
       hundreds place is increased to 9 and
       the digits to the right: 5 and 3, are
       replaced with zeros to yield 900.

*In Exercises 57-67, the procedure outlined in the Guide Problems is applied to round the given number to the specified place. 1) Locate the digit in the specified place. 2) Identify the digit to the right of the specified digit. 3) Increase the specified digit by one if the digit to the right is 5 or more. Leave the specified digit the same otherwise. 4) Replace all digits to the right of the specified digit with zeros.*

57. 4548 to the nearest ten
    The digit in the tens place is 4. The digit to the
    right is 8 which is more than 5. Increase the 4 to 5
    and replace the 8 with a zero to get  4550.

59. 590,341 to the nearest thousand
    The digit in the thousands place is 0. The digit to
    the right is 3 which  is less than 5. Leave the 0
    alone   and replace the 3,4 and 1 with zeros to get
    590,000.

61. 434,530 to the nearest ten thousand
    The 3 in the ten thousands place is  followed by a
    4 so the 3 is unchanged. The rounded number is
    430,000.

63. 4,970,001 to the nearest million
    The specified digit 4 is followed by 9. Round up
    to get 5,000,000.

65. 94,141,952 to the nearest ten million   The
    specified digit 9 is followed by 4.  Leave the 9
    alone and zero out the other digits to get the
    rounded 90,000,000.

67. 3,939,413 to the leftmost place value
    The leftmost place value is the millions place
    which contains a 3. The 9 in the place to the right
    indicates we increase 3 to 4 and use zeros for the
    remaining places to get 4,000,000.

69. "Seven hundred fifty-two thousand, one hundred twenty-eight" in standard notation is 752,128.

71. The word form "two trillion, nine hundred two billion dollars" indicates the trillions period consists of the digits 2 while the billions period consists of 902. The millions, thousands, and ones periods would consist of 000 since these periods do not appear in the word form. The standard notation is then 2,902,000,000,000.

73. The dollar amount $3497 has the digit 3 in the thousands period and digits 497 in the ones period. the word form is then "three thousand, four hundred ninety-seven dollars."

75. In word form, the amount spent, $30,000,000 is "thirty million dollars."

77. The length of 965 feet in word form is "nine hundred sixty-five feet."
The width of 106 feet in word form is "one hundred six feet."
The weight of 91,000 tons in word form is "ninety-one thousand tons."

79. In $1237, the digit in the tens place is 3. The digit to its right is 7 in the ones place. Since 7 is more than 5, increase the 3 to 4 and replace the 7 with 0 when rounding to the nearest ten. To the nearest ten, Ignacio spent $1240.

81. The number 46,847 rounded to the nearest thousand is 47,000 since the 6 in the thousands place is followed by and 8 in the hundreds place so 6 is increased to 7. The tax rules and regulations are 47,000 pages long to the nearest thousand.

83. From the chart, the average yearly salary with a bachelor's degree is $51,206. The periods here are the thousands and ones period containing the digit 51 and 206 respectively. The word form is then "fifty-one thousand, two hundred six dollars."

85. Workers with no high school diploma earn an average of $18,734 annually. To the nearest hundred this value is $18,700 since the digit to the right of the specified digit of 7 is 3.

87. Workers with a bachelor's degree earn an average of $51,206 annually. To the nearest thousand, their salary is $51,000 since the 1 in the thousands place is followed by a number less than 5, namely 2.

## Section 1.2 Adding Whole Numbers

## Concept Check

1. The mathematical process of combining two or more numbers to find the total is called _addition_.

3. When performing addition, we often format addends vertically so that their _place_ _values_ are aligned.

5. The commutative property of addition states that changing the _order_ of the addends does not change the sum.

7. When performing addition, if the sum of any column is a two-digit number, write the rightmost digit under the horizontal bar and _carry_ the leftmost digit to the top of the next column to the left.

## Guide Problems

9. $12 + 0 = \underline{\;12\;}$

This example demonstrates the addition property of  $\underline{\text{ zero }}$ .

The addition property of zero states that when 0 is added to any number the result is that number. Thus, $12 + 0 = 12$ .

13. Add $0 + 128$.
By the addition property of zero, $0 + 128 = 128$.

11. Show that $13 + 21 = 21 + 13$ .

$$13 + 21 = 21 + 13$$
$$34 = \underline{\;34\;}$$

Since both additions result in a sum of 34, the two are identical. This example demonstrates the  $\underline{\text{ commutative }}$  property of addition.

15. Show that $35 + 20 = 20 + 35$.

Since
$$35 + 20 = 55$$
$$20 + 35 = 35$$

both additions result in a sum of 55 and the two sums are identical. This demonstrates the  $\underline{\text{ commutative }}$  property of addition.

17. Show that $(3 + 10) + 8 = 3 + (10 + 8)$.

$$(3 + 10) + 8 = 13 + 8 = 21$$
$$3 + (10 + 8) = 3 + 18 = 21$$

Since the additions on both sides result in a sum of 21, the additions are identical. This demonstrates the  $\underline{\text{ associative }}$  property of addition.

## Guide Problems

19.
```
   79
 + 20
 ----
   99
```

21.
```
    1
  782
 + 55
 ----
  837
```

*In Exercises 23-53, the addition is performed as in the Guide Problems. Starting at the right, the digits in each column are added. If the sum of a column is 9 or less, write the result directly underneath that column. If the sum of a column is ten or more, write the units digits of the column sum underneath that column and carry any remaining digits to the top of the column to the left to be included in that column's sum. Continue this way to the left until all columns have been added.*

23.
```
   80
 + 12
 ----
   92
```

25.
```
   10
 + 61
 ----
   71
```

27.
```
   66
 + 22
 ----
   88
```

29.
```
   42
 + 70
 ----
  112
```

31.
```
    1
  371
 + 38
 ----
  409
```

33.
```
    1
  427
 +858
 ----
 1285
```

35.
```
  1256
 +1001
 -----
  2257
```

37.
```
    1 1 1
   5 735
 + 8 996
 -------
  14,73 1
```

39.
$$\overset{1}{8}31$$
523
+ 364
——
1718

41.
$$\overset{1\,1}{3}79$$
232
+ 536
——
1147

43.
$$\overset{2\,3\,2}{2}778$$
663
114
72
+ 398
——
4 025

45.
$$\overset{2\,3\,2}{5}472$$
4 126
850
58
+ 799
——
11,305

*In Exercises 47-53, the addition is first written vertically aligning the numbers so that the corresponding places align. The addition is then performed as in 23-45.*

47.
$$\overset{1\,1}{1}43$$
+ 89
——
232

49.
$$\overset{1\ \ 1}{2}656$$
+ 9519
——
12,175

51.
$$\overset{1\,2\,1}{5}98$$
1248
+ 1871
——
3717

53.
$$\overset{1\,1\,1}{2}39$$
1268
+ 1590
——
3097

55. The word "plus" indicates addition.

$$\overset{1\,1}{5}48$$
+ 556     The result is 1104.
——
1104

57. The phrase "more than" indicates addition.

$$\overset{1\,1}{9}91$$
+ 259     The result is 1250.
——
1250

59. The sum of the three numbers is

$$\overset{1\,2}{6}08$$
248     The result is 952.
+ 96
——
952

61. The airplane was at 23,400 feet and then climbed an additional 3500 feet so the new altitude is the sum of these two numbers or $23,400 + 3500 = 26,900$ feet.

63. The new sales figure is the sum of the previous figure and the increase

$$\overset{1}{\$}46,700$$
+ $12,300
——
$59,000

Sales this month will be $59,000.

65. Add the four given dollar amounts.

$$\overset{1\,1}{\$}1433$$
$231
$32
+ $78
——
$1774

Morley's total expenses were $1774.

67. The total number of meals per day is found by adding the numbers under each day in the table.

| Monday | Tuesday | Wednesday | Thursday | Friday |
|---|---|---|---|---|
| 2 | 1 1 | 3 2 | 1 1 | 2 1 |
| 215 | 238 | 197 | 184 | 258 |
| 326 | 310 | 349 | 308 | 280 |
| 429 | 432 | 375 | 381 | 402 |
| + 124 | + 129 | + 98 | + 103 | + 183 |
| 1094 | 1109 | 1019 | 976 | 1123 |

69. The total number of meals for the week can be found by adding the totals for each day given in Exercise 67. The total number is $1094 + 1109 + 1019 + 976 + 1123 = 5321$ meals.

71.
```
  1
  9 5   mg  (1 medium papaya)
  9 5   mg  (1 cup pepper)
+ 9 5   mg  (1 cup pepper)
─────────────────────────
2 8 5   mg  total
```

73. The perimeter of the figure is the sum of the lengths of the sides. The perimeter is given by the sum $34 + 34 + 29 = 97$ or 97 cm.

75. The amount of fencing needed is the sum of the lengths of the four sides. The sum is
```
  1 1
  114
  122
   65      The total number of fencing needed is 339 ft.
+  38
──────
  339
```

## Cumulative Skills Review

1. The word form "two hundred sixty-one thousand, eight hundred nine" in standard notation is 261,809.

3. In expanded notation , the number 14,739 can be written as
1 ten thousand + 4 thousands + 7 hundreds + 3 tens + 9 ones  or as  $10,000 + 4000 + 700 + 30 + 9$.

5. The number 6114 has the digit 6 in the thousands period and the digits 114 in the ones period. The word form is then "six thousand, one hundred fourteen."

7. From the right, the first three place values are the ones, tens, and hundreds so the 8 in 32,825 is in the hundreds place.

9. a. In standard form, "three hundred sixty-five thousand, five hundred twenty-nine" is 365,529.

   b. The thousands digit of 365,529 is 5 and the digit to the right is five or more so rounding to the nearest thousand gives 366,000.

## Section 1.3    Subtracting Whole Numbers

## Concept Check

1.  The mathematical process of taking away or deducting an amount from a given number is known as
    <u>subtraction</u>  .

3.  In subtraction, the number that is subtracted is called the <u>subtrahend</u>.

5.  In subtraction, we often format the digits of the whole numbers vertically so that the <u>place</u>  <u>values</u> are
    aligned.

## Guide Problems

*In problems 7 through 9, no borrowing is required.  The digits in each column can be subtracted starting with
the rightmost column and proceeding left.*

7.    $\begin{array}{r} 43 \\ -\ 3 \\ \hline \underline{40} \end{array}$

    since $3-3=0$
    and $4-0=4.$

9.    $\begin{array}{r} 67 \\ -\ \underline{5} \\ \hline 62 \end{array}$

    since $7-5=2$
    and $6-0=6.$

*In problems 11 through 13, the digit at the top of each column is smaller than the digit below it and so
borrowing is required.  For a given column, a 1 is borrowed from the digit to the left of the top digit and 10 is
added to the digit at the top of the column to perform the subtraction.*

11.    $\begin{array}{r} 82 \\ -17 \\ \hline \end{array}$   is computed as   $\begin{array}{r} \overset{7}{\cancel{8}}\ \overset{12}{\cancel{2}} \\ -1\ 7 \\ \hline 65 \end{array}$

    since $12-7=5$ and $7-1=6.$

13.    $\begin{array}{r} 145 \\ -58 \\ \hline \end{array}$   is computed as   $\begin{array}{r} \overset{0}{\cancel{1}}\ \overset{13}{\underset{}{\cancel{4}}}\ \overset{15}{\cancel{5}} \\ -5\ 8 \\ \hline 8\ 7 \end{array}$

    since $15-8=7$ and $13-5=8.$

*In problems 15 through 53, subtraction is performed as in the Guide Problems.  Subtraction is performed in
each column moving from right to left.  Borrowing is not needed in problems 15-25 and is employed when
necessary after problem 25.*

15.    $\begin{array}{r} 55 \\ -\ 3 \\ \hline 52 \end{array}$

17.    $\begin{array}{r} 39 \\ -\ 0 \\ \hline 39 \end{array}$

19.    $\begin{array}{r} 57 \\ -35 \\ \hline 22 \end{array}$

21.    $\begin{array}{r} 85 \\ -25 \\ \hline 60 \end{array}$

23.    $\begin{array}{r} 88 \\ -62 \\ \hline 26 \end{array}$

25.    $\begin{array}{r} 54 \\ -54 \\ \hline 0 \end{array}$

27.    $\begin{array}{r} \overset{5}{\cancel{6}}\ \overset{15}{\cancel{5}} \\ -\ 9 \\ \hline 5\ 6 \end{array}$

29.    $\begin{array}{r} \overset{4}{\cancel{5}}\ \overset{13}{\cancel{3}} \\ -\ 6 \\ \hline 47 \end{array}$

31.    $\begin{array}{r} \overset{8}{\cancel{9}}\ \overset{10}{\cancel{0}} \\ -3\ 9 \\ \hline 5\ 1 \end{array}$

33.    $\begin{array}{r} \overset{2}{\cancel{3}}\ \overset{15}{\cancel{5}} \\ -2\ 9 \\ \hline 6 \end{array}$

35.    $\begin{array}{r} 517 \\ -13 \\ \hline 504 \end{array}$

37.    $\begin{array}{r} 658 \\ -32 \\ \hline 626 \end{array}$

39.
$$
\begin{array}{r}
{}^{2}\,{}^{15}\\
8\,\cancel{3}\,\cancel{5}\\
-\,1\,2\,7\\
\hline
7\,0\,8
\end{array}
$$

41.
$$
\begin{array}{r}
{}^{6}\,{}^{11}\\
\cancel{7}\,\cancel{1}\,6\\
-\,3\,3\,0\\
\hline
3\,8\,6
\end{array}
$$

43.
$$
\begin{array}{r}
{}^{7}\,{}^{12}\,{}^{15}\\
\cancel{8}\,\cancel{3}\,\cancel{5}\,9\\
-\,4\,4\,8\,2\\
\hline
3\,8\,7\,7
\end{array}
$$

45.
$$
\begin{array}{r}
{}^{8}\,{}^{14}\\
76,\cancel{9}\,\cancel{4}\,7\\
-\,1\,5,8\,5\,0\\
\hline
6\,1,0\,9\,7
\end{array}
$$

In Problems 47 through 53, the subtraction is written horizontally. In each case, the first step is to rewrite the problem vertically aligning the corresponding place values in each number.

47.
$$
\begin{array}{r}
{}^{3}\,{}^{10}\,{}^{15}\\
\cancel{4}\,\cancel{1}\,\cancel{5}\\
-\,1\,8\\
\hline
3\,9\,7
\end{array}
$$

49.
$$
\begin{array}{r}
7\,5\,8\\
-\,6\,5\,4\\
\hline
1\,0\,4
\end{array}
$$

51.
$$
\begin{array}{r}
{}^{8}\,{}^{13}\\
1\,\cancel{9}\,\cancel{3}\,5\\
-\,8\,8\,5\\
\hline
1\,0\,5\,0
\end{array}
$$

53.
$$
\begin{array}{r}
{}^{8}\,{}^{13}\\
6\,8\,\cancel{9}\,\cancel{3}\\
-\,4\,8\,8\,7\\
\hline
2\,0\,0\,6
\end{array}
$$

55. "17 from 36" means $36-17$.
$$
\begin{array}{r}
{}^{2}\,{}^{16}\\
\cancel{3}\,\cancel{6}\\
-\,1\,7\\
\hline
1\,9
\end{array}
$$

57. "61 less 27" means $61-27$.
$$
\begin{array}{r}
{}^{5}\,{}^{11}\\
\cancel{6}\,\cancel{1}\\
-\,2\,7\\
\hline
3\,4
\end{array}
$$

59. "164 reduced by 48" means $164-48$.
$$
\begin{array}{r}
{}^{5}\,{}^{14}\\
1\,\cancel{6}\,\cancel{4}\\
-\,4\,8\\
\hline
1\,1\,6
\end{array}
$$

61. "2195 take away 1556" means $2195-1556$.
$$
\begin{array}{r}
{}^{1}\,{}^{11}\,{}^{8}\,{}^{15}\\
\cancel{2}\,\cancel{1}\,\cancel{9}\,\cancel{5}\\
-\,1\,5\,5\,6\\
\hline
6\,3\,9
\end{array}
$$

63. The temperature dropped from 86 to 69 degrees, a difference of $86-69=17$ degrees as shown below.
$$
\begin{array}{r}
{}^{7}\,{}^{16}\\
\cancel{8}\,\cancel{6}\\
-\,6\,9\\
\hline
1\,7
\end{array}
$$

65. The number of gallons dropped from 3400 to 1360, a difference of $3400-1360=2040$ gallons as shown below.
$$
\begin{array}{r}
{}^{3}\,{}^{10}\\
3\,\cancel{4}\,\cancel{0}\,0\\
-\,1\,3\,6\,0\\
\hline
2\,0\,4\,0
\end{array}
$$

67. The difference in inventory is
$$
\begin{array}{r}
{}^{11}\,{}^{16}\\
\cancel{1}\,\cancel{2}\,\cancel{6}\,0\\
-\,3\,8\,0\\
\hline
8\,8\,0
\end{array}
$$
so the store sold 880 ties.

69. a. The play exceeded budget by $\$2,560,000-\$2,370,000=\$190,000$.

   b. The profit is the difference between the revenue and the actual cost which is $\$7,450,000-\$2,560,000$ or $\$4,890,000$.

   c. In word form $\$4,890,000$ is "four million, eight hundred ninety thousand dollars."

71. a. Mike financed the difference between the total due and his down payment or $\$580-\$88=\$492$.

   b. At the start of the year Mike owed $492 while at the end of the year he owed $212. The amount he paid during the year is the difference or $\$492 - \$212 = \$280$.

73.  After paying rent and buying food, the Gordon family has *spent* $\$950 + \$350 = \$1300$.  The amount remaining in their $\$1800$ budget for other things is $\$1800 - \$1300 = \$500$.

75.  From the graph there were 14,321 DVDs in 2005 and 8723 DVDs available in 2004.  Computing the difference,

$$
\begin{array}{r}
\overset{3}{\,}\overset{12}{\,}\overset{11}{\,}\overset{11}{\,} \\
1\,4,3\,2\,1 \\
-\ 8\,7\,2\,3 \\
\hline
5\,5\,9\,8
\end{array}
$$
     There were 5598 more DVDs available in 2005 than in 2004.

77.  Given 34,500 DVDs in 2007 and 21,260 DVDs in 2006, compute the difference.

$$
\begin{array}{r}
\overset{4}{\,}\overset{10}{\,} \\
3\,4,5\,0\,0 \\
-\ 2\,1,2\,6\,0 \\
\hline
1\,3,2\,4\,0
\end{array}
$$
     The increase in DVDs from 2006 to 2007 was 13,240.

## Cumulative Skills Review

1.  In expanded notation, the number 2510 can be written as   2 thousands + 5 hundreds + 1 tens or as   $2000 + 500 + 10$.

3.  In the number 1,463,440 the digit 6 occupies the middle position in the thousands period and so is the ten thousands place value.

5.  The word form "two hundred nineteen thousand, eight hundred twelve" in standard notation is 219,812.

7.
$$
\begin{array}{r}
\overset{1}{2}32 \\
+\ 819 \\
\hline
1051
\end{array}
$$

9.  The number 35,429 has the digits 35 in the thousands period and the digits 429 in the ones period. The word form is then "thirty-five thousand, four hundred twenty-nine."

## Section 1.4    Multiplying Whole Numbers

## Concept Check

1.  The mathematical process of repeated addition of a value a specified number of times is called  multiplication  .

3.  The result of multiplying numbers is known as a   product  .

5.  The multiplication property of one states that the product of any number and  1, or one,  is the number itself.

7.  The associative property of addition states that the  grouping  of factors does not change the product.

## Guide Problems

9.  $4 \cdot 0 = \underline{\ 0\ }$

    This example demonstrates the multiplication property of __zero__. The multiplication property of zero states that the product of any number and 0 is 0.

11. Show that $7 \cdot 3 = 3 \cdot 7$.

    Perform the multiplication on both sides.

    $$7 \cdot 3 = 3 \cdot 7$$
    $$21 = \underline{\ 21\ }$$

    The results are the same so $7 \cdot 3 = 3 \cdot 7$. This example demonstrates the __commutative__ property of multiplication.

13. Show that $2(3 + 4) = 2 \cdot 3 + 2 \cdot 4$.

    Perform the operations on both sides. On the left side, evaluate within the parentheses first. On the right side, perform the multiplications before the addition.
    $$2(3 + 4) = 2 \cdot 3 + 2 \cdot 4$$
    $$2(\ \underline{7}\ ) = \underline{\ 6\ } + 8$$
    $$\underline{14} = \underline{14}$$

    The results are the same showing
    $$2(3 + 4) = 2 \cdot 3 + 2 \cdot 4$$
    This example demonstrates the __distributive__ property of multiplication over __addition__ .

15. By the multiplication property of zero, the product of any number and 0 is 0, so
    $$0 \cdot 215 = 0.$$

17. By the multiplication property of one, the product of any number and 1 is that number, so $82 \cdot 1 = 82$.

19. Show that $4 \cdot 8 = 8 \cdot 4$.
    Compute each side separately.
    $$4 \cdot 8 = 8 \cdot 4$$
    $$32 = 32$$
    The results are the same, showing
    $$4 \cdot 8 = 8 \cdot 4$$
    This illustrates the commutative property of multiplication.

21. Show that $3(2 \cdot 4) = (3 \cdot 2)4$.
    Compute each side separately.
    $$3(2 \cdot 4) = (3 \cdot 2)4$$
    $$3 \cdot 8 = 6 \cdot 4$$
    $$24 = 24$$
    The results are the same, showing
    $$3(2 \cdot 4) = (3 \cdot 2)4$$
    This illustrates the associative property of multiplication.

23. Show that $7(2 + 5) = 7 \cdot 2 + 7 \cdot 5$.
    Compute each side separately.
    $$7(2 + 5) = 7 \cdot 2 + 7 \cdot 5$$
    $$7 \cdot 7 = 14 + 35$$
    $$49 = 49$$
    The results are the same, showing
    $$7(2 + 5) = 7 \cdot 2 + 7 \cdot 5$$
    This illustrates the distributive property of multiplication over addition.

## Guide Problems

25.
$$\overset{2}{2}8$$
$$\underline{\times\ 3}$$
$$84$$

    In other words, "3 times 8 is 24. Write the 4 in the ones place of the answer and carry the 2 to the tens column. 3 times 2 is 6 plus the 2 carried over is 8. Write 8 in the tens place of the answer."

27.
$$\overset{2}{5}3$$
$$\underline{\times\ 9}$$
$$477$$

    Since $9 \times 3 = 27$ a 2 is carried to the tens column.

29.     16
     × 32
     ─────
       32
      480
     ─────
      512

31.     326
      × 75
     ──────
     1630
    22820
    ──────
    24,450

*In Problems 33 through 59, multiplication is performed using the techniques of this section. Numbers are not expressed in standard notation until the final answer.*

33.    23
     × 3
     ────
      69

35.    52
     × 4
     ────
     208

37.    ¹
       52
     × 9
     ────
     468

39.    ⁵
       59
     × 6
     ────
     354

41.     22
     × 78
     ─────
      176
     1540
     ─────
     1716

43.     94
     × 50
     ─────
     4700

45.     605
      × 40
     ──────
     24,200

47.     153
      × 33
     ──────
      459
     4590
     ──────
     5049

49.    6000
      × 74
     ───────
      24000
     420000
     ───────
     444000

     or 444,000

51.    4888
      × 23
     ───────
      14664
      97760
     ───────
     112424

     or 112,424

*In Problems 53 through 59, the multiplication problems are given horizontally. In each solution, the problem is first written vertically before multiplying.*

53.     4444
      × 270
     ───────
      311080
      888800
     ───────
     1199880

     or 1,199,880

55.     3342
      × 951
     ───────
        3342
      167100
     3007800
     ───────
     3178242

     or 3,178,242

57.     915
      × 25
     ──────
      4575
     18300
     ──────
     22875

     or 22,875

59.     282
      × 37
     ──────
     1974
     8460
     ──────
     10434

     or 10,434

61.  28 multiplied by 15 is
            28
          × 15
          ─────
           140
           280
          ─────
           420

63.  First multiply without dollars
            16
          × 12
          ─────
            32
           160
          ─────
           192

     12 calculators at $16 each cost $192.

65. First compute 25 times 42.
    Multiply the result by 4.

    $$
    \begin{array}{r}
    25 \\
    \times\ 42 \\
    \hline
    50 \\
    1000 \\
    \hline
    1050
    \end{array}
    \qquad
    \begin{array}{r}
    1050 \\
    \times\ 4 \\
    \hline
    4200
    \end{array}
    $$

    25 times 42 times 4 is 4200.

67. Multiply 26 by 30:

    $$
    \begin{array}{r}
    26 \\
    \times\ 30 \\
    \hline
    780
    \end{array}
    $$

    David saves $780.

69. If each airplane costs $5,660,000 dollars, the total cost of 13 planes is found by multiplying

    $$
    \begin{array}{r}
    5660000 \\
    \times\ \ \ \ \ \ 13 \\
    \hline
    16980000 \\
    56600000 \\
    \hline
    73580000
    \end{array}
    $$
    The total cost is $73,580,000.

71. In order to have a payment of $1149 per month, Maureen's monthly income must be $3 \times \$1149 = \$3447$.

73. Since 7 new people get on the Internet every second and there 60 seconds in a minute, 60 minutes in an hour, and 24 hours in a day, the total number of new Internet users each day is $7 \times 60 \times 60 \times 24 = 604{,}800$. Note, the multiplications involved can be done in any order. One example is $7 \times 60 = 420$, $420 \times 60 = 25{,}200$, and $25{,}200 \times 24 = 604{,}800$.

75. a. Since there are 60 minutes in one hour and the machine produces 74 toys per minute, the total produced in 1 hour is $60 \times 74 = 4440$ toys.

    b. From part a., one machine can produce 4440 toys in one hour. Therefore, in 8 hours, *one* machine can produce $8 \times 4440 = 35{,}520$ toys. A total of 9 machines can produce $9 \times 35{,}520 = 319{,}680$ toys per day.

77. a. The number of square feet in a doubles court is

    $$
    \begin{array}{r}
    78 \\
    \times\ 36 \\
    \hline
    468 \\
    2340 \\
    \hline
    2808
    \end{array}
    $$

    or 2808 square feet.

    b. The singles court is 27 feet (36 – 9) wide and has area

    $$
    \begin{array}{r}
    78 \\
    \times\ 27 \\
    \hline
    546 \\
    1560 \\
    \hline
    2106
    \end{array}
    $$

    or 2106 square feet.

    c. The difference between the area of a doubles and singles court is

    $$
    \begin{array}{r}
    2808 \\
    -\ 2106 \\
    \hline
    702
    \end{array}
    $$

    or 702 square feet.

79. From the graph, *one* acre was worth $1770 in 2004. The value of 300 acres is found by multiplying:

    $$
    \begin{array}{r}
    1770 \\
    \times\ 300 \\
    \hline
    531000
    \end{array}
    $$

    A 300-acre farm was worth $531,000.

81. In 2001, a 100-acre farm was worth $100 \times \$1580 = \$158{,}000$. In 1997, the same farm was worth $100 \times \$1270 = \$127{,}000$. So the 2001 value was $\$158{,}000 - \$127{,}000 = \$31{,}000$ more than in 1997.

## Cumulative Skills Review

1.  The number 82,184 has the digits 82 in the thousands period and the digits 184 in the ones period. The word form is then "eighty-two thousand, one hundred eighty-four."

3.
$$
\begin{array}{r}
\overset{1\ 1}{15}50 \\
122 \\
892 \\
+\ \ \ 30 \\
\hline
2594
\end{array}
$$

5.  The phrase "increased by" implies addition. The result is $422 + 110 = 532$.

7.  Adding the individual amounts
$$
\begin{array}{r}
\overset{2}{\$2}2 \\
\$39 \\
+\ \ \$9 \\
\hline
\$70
\end{array}
$$
$\ \ \ \ $ You spent \$70 total on your car.

9.  Subtracting the number of tickets sold from the original number of tickets gives
$$
\begin{array}{r}
\overset{6\ \ 13\ 18}{7\,4\,8\,5} \\
-\ 5\ 7\ 9\ 5 \\
\hline
1\ 6\ 9\ 0
\end{array}
$$
so 1690 tickets remain.

## Section 1.5    Dividing Whole Numbers

## Concept Check

1.  The mathematical process of repeated subtraction of a value is called __division__ .

3.  The number by which the dividend is divided is known as the __divisor__ .

5.  Show three ways to express 8 divided by 4.  $8 \div 4$,  $4\overline{)8}$,  $\dfrac{8}{4}$

7.  The quotient of any nonzero number and itself is __one__ .

9.  The quotient of any number and zero is __undefined__ .

## Guide Problems

11. $\dfrac{213}{1} = \underline{\ 213\ }$

The result is 213 because the quotient of any number and 1 is that number.

13. $23\overline{)0}$ with quotient $0$

The result is 0 because the quotient of 0 and any nonzero number is 0.

15. $18 \div 1 = 18$
Any number divided by one results in that number.

17. $51 \div 51 = 1$
Any number divided by itself results in 1.

19. $0 \div 3 = 0$
Zero divided by any nonzero number is 0.

21. $54 \div 0$
Division by 0 is *undefined*.

23. $\dfrac{67}{0}$
Division by 0 is *undefined*.

25. $\dfrac{0}{12} = 0$
Zero divided by any nonzero number is 0.

27. $76\overline{)0}$ with quotient $0$  Zero divided by any nonzero number is 0.

29. $0\overline{)210}$  Division by 0 is *undefined*.

## Guide Problems

31. $5\overline{)40}$ with quotient $8$

because there are eight 5s in 40.

33. $8\overline{)368}$ with quotient $4$

$\underline{32}$

$48$

After subtracting 32 from 36, bring down the 8. Note the problem illustrates one step in the division process and is not a complete division.

*In Problems 35 through 65 division problems that are stated vertically or with a fraction bar are rewritten as long division problems for solving. If the division is straightforward as in problem 35, then the long division format is not used.*

35. $16 \div 2 = 8$

37. $8\overline{)48}$ with quotient $6$

39. $55\overline{)495}$ with quotient $9$

$\underline{495}$

41. $15\overline{)675}$ with quotient $45$

$\underline{60}$

$75$

$\underline{75}$

43. $\dfrac{131}{16)\overline{2096}}$

$\underline{16}$

49

$\underline{48}$

16

$\underline{16}$

45. $\dfrac{114}{23)\overline{2622}}$

$\underline{23}$

32

$\underline{23}$

92

$\underline{92}$

47. $\dfrac{234}{100)\overline{23400}}$

$\underline{200}$

340

$\underline{300}$

400

$\underline{400}$

49. $\dfrac{5 \ R \ 4}{6)\overline{34}}$

$\underline{30}$

4

51. $\dfrac{8 \ R \ 6}{9)\overline{78}}$

$\underline{72}$

6

53. $\dfrac{10 \ R \ 56}{71)\overline{766}}$

$\underline{71}$

56

55. $\dfrac{17 \ R \ 18}{26)\overline{460}}$

$\underline{26}$

200

$\underline{182}$

18

57. $\dfrac{47 \ R \ 17}{19)\overline{910}}$

$\underline{76}$

150

$\underline{133}$

17

59. $\dfrac{11 \ R \ 2}{85)\overline{937}}$

$\underline{85}$

87

$\underline{85}$

2

61. $\dfrac{52 \ R \ 12}{13)\overline{688}}$

$\underline{65}$

38

$\underline{26}$

12

63. $\dfrac{27 \ R \ 1}{343)\overline{9262}}$

$\underline{686}$

2402

$\underline{2401}$

1

65. $\dfrac{28 \ R \ 98}{169)\overline{4830}}$

$\underline{338}$

1450

$\underline{1352}$

98

67. $\dfrac{105}{7)\overline{735}}$

$\underline{7}$

35

$\underline{35}$

69. $\dfrac{426 \ R \ 5}{8)\overline{3413}}$

$\underline{32}$

21

$\underline{16}$

53

$\underline{48}$

5

71. Use long division.

$\dfrac{12}{88)\overline{1056}}$

$\underline{88}$

176

$\underline{176}$

88 goes into 1056 exactly 12 times.

73. 200 minutes is divided among 5 tests
Dividing

$\dfrac{40}{5)\overline{200}}$

$\underline{20}$

gives 40 minutes per test.

75. Divide 34,300 by 2450 to find the average amount per home.

$$
\begin{array}{r}
14 \\
2450\overline{)34300} \\
2450 \\
\hline
9800 \\
9800 \\
\hline
\end{array}
$$

The average home recycled 14 pounds.

77. a. Determine the number of times 22 goes into 1760.

$$
\begin{array}{r}
80 \\
22\overline{)1760} \\
176 \\
\hline
\end{array}
$$

80 pieces can be cut.

b. Determine the number of times 5 goes into 80 (from part a.)

$$
\begin{array}{r}
16 \\
5\overline{)80} \\
5 \\
\hline
30 \\
30 \\
\hline
\end{array}
$$

16 boats can be furnished with 5 lines.

79. The insurance rates for 1996 and 1997 were $691 and $707 dollars respectively. The average for the two years is found by adding these amounts, $691 + $707 = $1398 and then dividing by 2, $1398 ÷ 2 = $699.

81. The rates for 2001, 2002, and 2003 were $723, $784, and $855 respectively. When rounded to the nearest hundred, these three values become $700, $800, and $900 respectively. Find their sum $700 + $800 + $900 = $2400 and divide by three, $2400 ÷ 3 = $800. The average is $800.

## Cumulative Skills Review

1.
$$
\begin{array}{r}
516 \\
\times\ 200 \\
\hline
103,200
\end{array}
$$

3. The phrase "increased by" means addition. Adding vertically gives

$$
\begin{array}{r}
\overset{1}{9}54 \\
+\ 181 \\
\hline
1135
\end{array}
$$

5. In expanded form 22,185 is "2 ten thousands + 2 thousands + 1 hundred + 8 tens + 5 ones" or $20,000 + 2000 + 100 + 80 + 5$.

7.
$$
\begin{array}{r}
\overset{7\ \ 13\ 15}{\cancel{8}\cancel{4}\cancel{5}} \\
-\ 2\ 6\ 8 \\
\hline
5\ 7\ 7
\end{array}
$$
845 less 268 is 577

9. Tom worked 6 days at 7 hours per day, so worked a total of $6 \times 7 = 42$ hours.

## Section 1.6     Evaluating Exponential Expressions and Applying Order of Operations

## Concept Check

1.   _Exponential_ notation is a shorthand way of expressing repeated multiplication.

3.   In exponential notation, the _exponent or power_ is the number that indicates how many times the base is used as a factor.

5.   In the expression $4^3$, the number 3 is known as the _exponent_.

7.   Write the number "15 cubed" using exponential notation.  Here 15 is the base and "cubed" means an exponent of 3 so "15 cubed" in exponential notation is $15^3$.

9.   The result of raising a nonzero number to the zero power is _one_.

## Guide Problems

11.  $3 \cdot 3 \cdot 3 \cdot 3 = \underline{3^4}$

The factor being multiplied repeatedly is 3 so 3 is the base in exponential notation. There are four factors of 3 so the exponent is 4.

13.  $9 \cdot 9 \cdot 9 \cdot 9 \cdot 9 = 9^{\underline{5}}$

The base is 9.  There are five factors of 9 in the product so the exponent is 5.

15.  $\underline{8} \cdot \underline{8} \cdot \underline{8} = 8^3$

The exponential notation $8^3$ means a product consisting of three factors of 8.

17.  $5 \cdot 5 \cdot 3 \cdot 3 \cdot 3 = \underline{5}^{\underline{2}} \cdot \underline{3}^{\underline{3}}$

The product on the left consists of two factors of 5 and three factors of 3. Thus a base of 5 goes with the exponent 2 and a base of 3 goes with the exponent 3.

19.  $3 \cdot 3 \cdot 3$
There are three factors of 3. So 3 is the base and 3 is the exponent.
Exponential notation:  $3^3$
Word form: "three to the third power" or "three cubed"

21.  $5 \cdot 5 \cdot 5 \cdot 5$
Exponential notation:  $5^4$
Word form: "five to the fourth power"

23.  9
Exponential notation:  $9^1$
(note there is one factor of nine)
Word form: "nine to the first power"

25.  $1 \cdot 1 \cdot 1$
Exponential notation:  $1^3$
Word form: "one to the third power" or "one cubed"

27.  $4 \cdot 4 \cdot 4 \cdot 9 \cdot 9$
There are three factors of 4 forming $4^3$ and two factors of 9 forming $9^2$.
The product can be written as $4^3 \cdot 9^2$.

29.  $3 \cdot 3 \cdot 4 \cdot 4 \cdot 4 \cdot 5$ or $3^2 \cdot 4^3 \cdot 5$　　31.  $5 \cdot 5 \cdot 8 \cdot 8 \cdot 8 \cdot 12$ or $5^2 \cdot 8^3 \cdot 12$　　33.  $2 \cdot 5 \cdot 5 \cdot 7 \cdot 9 \cdot 9 \cdot 9$ or $2 \cdot 5^2 \cdot 7 \cdot 9^3$

## Guide Problems

35.  $5^0 = 1$  because an exponent of 0
　　results in 1 (if the base is not zero.)

37.  $3^3 = 3 \cdot 3 \cdot 3 = \underline{27}$  The exponent is
　　3 because there are 3 factors being
　　multiplied. The result is 27 since
　　$3 \cdot 3 \cdot 3 = 9 \cdot 3 = 27$.

39.  $2^0 = 1$

41.  $14^1 = 14$

43.  $1^4 = 1 \cdot 1 \cdot 1 \cdot 1 = 1$

45.  $5^2 = 5 \cdot 5 = 25$

47.  $12^2 = 12 \cdot 12 = 144$

49.  $7^3 = 7 \cdot 7 \cdot 7 = 49 \cdot 7$
　　　　$= 343$

51.  $2^6 = 2 \cdot 2 \cdot 2 \cdot 2 \cdot 2 \cdot 2 = 64$

53.  $10^3 = 10 \cdot 10 \cdot 10 = 100 \cdot 10$
　　　　$= 1000$

55.  $86^1 = 86$

57.  $4^4 = 4 \cdot 4 \cdot 4 \cdot 4 = 16 \cdot 4 \cdot 4$
　　　　$= 64 \cdot 4 = 256$

59.  $132^0 = 1$

61.  $15^2 = 15 \cdot 15 = 225$

## Guide Problems

63.  Simplify  $67 + 3 \cdot 18$.
　　$67 + 3 \cdot 18$　No grouping symbols and
　　$67 + \underline{54}$　　no exponents. Multiply first.
　　$\underline{121}$　　　Perform addition last.

65.  Simplify  $19 \cdot (4 + 21) - 2$.
　　$19 \cdot (4 + 21) - 2$
　　$19 \cdot \underline{25} - 2$　Evaluate within parentheses
　　first.
　　$\underline{475} - 2$　　No exponents so perform
　　multiplication next.
　　$\underline{473}$　　　Lastly, subtract.

*In problems 67 through 93, the order of operations is applied to evaluate each given expression. First,
expressions within grouping symbols such as ( ) and [ ] are evaluated. Remember, when a fraction bar appears
in an expression, as in problem 79, treat the numerator as if it was contained between grouping
symbols and similarly with the denominator. Second, evaluate any exponential expressions. Third, evaluate any
multiplications and division as they appear from left to right. Fourth, evaluate any additions and subtractions
as they appear from left to right.*

67.  $22 - (4 + 5)$
　　$22 - 9$
　　$13$

69.  $5 \cdot 4 + 2 \cdot 3$
　　$20 + 6$
　　$26$

71.  $144 \div (100 - 76)$
　　$144 \div 24$
　　$6$

73.  $81 + 0 \div 3 + 9^2$
　　$81 + 0 \div 3 + 81$
　　$81 + 0 + 81$
　　$81 + 81$
　　$162$

75.  $20 \div 5 + 3^3$
　　$20 \div 5 + 27$
　　$4 + 27$
　　$31$

77.  $48 \div 2 - 11 \cdot 2$
　　$24 - 11 \cdot 2$
　　$24 - 22$
　　$2$

79.  $\dfrac{15+13}{9-2}$

$\dfrac{28}{7}$

$4$

81.  $42+3(6-2)+5^2$

$42+3(4)+5^2$

$42+3(4)+25$

$42+12+25$

$54+25$

$79$

83.  $45\cdot2-(6+1)^2$

$45\cdot2-7^2$

$45\cdot2-49$

$90-49$

$41$

85.  $18-6\cdot2+(2+4)^2-10$

$18-6\cdot2+6^2-10$

$18-6\cdot2+36-10$

$18-12+36-10$

$6+36-10$

$42-10$

$32$

87.  $2^4\cdot5+5-5^0$

$16\cdot5+5-1$

$80+5-1$

$85-1$

$84$

89.  $23+(15-12)^3\div(134-5^3)$

$23+3^3\div(134-125)$

$23+3^3\div9$

$23+27\div9$

$23+3$

$26$

91.  $12\div4\cdot4[(6-2)+(5-3)]$

$12\div4\cdot4[4+2]$

$12\div4\cdot4\cdot6$

$3\cdot4\cdot6$

$12\cdot6$

$72$

93.  $\dfrac{3(14+3)-(20-5)}{6^2}+1$

$\dfrac{3\cdot17-15}{6^2}+1$

$\dfrac{51-15}{36}+1$

$\dfrac{36}{36}+1$

$1+1$

$2$

95.  The area is given by $A=s^2=(15\text{ inches})^2=15\cdot15$ square inches $=225$ square inches.

97.  a. The floor is square with a side of 20 feet. The area is $A=s^2=(20\text{ ft})^2=400$ square feet.

   b. Using part a., 400 square feet at \$9 per square foot yields a total cost of $400\cdot\$9=\$3600$.

   c. Floormasters, Inc. will charge a total of $400\cdot\$7=\$2800$. Subtracting this from the WoodWorks, Inc. estimate from part b., $\$3600-\$2800=\$800$, we see \$800 can be saved by taking the cheaper bid.

## Cumulative Skills Review

1.  Subtract vertically:

$$\begin{array}{r} {}^{8}\ {}^{11} \\ 83,2\cancel{9}\,\cancel{1} \\ -\ 12,2\,6\,9 \\ \hline 71,0\,2\,2 \end{array}$$

3.  Add vertically:

$$\begin{array}{r} {}^{1}\ {}^{1} \\ 16,824 \\ +\ 9\,542 \\ \hline 26,366 \end{array}$$

5.  $\begin{array}{r} 1300 \\ \times\ 200 \\ \hline 260,000 \end{array}$

7.  $0 \div 376 = 0$ since 0 divided by any nonzero number is 0.

9.  The total amount is found by multiplying

$\begin{array}{r} 470 \\ \times\ 21 \\ \hline 470 \\ 9400 \\ \hline 9870 \end{array}$

The total of the bonuses was $9870.

## Section 1.7    Solving Application Problems

## Concept Check

1.  The keyword is used to indicate the operation of __addition__ in an application problem.

3.  *Of* and *at* are words used to indicate __multiplication__ .

5.  The words *is* and *are* indicate __equality or the equal sign__ .

7.  *Understand the situation* You are given former horsepower information for a car and asked to find the current horsepower. *Take inventory* The knowns are the old horsepower and the increase in horsepower. The unknown is the new horsepower. *Translate the problem* The word "increased" indicates we add the change in horsepower to the old horsepower. *Solve the problem* Add $460 + 115 = 575$, the engine is now rated at 575 horsepower.

9.  You know the total memory of the camera, 512 megabytes, and the amount of memory each photo requires, 4 megabytes. The keyword *each* implies division is involved. The number of photos the card can hold is $512 \div 4 = 128$.

11. Jan's rent is $385 per month. her total rent for a year is $385 \times 12 = \$4620$.

13. a. The keyword *total* indicates addition. The total number of jewelry pieces is $35 + 19 + 12 = 66$.

    b. From part a., Julie sold 66 pieces. At $30 apiece the total is revenue is found by multiplying $\$30 \times 66 = \$1980$.

    c. From part b., her monthly revenue is $1980. If she makes this for each of 12 months, her total revenue would be $12 \times \$1980 = \$23,760$.

15. The keyword *per* in price *per square foot* indicates division of price by square feet. For the first townhouse, the price per square foot is $\$189,000 \div 1350 = \$140$ per square foot. For the second townhouse, the cost per square foot is $\$166,250 \div 950 = \$175$ per square foot. The first townhouse is the better buy being cheaper by the square foot.

17. a. The ten thousands digit of 34,646,419 is 4 and it followed by 6. So round 4 up to 5 and zero out 6,419 to get 34,650,000.

    b. The phrase how much closer indicates we need to compute a difference. The difference in distances is $60,000,000 - 34,650,00 = 25,350,00$ miles. Mars was approximately 25,350,000 miles closer in 2003.

19. For every 30 minutes, tennis burns 275 calories while gardening burns 170. The phrase *how many more* says we need to find a difference and so must subtract. Tennis burns $275 - 170 = 105$ calories more than gardening each half hour.

21. First consider stair climbing. There are 4 thirty minute periods in 2 hours, so stair climbing will burn $4 \times 306 = 1224$ calories per week. Next, consider weightlifting. There are 8 thirty minute periods in 4 hours, so weightlifting will burn $8 \times 234 = 1872$ calories per week. In total, the two activities will burn $1224 + 1872 = 3096$ calories each week.

23. There are two differently priced computers involved. First determine the cost of the computers at each price: 26 computers *at* \$1590 cost $26 \times \$1590 = \$41,340$ while 6 computers *at* \$1850 cost $6 \times \$1850 = \$11,100$. The *total* amount spent was $\$41,340 + \$11,100 = \$52,440$.

25. First consider the black and white pages. At 21 pages per minute, 252 pages will take $252 \div 21 = 12$ minutes. The color pages will take $104 \div 8 = 13$ minutes. The total time will be $12 + 13 = 25$ minutes.

27. First. compute the cost for one night 4 years ago. The current rate of \$495 is triple, or three times, what is what then so dividing by three gives $\$495 \div 3 = \$165$ the cost for one night 4 years ago. Next, compute the cost of a five night stay both now and then. Currently, 5 nights *at* \$495 per night is $\$495 \times 5 = \$2475$ while four years ago the cost was $\$165 \times 5 = \$825$. The difference is $\$2475 - \$825 = \$1650$.

29. As a first step, convert the numbers involved to have the same units, inches. A roll of floss is 16 yards long. Given 3 feet per yard, this is $16 \cdot 3 = 48$ feet which is $48 \cdot 12 = 576$ inches. The number of 16 inch pieces is found by dividing the total length by 16 resulting in $576 \div 16 = 36$. There are 36 uses in a roll.

31. First, compute the revenue from each type of book.

| Subject | Books sold | Price per book | Revenue |
|---------|-----------|----------------|---------|
| algebra | 463 | \$85 | $463 \times \$85 = \$39,355$ |
| English | 328 | \$67 | $328 \times \$67 = \$21,976$ |
| economics | 139 | \$45 | $139 \times \$45 = \$6255$ |

Now add the separate revenues to find a total revenue of $\$39,355 + \$21,976 + \$6255 = \$67,586$.

33. First. compute the total increase in population which means the difference between the current and previous populations: $35,418 - 18,408 = 17,010$. The average increase per year over 5 years is found by dividing this difference by 5. The average increase was $17,010 \div 5 = 3402$ people per year.

35. a. The estimated weight of 12 buses *at* 15,000 pounds each is $12 \times 15,000 = 180,000$ pounds. The estimated weight of 40 cars *at* 2500 pounds each is $40 \times 2500 = 100,000$ pounds. The total estimated weight is $180,000 + 100,000 = 280,000$ pounds. This is within the 300,000 pound limit.

    b. From part a., the load is $300,000 - 280,000 = 20,000$ short of the limit. Since, a bus weighs 15,000 pounds there is room for one more bus and still the load would be $20,000 - 15,000 = 5000$ pounds under the limit. Since a car weighs 2500 pounds, there would be room for two more cars. Alternatively, if no bus were added to the load, there would be room for $20,000 \div 2500 = 8$ cars could be added.

37. First, compute the total cost of leasing a Supercrew XLT and leasing an F-150 Supercab XLT. The Supercrew will cost $36 \times \$247 = \$8892$ while the Supercab will cost $36 \times \$219 = \$7884$. The difference is $\$8892 - \$7884 = \$1008$ so the Supercrew XLT would cost $\$1108$ more to lease than the Supercab XLT.

39. The amount you save is the difference between the original price and the sale price or $\$15,165 - \$11,495 = \$3670$.

## Cumulative Skills Review

1.
$$
\begin{array}{r}
163 \\
3\overline{)489} \\
3 \\
\hline
18 \\
18 \\
\hline
9 \\
9
\end{array}
$$

3.
$$
\begin{array}{r}
{}^{7}\cancel{8}\,{}^{13}\cancel{3} \\
-\ 5\ 9 \\
\hline
2\ 4
\end{array}
$$

5. The phrase "more than" means addition.
   The sum is
$$
\begin{array}{r}
{}^{1\ 1} \\
1563 \\
+\ 982 \\
\hline
2545
\end{array}
$$

7. $48 + (3 + 2)^3$
   $48 + 5^3$
   $48 + 125$
   $173$

9. In expanded notation 1549 is "1 thousand + 5 hundreds + 4 tens + 9 ones" or "$1000 + 500 + 40 + 9$"

## Chapter 1 Numerical Facts of Life

*The personal balance sheet for Todd and Claudia is filled in below. Next to each entry, in the rightmost column, is either the word "given" or the arithmetic operation used to arrive at the entry. The word "given" indicates that the value entered was given in the statement of the problem as one of the couple's assets or liabilities. All values represent amounts in dollars.*

| ASSETS | | | |
|---|---|---|---|
| **CURRENT ASSETS** | | | |
| Checking account | 3640 | | Given |
| Savings account | 4720 | | Given |
| Certificates of deposit | 18,640 | | Given |
| **Total Current Assets** | | 27,000 | $= 3640 + 4720 + 18,640$ |
| | | | |
| **LONG-TERM ASSETS** | | | |
| **Investments** | | | |
| Retirement plans | 67,880 | | Given |
| Stocks and bonds | 25,550 | | Given |
| Mutual funds | 15,960 | | Given |
| **Personal** | | | |
| Home | 225,500 | | Given |
| Automobiles | 32,300 | | Given |
| Personal property | 6400 | | Given |
| Other | 12,100 | | Given (sailboat) |
| Other | 7630 | | Given (electronics) |
| **Total Long-Term Assets** | | 393,320 | $= 67,880 + 25,550 + 15,960 + 225,500 + 32,300 + 6400 + 12,100 + 7630$ |
| **TOTAL ASSETS** | | $420,320 | $=$ Total Current Assets $+$ Total Long-Term Assets $= 27,000 + 393,320$ |

| LIABILITIES | | | |
|---|---|---|---|
| **CURRENT LIABILITIES** | | | |
| Store charge accounts | 1940 | | Given |
| Credit card accounts | 8660 | | Given |
| Other current debt | 0 | | Given |
| **Total Current Liabilities** | | 10,600 | $= 1940 + 8660 + 0$ |
| | | | |
| **LONG-TERM LIABILITIES** | | | |
| Home mortgage | 165,410 | | Given |
| Automobile loans | 13,200 | | Given |
| Other loans | 4580 | | Given (boat loan) |
| **Total Long-Term Liabilities** | | 183,190 | $= 165,410 + 13,200 + 4580$ |
| **TOTAL LIABILITIES** | | $193,790 | $=$ Total Current Liabilities $+$ Total Long-Term Liabilities $= 10,600 + 183,190$ |

| NET WORTH | | | |
|---|---|---|---|
| **Total Assets** | 420,320 | | From the assets table |
| **Total Liabilities** | − 193,790 | | From the liabilities table |
| **NET WORTH** | | $226,430 | = Total Assets − Total Liabilities |
| | | | = 420,320 − 193,790 |

## Chapter 1 Review Exercises

1.  In the number 5<u>4</u>,220, the indicated digit 4 is in the *thousands* place (the rightmost digit in the thousands period.)

2.  In the number <u>7</u>27, the indicated digit 7 is in the *hundreds* place (the leftmost digit in the ones period.)

3.  In the number 7<u>8</u>,414,645, the indicated digit 8 is in the *millions* place (the rightmost digit in the millions period.)

4.  In the number 334<u>1</u>, the indicated digit 1 is in the *ones* place (the rightmost digit in the ones period.)

5.  In the number 35,6<u>8</u>6, the indicated digit 8 is in the *tens* place (the middle digit in the ones period.)

6.  In the number 18,2<u>8</u>6,719, the indicated digit 8 is in the *ten thousands* place (the middle digit in the thousands period.)

7.  336
    standard notation: 336
    word form: three hundred thirty-six

8.  8475
    standard notation: 8,475 or 8475
    word form: eight thousand, four hundred seventy-five

9.  784341
    standard notation: 784,341
    word form: seven hundred eighty-four thousand, three hundred forty-one

10. 380633
    standard notation: 380,633
    word form: three hundred eighty thousand, six hundred thirty-three

11. 62646
    standard notation: 62,646
    word form: sixty-two thousand, six hundred forty-six

12. 1326554
    standard notation: 1,326,554
    word form: one million, three hundred twenty-six thousand, five hundred fifty-four

13. 10102
    standard notation: 10,102
    word form: ten thousand, one hundred two

14. 6653634
    standard notation: 6,653,634
    word form: six million, six hundred fifty-three thousand, six hundred thirty-four

15. 4022407508
    standard notation: 4,022,407,508
    word form: four billion, twenty-two million, four hundred seven thousand, five hundred eight

16. The expanded form of 23 is $20 + 3$ or 2 tens + 3 ones.

17. The expanded form of 532 is $500 + 30 + 2$ or 5 hundreds + 3 tens + 2 ones.

18. The expanded form of 109 is $100+9$ or 1 hundred + 9 ones.

19. The expanded form of 26,385 is  or $20,000+6000+300+80+5$ or 2 ten thousands + 6 thousands + 3 hundreds + 8 tens + 5 ones

20. The expanded form of 2,148 is $2000+100+40+8$ or 2 thousands + 1 hundred + 4 tens + 8 ones

21. The expanded form of 1,928,365 is $1,000,000+900,000+20,000+8000+300+60+5$     or 1 million + 9 hundred thousands + 2 ten thousands + 8 thousands + 3 hundreds + 6 tens + 5 ones

22. 36<u>3</u>,484

The indicated place, thousands, contains a 3. Since a 4 is to the right, round down to 363,000.

23. 18,1<u>3</u>6

The indicated place, tens, contains a 3. Since a 6 is to the right, round up to 18,140.

24. <u>8</u>6,614

The indicated place, ten thousands, contains an 8. Since a 6 is to the right, round up to 90,000.

25. 601,<u>9</u>27

The indicated place, hundreds, contains a 9. Since a 2 is to the right, round down to 601,900.

26. <u>4</u>,829,387

The indicated place, millions, contains a 4. Since an 8 is to the right, round up to 5,000,000.

27. 3,<u>1</u>46,844

The indicated place, hundred thousands, contains a 1. Since a 4 is to the right, round down to 3,100,000.

28. 81,0<u>8</u>4

The indicated place, tens, contains an 8. Since a 4 is to the right, round down to 81,080.

29. 19<u>6</u>,140

The indicated place, thousands, contains a 6. Since a 1 is to the right, round down to 196,000.

30. <u>4</u>2,862,785

The indicated place, ten millions, contains a 4. Since a 2 is to the right, round down to 40,000,000.

31.
```
   30
 + 59
 ----
   89
```

32.
```
   ¹
   45
 + 68
 ----
  113
```

33.
```
  319
 + 60
 ----
  379
```

34.
```
    ¹
  916
 + 35
 ----
  951
```

35.
```
   414
 + 181
 -----
   595
```

36.
```
    ¹
   360
   971
 + 964
 -----
  2295
```

37.
```
   ¹  ¹
  43,814
 + 71,658
 -------
 115,472
```

38.
```
   ¹ ¹
  1700
   130
   421
    81
 + 237
 -----
  2569
```

39.
$$
\begin{array}{r}
^{1\ 1}59 \\
294 \\
1100 \\
+\ \ \ 10 \\
\hline
1463
\end{array}
$$

40.
$$
\begin{array}{r}
^{1}853 \\
121 \\
0 \\
+\ 2912 \\
\hline
3886
\end{array}
$$

41.
$$
\begin{array}{r}
^{1}25 \\
0 \\
53 \\
180 \\
+\ \ \ 0 \\
\hline
258
\end{array}
$$

42.
$$
\begin{array}{r}
^{1}9 \\
0 \\
71 \\
0 \\
+\ 312 \\
\hline
392
\end{array}
$$

43.
$$
\begin{array}{r}
67 \\
-\ \ 3 \\
\hline
64
\end{array}
$$

44.
$$
\begin{array}{r}
16 \\
-\ 7 \\
\hline
9
\end{array}
$$

45.
$$
\begin{array}{r}
89 \\
-\ 62 \\
\hline
27
\end{array}
$$

46.
$$
\begin{array}{r}
55 \\
-\ 32 \\
\hline
23
\end{array}
$$

47.
$$
\begin{array}{r}
695 \\
-\ 12 \\
\hline
683
\end{array}
$$

48.
$$
\begin{array}{r}
386 \\
-\ 24 \\
\hline
362
\end{array}
$$

49.
$$
\begin{array}{r}
649 \\
-\ 226 \\
\hline
423
\end{array}
$$

50.
$$
\begin{array}{r}
867 \\
-\ 253 \\
\hline
614
\end{array}
$$

51.
$$
\begin{array}{r}
^{6\ \ 12\ 12\ 12}\cancel{7}\cancel{3}\cancel{3}\cancel{2} \\
-\ 4\ 9\ 9 \\
\hline
6\ 8\ 3\ 3
\end{array}
$$

52.
$$
\begin{array}{r}
1565 \\
-\ 360 \\
\hline
1205
\end{array}
$$

53.
$$
\begin{array}{r}
^{2\ \ ^{9}\cancel{10}\ 10}4\cancel{3}\cancel{0}\cancel{0} \\
-\ \ \ 3\ 1 \\
\hline
4\ 2\ 6\ 9
\end{array}
$$

54.
$$
\begin{array}{r}
^{6\ \ 14\ ^{9}\cancel{10}\ 10}\cancel{7}\cancel{5}\cancel{0}\cancel{0} \\
-\ \ 9\ 7\ 3 \\
\hline
6\ 5\ 2\ 7
\end{array}
$$

55.
$$
\begin{array}{r}
64 \\
\times\ \ 1 \\
\hline
64
\end{array}
$$

56.
$$
\begin{array}{r}
72 \\
\times\ \ 0 \\
\hline
0
\end{array}
$$

57.
$$
\begin{array}{r}
63 \\
\times\ 25 \\
\hline
315 \\
1260 \\
\hline
1575
\end{array}
$$

58.
$$
\begin{array}{r}
78 \\
\times\ 55 \\
\hline
390 \\
3900 \\
\hline
4290
\end{array}
$$

59.
$$
\begin{array}{r}
39 \\
\times\ 95 \\
\hline
195 \\
3510 \\
\hline
3705
\end{array}
$$

60.
$$
\begin{array}{r}
342 \\
\times\ 37 \\
\hline
2394 \\
10260 \\
\hline
12654
\end{array}
$$
or 12,654

61.
$$
\begin{array}{r}
318 \\
\times\ 40 \\
\hline
12{,}720
\end{array}
$$

62.
$$
\begin{array}{r}
111 \\
\times\ 55 \\
\hline
555 \\
5550 \\
\hline
6105
\end{array}
$$

63.
$$
\begin{array}{r}
18 \\
\times\ 45 \\
\hline
90 \\
720 \\
\hline
810
\end{array}
$$

64.
$$
\begin{array}{r}
270 \\
\times\ 64 \\
\hline
1\ 080 \\
16{,}200 \\
\hline
17{,}280
\end{array}
$$

65.
$$
\begin{array}{r}
815 \\
\times\ 60 \\
\hline
48{,}900
\end{array}
$$

66.
$$
\begin{array}{r}
2900 \\
\times\ 328 \\
\hline
23{,}200 \\
58{,}000 \\
870{,}000 \\
\hline
951{,}200
\end{array}
$$

67. $48 \div 0$ undefined

   Division by 0 is always undefined.

68. $0 \div 63 = 0$

   0 divided by any nonzero number is 0.

69. $\dfrac{46}{1} = 46$

Dividing any number by 1 gives the number.

70. $\dfrac{79}{79} = 1$

Dividing any nonzero number by itself gives 1.

71.
$$\begin{array}{r} 16 \\ 49\overline{)784} \\ 49 \\ \hline 294 \\ 294 \\ \hline \end{array}$$

72.
$$\begin{array}{r} 21 \\ 42\overline{)882} \\ 84 \\ \hline 42 \\ 42 \\ \hline \end{array}$$

73.
$$\begin{array}{r} 11 \ R\ 2 \\ 4\overline{)46} \\ 4 \\ \hline 6 \\ 4 \\ \hline 2 \end{array}$$

74.
$$\begin{array}{r} 21 \ R\ 26 \\ 29\overline{)635} \\ 58 \\ \hline 55 \\ 29 \\ \hline 26 \end{array}$$

75.
$$\begin{array}{r} 20 \ R\ 4 \\ 14\overline{)284} \\ 28 \\ \hline 4 \end{array}$$

$\dfrac{284}{14}$ is 20 $R$ 4

76.
$$\begin{array}{r} 34 \ R\ 4 \\ 182\overline{)6192} \\ 546 \\ \hline 732 \\ 728 \\ \hline 4 \end{array}$$

$6192 \div 182$ is 34 $R$ 4

77.
$$\begin{array}{r} 4 \ R\ 2 \\ 11\overline{)46} \\ 44 \\ \hline 2 \end{array}$$

78.
$$\begin{array}{r} 31 \ R\ 2 \\ 3\overline{)95} \\ 9 \\ \hline 5 \\ 3 \\ \hline 2 \end{array}$$

79. $7 \cdot 7 \cdot 7 \cdot 7 = 7^4$

Since there are 4 factors of 7.

80. $13 \cdot 13 \cdot 13 = 13^3$

Since there are 3 factors of 13.

81. $17 = 17^1$

82. $5 \cdot 5 \cdot 5 \cdot 5 \cdot 5 \cdot 5$

$= 5^6$

83. $3 \cdot 3 \cdot 5 \cdot 5 \cdot 5 \cdot 11 \cdot 11$

$= 3^2 \cdot 5^3 \cdot 11^2$

84. $5 \cdot 5 \cdot 7 \cdot 17 \cdot 17 \cdot 19$

$= 5^2 \cdot 7 \cdot 17^2 \cdot 19$

85. $2 \cdot 2 \cdot 2 \cdot 2 \cdot 23 \cdot 23 \cdot 29$

$= 2^4 \cdot 23^2 \cdot 29$

86. $11 \cdot 11 \cdot 11 \cdot 19 \cdot 19$

$= 11^3 \cdot 19^2$

87. $7^2 = 7 \cdot 7 = 49$

88. $2^4 = 2 \cdot 2 \cdot 2 \cdot 2$

$= 4 \cdot 2 \cdot 2$

$= 8 \cdot 2 = 16$

89. $39^1 = 39$

90. $3^5 = 3 \cdot 3 \cdot 3 \cdot 3 \cdot 3$

$= 9 \cdot 3 \cdot 3 \cdot 3$

$= 27 \cdot 3 \cdot 3$

$= 81 \cdot 3$

$= 343$

91. $10^3 = 10 \cdot 10 \cdot 10$

$= 100 \cdot 10$

$= 1000$

92. $66^0 = 1$

Any nonzero number raised to the 0 power is 1.

93. $19^2 = 19 \cdot 19$

$= 361$

94. $1^{20} = 1$

1 raised to any power is 1.

95. $10^6 = 10 \cdot 10 \cdot 10 \cdot 10 \cdot 10 \cdot 10$

$= 1,000,000$

96. $6^3 = 6 \cdot 6 \cdot 6$
    $= 36 \cdot 6$
    $= 216$

97. $0^7 = 0$
    0 raised to any
    power (except 0)
    is 0.

98. $2^9 = 2 \cdot 2 \cdot 2 \cdot 2 \cdot 2 \cdot 2 \cdot 2 \cdot 2 \cdot 2$
    $= 512$

99. $9 + 17 \cdot 20$
    $9 + 340$
    $349$

100. $34 \div 2 + 9(20 + 5)$
     $34 \div 2 + 9 \cdot 25$
     $17 + 225$
     $242$

101. $20 \div 2^2 + (5 \cdot 4)$
     $20 \div 2^2 + 20$
     $20 \div 4 + 20$
     $5 + 20$
     $25$

102. $\dfrac{360 \div 6^2}{35 - 25} + 5^3$
     $\dfrac{360 \div 36}{10} + 5^3$
     $\dfrac{10}{10} + 5^3$
     $\dfrac{10}{10} + 125$
     $1 + 125$
     $126$

103. $8^2 - (8 - 4)^3$
     $8^2 - 4^3$
     $64 - 64$
     $0$

104. $5 + \left(\dfrac{300}{12} - 4^2\right)^2$
     $5 + \left(\dfrac{300}{12} - 16\right)^2$
     $5 + (25 - 16)^2$
     $5 + 9^2$
     $5 + 81$
     $86$

105. $50 \cdot 8^2 \div (10 + 30)^2$
     $50 \cdot 8^2 \div 40^2$
     $50 \cdot 64 \div 1600$
     $3200 \div 1600$
     $2$

106. $12 + 2[6 - (5 - 2)]$
     $12 + 2[6 - 3]$
     $12 + 2 \cdot 3$
     $12 + 6$
     $18$

107. $\dfrac{(4 \cdot 3^2)}{18 - 12} \cdot 10$
     $\dfrac{(4 \cdot 9)}{18 - 12} \cdot 10$
     $\dfrac{36}{18 - 12} \cdot 10$
     $\dfrac{36}{6} \cdot 10$
     $6 \cdot 10$
     $60$

108. $111{,}000 - 500(12 - 6)^3$
     $111{,}000 - 500 \cdot 6^3$
     $111{,}000 - 500 \cdot 216$
     $111{,}000 - 108{,}000$
     $3000$

109. $3[100 - 8(4) + 9 \div 3 + 7]$
     $3[100 - 32 + 9 \div 3 + 7]$
     $3[100 - 32 + 3 + 7]$
     $3[68 + 3 + 7]$
     $3[71 + 7]$
     $3[78]$
     $234$

110. $\dfrac{7^2 - 6^2}{(7 + 6)} + 19$
     $\dfrac{7^2 - 6^2}{13} + 19$
     $\dfrac{49 - 36}{13} + 19$
     $\dfrac{13}{13} + 19$
     $1 + 19$
     $20$

111. a. The hundreds digit of 453,229 is 2 and it is followed by a 2 so round down to get 453,200 gallons.

b. The number 453,200 has the digits 453 in the thousands period and 200 in the ones period. The word form is "four hundred fifty-three thousand, two hundred."

112. a. Adding the acreage for each crop, the pasture, and the buildings gives a total of
$450 + 259 + 812 + 18 + 22 + 6 = 1567$ acres.

   b. Using the total of 1567 acres from part a., if 329 acres are sold the number of acres remaining would be $1567 - 329 = 1238$.

113. First compute the total yardage for the holes having a given par.
Par-3: $6 \times 175 = 1050$ yards    Par-4: $7 \times 228 = 1596$ yards    Par-5: $5 \times 340 = 1700$ yards
The total yardage for all eighteen holes is $1050 + 1596 + 1700 = 4346$ yards.

114. First, compute the total sales for each type of real estate sold.
Land: $4 \times \$32,500 = \$130,000$        Condominiums: $1 \times \$55,600 = \$55,600$
Homes: $\$79,200 + \$96,200 = \$175,400$
Next, add these amounts $\$130,000 + \$55,600 + \$175,400 = \$361,000$. Sales totaled $361,000.

115. a. The painting has two sides measuring 22 inches and two sides measuring 14 inches. The perimeter is $22 + 22 + 14 + 14 = 72$ inches. At $3 per inch, the cost of a standard frame would be $72 \times \$3 = \$216$.

   b. From part a., the perimeter is 72 inches. At $4 per inch, the cost of a deluxe frame would be $72 \times \$4 = \$288$. This is $\$288 - \$216 = \$72$ more than the standard frame.

116. a. First note that there are two types of costs. The food cost depends on the number of people who attend while the remaining costs are fixed regardless of the number of people who attend. The fixed costs total $\$1250 + \$700 + \$328 + \$382 + \$1500 = \$4160$. If 160 people attend at a cost of $16 per person then the total cost of food is $160 \times \$16 = \$2560$. Total cost of the party is $\$4160 + \$2560 = \$6720$.

   b. From part a. the total cost is $6720, the average cost over 160 people is $\$6720 \div 160 = \$42$ per person.

117. Girls work 8 eight hours and do homework 12 hours for a total of $8 + 12 = 20$ hours per week.

118. Boys exercise 7 hours and work 10 hours for a total of $7 + 10 = 17$ hours per week.

119. In one year girls surf and email a total of $52 \times 17 = 884$ hours while boys use $52 \times 16 = 832$ hours. So girls surf and email $884 - 832 = 52$ hours more than boys per year. Another way to see this is to note that in any given week girls surf and write emails $17 - 16 = 1$ hour more than boys. Over a year (52 weeks) girls surf and write emails 52 hours more than boys.

120. Over 52 weeks girls work a total $52 \times 8 = 416$ hours wile boys work $52 \times 10 = 520$ hours. Thus girls work $520 - 416 = 104$ fewer hours than boys.

121. Boys and girls together surf and write emails a total of $16 + 17 = 33$ hours per week. Over a year, the total time spent surfing and writing emails would be $52 \times 33 = 1716$. The average among boys and girls is then $1716 \div 2 = 858$.

122. The total time per week spent by boys and girls doing homework is $8 + 12 = 20$ hours. The average among boys and girls is then $20 \div 2 = 10$.

## Chapter 1 Assessment Test

1.  In the number 6877, the indicated digit 8 is in the *hundreds* place (the leftmost digit in the hundreds period.)

2.  In the number 2,336,029, the indicated digit 3 is in the *ten thousands* place(the middle digit in the thousands period.)

3.  $10,000 + 5000 + 800 + 60 + 2$
    standard notation:  15,862
    word form:  fifteen thousand,
         eight hundred sixty-two

4.  $100,000 + 20,000 + 3000 + 500 + 9$
    standard notation:  123,509
    word form:  one hundred twenty-three
         thousand, five hundred nine

5.  The expanded form of 475 is $400 + 70 + 5$
    or 4 hundreds + 7 tens + 5 ones.

6.  The expanded form of 1397 is $1000 + 300 + 90 + 7$
    or 1 thousand + 3 hundreds + 9 tens + 7 ones.

7.  34,771 rounded to the nearest thousand is 35, 000. The digit in the thousands place is 4 and the digit to its right is 7 so the thousands place is rounded up to 5 and then by zeros.

8.  6,529,398 rounded to the nearest hundred thousands is 6,500,000.  The digit in the hundred thousands place is 5 and the digit to its right is 2 so we round down, leaving the thousands place alone and following it with zeros.

9.
$$
\begin{array}{r}
463 \\
+\ 25 \\
\hline
488
\end{array}
$$
Simply add column wise.

10. Write the addition vertically.
$$
\begin{array}{r}
{}^{1\ 1} \\
652 \\
0 \\
257 \\
+\ 576 \\
\hline
1485
\end{array}
$$

11.
$$
\begin{array}{r}
{}^{4\ 11} \\
\cancel{5}\,\cancel{1} \\
-\ 3\ 4 \\
\hline
1\ 7
\end{array}
$$

12. Write vertically and subtract columnwise.
$$
\begin{array}{r}
782 \\
-\ 41 \\
\hline
741
\end{array}
$$

13.
$$
\begin{array}{r}
318 \\
\times\ 36 \\
\hline
1908 \\
9540 \\
\hline
11,448
\end{array}
$$

14. Write vertically and multiply.
$$
\begin{array}{r}
3132 \\
\times\ 58 \\
\hline
25,056 \\
156,600 \\
\hline
181,656
\end{array}
$$

15.
$$
\begin{array}{r}
63 \\
34\overline{)2142} \\
\underline{204\phantom{2}} \\
102 \\
102
\end{array}
$$

16.  $\begin{array}{r} 14 \ R \ 1 \\ 7\overline{)99} \\ \underline{7} \\ 29 \\ \underline{28} \\ 1 \end{array}$

17.  $13 \cdot 13 \cdot 13$

Each factor is 13 so
13 is the base. There
are three factors so the
exponent is 3.
Exponential notation is $13^3$.

18.  $3 \cdot 3 \cdot 5 \cdot 5 \cdot 5 \cdot 5 \cdot 7$

$3^2 \cdot 5^4 \cdot 7$

19.  $6^2 = 6 \cdot 6 = 36$

20.  $2^4 = 2 \cdot 2 \cdot 2 \cdot 2 = 4 \cdot 2 \cdot 2 = 8 \cdot 2 = 16$

21.  $6 + 7 \cdot 2$

$6 + 14$

$20$

22.  $64 \div 2^3 + (5 \cdot 7)$

$64 \div 2^3 + 35$

$64 \div 8 + 35$

$8 + 35$

$43$

23.  First determine the number of miles Alice traveled in the two days: $238 + 287 = 525$ miles. The odometer reading would have increased by this amount so add 525 to the original odometer reading to get the new reading.  $23,414 + 525 = 23,939$ is the new reading.

24.  Since the amount is being split, use division. Since  $2,520,000 \div 14 = 180,000,$  each person receives $180,000.

25.  a. Determine how many times 4 *goes into* 420, that is, divide. Since  $420 \div 4 = 105$ , there are 105 portions.

    b. 225 meals a night at 7 nights a week is  $7 \times 225 = 1575$  meals. Since each meal uses 4 ounces of pasta, this is a total of  $4 \times 1575 = 6300$  ounces of pasta. A carton consists of 420 ounces so the number of cartons is  $6300 \div 420 = 15$ .

26.  a. The rectangular section has two sides of length 150 feet and two sides of 110 feet. The perimeter is  $150 + 150 + 110 + 110 = 520$  feet.

    b. The area of the rectangular section is  $A = lw = (150 \text{ feet})(110 \text{ feet}) = 150 \cdot 110$  square feet or 16,500 square feet.

# Chapter 2 Fractions

## Section 2.1 Factors, Prime Factorizations, and Least Common Multiples

### Concept Check

1. A _prime_ number is a natural number greater than 1 that has only two factors (divisors), namely, 1 and itself.

3. To _factor_ a quantity is to express it as a product of factors.

5. An illustration used to determine the prime factorization of a composite number is known as a _factor tree_.

7. A nonzero multiple that is shared by a set of two or more whole numbers is called a _common_ multiple.

### Guide Problems

9. The following quotients with 40 as the dividend result in a natural number: $40 \div 1 = 40$, $40 \div 2 = 20$, $40 \div 4 = 10$, $40 \div 5 = 8$ before factors begin to repeat. The factors of 40 are 1, 2, 4, 5, 8, 10, 20 and 40.

11. Among 1, 2, 3, 4 and 5 only 1 and 5 divide 5 evenly so the factors of 5 are 1 and 5.

13. Find the factors of 6: $6 \div 1 = 6$, $6 \div 2 = 3$, $6 \div 3 = 2$. Stop with the last division since factors are repeating. The factors of 6 are 1, 2, 3, and 6.

15. The following quotients with 49 as the dividend result in a natural number: $49 \div 1 = 49$, $49 \div 7 = 7$, where factors begin to repeat. The factors of 49 are 1, 7, and 49.

17. The factors of 29 are 1 and 29 as these are the only natural numbers that divide 29 evenly.

19. The following quotients with 77 as the dividend result in a natural number: $77 \div 1 = 77$, $77 \div 7 = 11$, before factors begin to repeat. The factors of 77 are 1, 7, 11, and 77.

21. Find the factors of 28: $28 \div 1 = 28$, $28 \div 2 = 14$, $28 \div 4 = 7$, $28 \div 7 = 4$ are the quotients that divide evenly. Note the factors begin to repeat with the last division. The factors of 28 are 1, 2, 4, 7, 14, and 28.

23. The factors of 61 are 1 and 61 as these are the only natural numbers that divide 61 evenly.

25. Find the factors of 75: $75 \div 1 = 75$, $75 \div 3 = 25$, $75 \div 5 = 15$, $75 \div 15 = 5$ are the quotients that divide evenly. Note the factors begin to repeat with the last division. The factors of 75 are 1, 3, 5, 15, 25, and 75.

27. Find the factors of 54.

$54 \div 1 = 54$      $54 \div 2 = 27$
$54 \div 3 = 18$
$54 \div 4$ Does not divide evenly.
$54 \div 5$ Does not divide evenly.
$54 \div 6 = 9$
$54 \div 7$ Does not divide evenly.
$54 \div 8$ Does not divide evenly.
$54 \div 9 = 6$ Factors are repeating. Stop.

The factors of 54 are 1, 2, 3, 6, 9, 18, 27, and 54.

29. Find the factors of 100.

$100 \div 1 = 100$      $100 \div 2 = 50$
$100 \div 3$ Does not divide evenly.
$100 \div 4 = 25$      $100 \div 5 = 20$
$100 \div 6$ Does not divide evenly.
$100 \div 7$ Does not divide evenly.
$100 \div 8$ Does not divide evenly.
$100 \div 9$ Does not divide evenly.
$100 \div 10 = 10$ Factors are repeating. Stop.

The factors of 100 are 1, 2, 4, 5, 10, 20, 25, 50 and 100.

31. You can demonstrate a natural number is composite by exhibiting one factor of the number different from itself or one. The number 40 is even so 2 is a factor ( $40 \div 2 = 20$ ) and so 40 is a composite number.

33. The only factors of 43 are 1 and 43, so 43 is a prime number.

35. By definition, prime and composite numbers are greater than one. Hence, 1 is neither prime nor composite.

37. By definition, prime and composite numbers are greater than one. Hence, 0 is neither prime nor composite.

39. Since 2 divides 28 evenly, $28 \div 2 = 14$ , the number 28 is composite. The factors of 28 are 1, 2, 4, 7, 14, and 28.

41. The number 165 is divisible by 5 ( $165 \div 5 = 33$ ) and so 165 is composite. The factors of 165 are 1, 3, 5, 11, 15, 33, 55, and 165.

43. The only factors of 17 are 1 and 17, so 17 is a prime number.

45. The number 81 is divisible by 9 ( $81 \div 9 = 9$ ) and so 81 is composite. The factors of 81 are 1, 3, 9, 27, and 81.

47. The only factors of 83 are 1 and 83, so 83 is a prime number.

## Guide Problems

49. Find the prime factorization of 30.
Construct a factor tree for the number 30. The two branches from a particular number are factors whose product is the given number. Continue in this fashion until all branches end with a prime number. These primes constitute the prime factorization of the starting value.

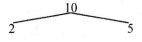

$30 = \underline{2} \cdot \underline{3} \cdot \underline{5}$

*In problems 51 through 73, a prime factorization is determined using a factor tree. A factor tree is typically not unique. For example, a factor tree for 30 might begin with the factors 5 and 6 or the factors 3 and 10, and so solutions may vary. However any factor tree must lead to the same prime factorization.*

51.

$10 = 2 \cdot 5$

53.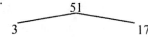

$51 = 3 \cdot 17$

55.

$42 = 2 \cdot 3 \cdot 7$

57.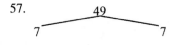

$49 = 7 \cdot 7 = 7^2$

59.

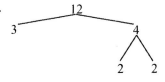

$$12 = 2 \cdot 2 \cdot 3 = 2^2 \cdot 3$$

61.

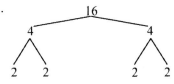

$$16 = 2 \cdot 2 \cdot 2 \cdot 2 = 2^4$$

63.

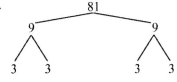

$$81 = 3 \cdot 3 \cdot 3 \cdot 3 = 3^4$$

65.

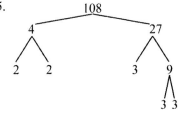

$$108 = 2 \cdot 2 \cdot 3 \cdot 3 \cdot 3 = 2^2 \cdot 3^3$$

67.

$$175 = 5 \cdot 5 \cdot 7 = 5^2 \cdot 7$$

69.

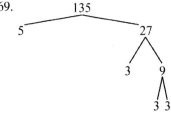

$$135 = 3 \cdot 3 \cdot 3 \cdot 5 = 3^3 \cdot 5$$

71.

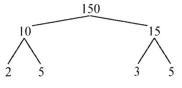

$$150 = 2 \cdot 3 \cdot 5 \cdot 5 = 2 \cdot 3 \cdot 5^2$$

73.

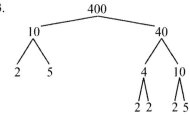

$$400 = 2 \cdot 2 \cdot 2 \cdot 2 \cdot 5 \cdot 5 = 2^4 \cdot 5^2$$

## Guide Problems

75.  Find the LCM of 4 and 6 by listing multiples.

a. List the first few multiples of each number.
   4:  4, 8, <u>12</u>, <u>16</u>, <u>20</u>, <u>24</u>, …
   6:  6, 12, <u>18</u>, <u>24</u>, <u>30</u>, <u>36</u>, …

b. The LCM is the smallest multiple in common to each list.  The multiples 12 and 24 are common to both lists but 12 is the smallest or least common multiple.  What is the LCM?

The LCM of 4 and 6 is <u>12</u>.

77.  Find the LCM of 8 and 18 using prime factorization.

a. Find the prime factorization of each
   number.

$$8 = 2 \cdot 2 \cdot 2 = 2^3$$
$$18 = \underline{2} \cdot \underline{3} \cdot \underline{3} = \underline{2} \cdot \underline{3^2}$$

b. The LCM is the product of those prime
   factors occurring the greatest number of
   times in any one factorization.  What is
   the LCM?

   The greatest number of times that 2 appears
   in either factorization is __three__ times.
   The greatest number of times that 3 appears
   in either factorization is __two__ times.

   The LCM of 8 and 18 is
   $\underline{2} \cdot \underline{2} \cdot \underline{2} \cdot \underline{3} \cdot \underline{3} = 72.$

*In problems 79 through 97, either the prime factorization method or the method of listing multiples is used to
find the LCM.  All problems could be worked by either method although, on occasion, the method of listing
multiples may require long lists before a common multiple is found.*

79.  Prime factorizations:
     $$2 = 2, \quad 9 = 3 \cdot 3 = 3^2$$

     The factor 2 appears at most once.
     The factor 3 appears at most twice.

     LCM of 2 and 9 is $2 \cdot 3^2 = 18$.

81.  List multiples:
     4:  4, 8, 12, 16, 20, 24, ...
     5:  5, 10, 15, 20, 25, ...
     The multiple 20 is common to both lists
     and is the smallest number common to
     both lists.  The LCM of 4 and 5 is 20.

83.  List multiples:
     6:  6, 12, 18, 24, 30, ...
     9:  9, 18, 27, 36, ...
     The LCM of 6 and 9 is 18.

85.  List multiples:
     8:  8, 16, 24, 32, 40, 48, 56, 64, 72, ...
     9:  9, 18, 27, 36, 45, 54, 63, 72, ...
     The LCM of 8 and 9 is 72.

87.  Prime factorizations:
     $$8 = 2 \cdot 2 \cdot 2 = 2^3$$
     $$16 = 2 \cdot 2 \cdot 2 \cdot 2 = 2^4$$
     The LCM of 8 and 16 is $2 \cdot 2 \cdot 2 \cdot 2 = 16$.

89.  Prime factorizations:
     $$15 = 3 \cdot 5$$
     $$20 = 2 \cdot 2 \cdot 5 = 2^2 \cdot 5$$
     The LCM of 15 and 20 is $2 \cdot 2 \cdot 3 \cdot 5 = 60$.

91.  Prime factorizations:
     $$2 = 2, \quad 3 = 3$$
     $$16 = 2 \cdot 2 \cdot 2 \cdot 2 = 2^4$$
     The LCM of 2, 3, and 16 is $2 \cdot 2 \cdot 2 \cdot 2 \cdot 3 = 48$.

93.  List multiples:
     12:  12, 24, 36, 48, 60, 72, 84, 96, ...
     14:  14, 28, 42, 56, 70, 84, 98, 112, ...
     21:  21, 42, 63, 84, 105, ...
     The LCM of 12, 14, and 21 is 84.

95.  Prime factorizations:
     $$2 = 2, \quad 6 = 2 \cdot 3, \quad 33 = 3 \cdot 11$$
     The LCM of 2, 6, and 33 is $2 \cdot 3 \cdot 11 = 66$.

97.  Prime factorizations:
     $$8 = 2 \cdot 2 \cdot 2 = 2^3$$
     $$9 = 3 \cdot 3 = 3^2$$
     $$12 = 2 \cdot 2 \cdot 3 = 2^2 \cdot 3$$
     $$18 = 2 \cdot 3 \cdot 3 = 2 \cdot 3^2$$
     The LCM of 8, 9, 12, and 18 is $2 \cdot 2 \cdot 2 \cdot 3 \cdot 3 = 72$.

99.  Since Tony goes to the beach every fourth day and Kathy goes every sixth day, they will meet on days
     corresponding to the least common multiple of 4 and 6.  Listing multiples of each number,
            4: 4, 8, 12, 16, 20, 24, ...   and  6: 6, 12, 18, 24, 30, ...

it is easy to see the least common multiple on the two lists is 12. Tony and Kathy will meet every 12 days.

101. If Thomas and Elaine start at the same time, then Thomas will take a break after 20 minutes, 40 minutes, 60 minutes, 80 minutes, and so on while Elaine takes a break after 30 minutes, 60 minutes, 90 minutes, etc. The least common multiple on these lists is 60 minutes, the LCM of 20 and 30. Thomas and Elaine will share a break every 60 minutes.

## Cumulative Skills Review

1. $30 - (4-2)^2 \cdot 7 = 30 - (2)^2 \cdot 7$    Simplify within parentheses first.

                $= 30 - 4 \cdot 7$    Simplify exponents.

                $= 30 - 28$    Perform multiplications from left to right.

                $= 2$    Subtract.

3.    18,255

      1399

  +   415

   20,069

5.    $105 - 26 = 79$

7.    $6 \cdot 6 \cdot 7 \cdot 7 \cdot 7 = (6 \cdot 6) \cdot (7 \cdot 7 \cdot 7) = 6^2 \cdot 7^3$

9.    In the number \$34,232,900 the digit to the right of the thousands digit is 9 so to round to the thousands place we round up giving \$34,233,000 .

## Section 2.2    Introduction to Fractions and Mixed Numbers

### Concept Check

1. A  _fraction_  is a number written in the form $\dfrac{a}{b}$ where $a$ and $b$ are whole numbers and $b$ is not zero.

3. The bottom number in a fraction is called the  _denominator_ .

5. A fraction in which the numerator is less than the denominator is known as a  _proper_  fraction.

7. A  _mixed_   _number_  is a number that combines a whole number with a fraction.

### Guide Problems

9. a. The numerator, 5, is less than the denominator, 19, so $\dfrac{5}{19}$ is a proper fraction.

    b. The numerator, 13, is greater than the denominator, 8, so $\dfrac{13}{8}$ is an improper fraction.

    c. The number $3\dfrac{1}{7}$ combines a whole number, 3, with a fraction, $\dfrac{1}{7}$, and so is a mixed number.

11. The fraction $\dfrac{3}{8}$ is a proper fraction since 3 is less than 8.

13. The fraction $\dfrac{9}{9}$ is an improper fraction since 9 is equal to 9.

15. The number $4\dfrac{1}{5}$ is a mixed number since it is a whole number combined with a fraction.

17. The fraction $\dfrac{33}{6}$ is an improper fraction since 33 is greater than 6.

19. The fraction $\dfrac{7}{7}$ is an improper fraction since 7 is equal to 7.

21. The number $5\dfrac{2}{3}$ is a mixed number since it is a whole number combined with a fraction.

23. The rectangle is divided into 9 squares. Of these 9 squares, 5 squares are shaded. The shaded portion is then represented by the fraction $\dfrac{5}{9}$.

25. The cake has been cut into eight equal pieces. Of these five pieces are shaded. The shaded pieces represent the fraction $\dfrac{5}{8}$.

27. The rectangle is divided into 18 squares of the same size. Seven of these squares have been shaded. The fraction shaded is $\dfrac{7}{18}$.

29. There are seven triangles, of which four have been shaded. The figure represents the fraction $\dfrac{4}{7}$.

31. There are three circles shaded completely. The fourth circle is divided into four parts, three of which are shaded, representing the fraction $\dfrac{3}{4}$. Thus we have three wholes and three-fourths of a whole together representing the mixed number $3\dfrac{3}{4}$.

33. The first three trapezoids are shaded completely. The last trapezoid has two of its three triangle shaded representing the fraction $\dfrac{2}{3}$. Thus we have three wholes and two-thirds of a whole together representing the mixed number $3\dfrac{2}{3}$. Alternatively, think of each whole trapezoid as being divided into three triangles. There are 11 triangles shaded so the fraction represented is $\dfrac{11}{3}$.

35. There are a total of 11 animals, 6 fish and 5 frogs. The fish are represented by the fraction $\dfrac{6}{11}$.

## Guide Problems

37. Write $\dfrac{54}{9}$ as a mixed or whole number.

Divide the denominator into the numerator

$$9\overline{)54} \phantom{x} \begin{array}{r} 6 \\ \hline \end{array}$$

$$\underline{54}$$
$$0$$

The quotient is 6 and the remainder is 0. The quotient is the whole number part of the equivalent mixed number. The fraction part of the mixed number has the remainder as its numerator and the denominator is that of the original fraction. Thus

$$\frac{54}{9} = 6$$

When the remainder is 0, the given fraction is equivalent to a whole number and there is no fraction part.

39. Write $8\dfrac{3}{5}$ as an improper fraction.

To convert a mixed number to an improper fraction use the denominator of the fraction part of the mixed number as the denominator of the improper fraction. To find the numerator of the improper fraction, multiply the denominator by the whole number part of the mixed number and add the numerator of the fraction part.

$$8\frac{3}{5} = \frac{8 \cdot 5 + 3}{5} = \frac{43}{5}$$

41. 
$$9\overline{)90} \quad \begin{array}{r} 10 \\ \hline \end{array}$$
$$\underline{90}$$
$$0$$

$$\frac{90}{9} = 10$$

43. 
$$2\overline{)13} \quad \begin{array}{r} 6 \\ \hline \end{array}$$
$$\underline{12}$$
$$1$$

$$\frac{13}{2} = 6\frac{1}{2}$$

45. 
$$9\overline{)88} \quad \begin{array}{r} 9 \\ \hline \end{array}$$
$$\underline{81}$$
$$7$$

$$\frac{88}{9} = 9\frac{7}{9}$$

47. 
$$10\overline{)130} \quad \begin{array}{r} 13 \\ \hline \end{array}$$
$$\underline{130}$$
$$0$$

$$\frac{130}{10} = 13$$

49. 
$$8\overline{)15} \quad \begin{array}{r} 1 \\ \hline \end{array}$$
$$\underline{8}$$
$$7$$

$$\frac{15}{8} = 1\frac{7}{8}$$

51. 
$$3\overline{)40} \quad \begin{array}{r} 13 \\ \hline \end{array}$$
$$\underline{3}$$
$$10$$
$$\underline{9}$$
$$1$$

$$\frac{40}{3} = 13\frac{1}{3}$$

53.  $\begin{array}{r} 10 \\ 3\overline{)31} \\ \underline{3} \\ 1 \end{array}$

55.  $\begin{array}{r} 10 \\ 12\overline{)131} \\ \underline{120} \\ 11 \end{array}$

$\dfrac{31}{3} = 10\dfrac{1}{3}$

$\dfrac{131}{12} = 10\dfrac{11}{12}$

57.  $10\dfrac{3}{7} = \dfrac{10 \cdot 7 + 3}{7} = \dfrac{73}{7}$

59.  $11\dfrac{3}{5} = \dfrac{11 \cdot 5 + 3}{5} = \dfrac{58}{5}$

61.  $7\dfrac{3}{10} = \dfrac{7 \cdot 10 + 3}{10} = \dfrac{73}{10}$

63.  $8\dfrac{4}{5} = \dfrac{8 \cdot 5 + 4}{5} = \dfrac{44}{5}$

65.  $12\dfrac{2}{3} = \dfrac{12 \cdot 3 + 2}{3} = \dfrac{38}{3}$

67.  $12\dfrac{1}{10} = \dfrac{12 \cdot 10 + 1}{10} = \dfrac{121}{10}$

69.  $10\dfrac{8}{13} = \dfrac{10 \cdot 13 + 8}{13} = \dfrac{138}{13}$

71.  $9\dfrac{1}{7} = \dfrac{9 \cdot 7 + 1}{7} = \dfrac{64}{7}$

73.  a. There are 37 students total, 21 of which are females.  The fraction of the students represented by the females will have the total number of students as denominator and the number of females as its numerator, that is $\dfrac{21}{37}$.

b. If there are 37 students and 21 are female then $37 - 21 = 16$ must be male.  The males represent $\dfrac{16}{37}$ of the students.

75.  There is a total of 55 players, 14 of which are freshmen.  The fraction of players represented by freshmen is $\dfrac{\text{no. of freshmen players}}{\text{no. of players}} = \dfrac{14}{55}$.

77.  We are looking for the fraction of a week a given number of days represents.  Thus the denominator will be the total number of days in a week, 7, and the numerator will be the given number of days.

a. 3 days is $\dfrac{3}{7}$ of a week          b. 11 days is $\dfrac{11}{7}$ of a week

79.  a. You spent a total of $\$24 + \$17 + \$42 = \$83$ out of your budgeted $\$200$.  The fraction of the budget your spending represents is $\dfrac{83}{200}$.

b. The amount of your budget you did not spend is $\$200 - \$83 = \$117$.  The fraction of your budget this amount represents is $\dfrac{117}{200}$.

81.  a. The amount the charity still needs to raise is

$\begin{array}{r} \$30,000 \\ - \ \$11,559 \\ \hline \$18,441 \end{array}$

and so the fraction representing the amount left to raise is $\dfrac{\$18,441}{\$30,000}$ since $\$30,000$ is the desired total.

b. If the goal were, in fact, $36,000 and $11,559 has been raised then the fraction of the goal raised is represented by the fraction $\dfrac{\$11,559}{\$36,000}$ .

83. At the end of the day Thursday you have given performances on Monday, Tuesday, Wednesday and Thursday for a total of 4 shows. Over the course of the seven day run, you will give one performance in the evening of each day and one additional matinee performance on both Friday and Saturday for a total of 9 performances. The fraction of total performances given after Thursday's show is then $\dfrac{4}{9}$ .

## Cumulative Skills Review

1. Find the factors of 16: $16 \div 1 = 16$, $16 \div 2 = 8$, $16 \div 3$ does not divide evenly, $16 \div 4 = 4$. Stop with the last division since there is a repeat factor. The factors of 16 are 1, 2, 4, 8 and 16.

3. $48 + (3 + 2)3 = 48 + (5)3$    Simplify within parentheses first.

$\qquad\qquad\quad = 48 + 15$    Perform multiplications from left to right.

$\qquad\qquad\quad = 63$    Add.

5. In the number 132,596 the digit to the right of the thousands digit is 5 so to round to the thousands place we round up giving 133,000.

7. Since 5 divides 65, that is, $65 \div 5 = 13$, the number 65 is composite.

9. In standard notation, the number 125662 is 125,662. The word form for 125662 is "one hundred twenty-five thousand, six hundred sixty-two."

## Section 2.3    Equivalent Fractions

## Concept Check

1. Fractions that represent the same quantity are called  equivalent  fractions.

3. The largest factor common to two or more numbers is known as the  greatest   common   factor  or GCF .

5. When we multiply the numerator and denominator of a fraction by the same nonzero number, we are actually multiplying the fraction by  one  .

7. A  common   denominator  is a common multiple of all the denominators for a set of fractions.

9. To list fractions in order of value, first find the  LCD  of the fractions. Then, write each fraction as an equivalent  fraction with this denominator. Then, list the fractions in order by comparing their numerators  .

## Guide Problems

11. Consider the fraction $\dfrac{20}{24}$.

    a. Write the prime factorization of the numerator. Use the methods of Section 2.1 such as factor trees.

$$20 = \underline{2} \cdot \underline{2} \cdot \underline{5}$$

    b. Write the prime factorization of the denominator.

$$24 = \underline{2} \cdot \underline{2} \cdot \underline{2} \cdot \underline{3}$$

    c. Simplify by dividing out the factors common to the numerator and denominator.

$$\frac{20}{24} = \frac{\overset{1}{\cancel{2}} \cdot \overset{1}{\cancel{2}} \cdot 5}{\underset{1}{\cancel{2}} \cdot \underset{1}{\cancel{2}} \cdot 2 \cdot 3} = \frac{5}{2 \cdot 3} = \frac{5}{6}$$

    So $\dfrac{20}{24} = \dfrac{5}{6}$.

13. Consider the fraction $\dfrac{12}{20}$.

    a. What is the greatest common factor (GCF) of the numerator and denominator of $\dfrac{12}{20}$.

    The greatest common factor of 12 and 20 is 4.

    b. Simplify $\dfrac{12}{20}$ by diving out the GCF.

    Divide both the numerator and denominator by the GCF, 4.

$$\frac{12}{20} = \frac{\overset{3}{\cancel{12}}}{\underset{5}{\cancel{20}}} = \frac{3}{5}$$

*Problems 15 through 49 employ the following strategy to reduce a given fraction. If the GCF of the numerator and denominator is obvious, then the GCF method is used. If not, the prime factorization method is applied.*

15. The GCF of 6 and 14 is 2.

$$\frac{6}{14} = \frac{\overset{3}{\cancel{6}}}{\underset{7}{\cancel{14}}} = \frac{3}{7}$$

17. The GCF of 6 and 36 is 6.

$$\frac{6}{36} = \frac{\overset{1}{\cancel{6}}}{\underset{6}{\cancel{36}}} = \frac{1}{6}$$

19. The GCF of 8 and 36 is 4.

$$\frac{8}{36} = \frac{\overset{2}{\cancel{8}}}{\underset{9}{\cancel{36}}} = \frac{2}{9}$$

21. The GCF of 2 and 22 is 2.

$$\frac{2}{22} = \frac{\overset{1}{\cancel{2}}}{\underset{11}{\cancel{22}}} = \frac{1}{11}$$

23. The GCF of 16 and 20 is 4.

$$\frac{16}{20} = \frac{\overset{4}{\cancel{16}}}{\underset{5}{\cancel{20}}} = \frac{4}{5}$$

25. 7 and 19 are both primes and so have no common factors.

    $\dfrac{7}{19}$ is in reduced form.

27. The GCF of 9 and 90 is 9.

$$\frac{9}{90} = \frac{\overset{1}{\cancel{9}}}{\underset{10}{\cancel{90}}} = \frac{1}{10}$$

29. $32 = 2 \cdot 2 \cdot 2 \cdot 2 \cdot 2$
    $56 = 2 \cdot 2 \cdot 2 \cdot 7$

$$\frac{32}{56} = \frac{\cancel{2} \cdot \cancel{2} \cdot \cancel{2} \cdot 2 \cdot 2}{\cancel{2} \cdot \cancel{2} \cdot \cancel{2} \cdot 7} = \frac{2 \cdot 2}{7} = \frac{4}{7}$$

31. $48 = 2 \cdot 2 \cdot 2 \cdot 2 \cdot 3$
    $80 = 2 \cdot 2 \cdot 2 \cdot 2 \cdot 5$

$$\frac{48}{80} = \frac{\cancel{2} \cdot \cancel{2} \cdot \cancel{2} \cdot \cancel{2} \cdot 3}{\cancel{2} \cdot \cancel{2} \cdot \cancel{2} \cdot \cancel{2} \cdot 5} = \frac{3}{5}$$

33. $7 = 7$
$24 = 2 \cdot 2 \cdot 2 \cdot 3$

$\dfrac{7}{24}$ cannot be simplified
further since 7 and 24 have
no factors in common.

35. $9 = 3 \cdot 3$
$75 = 3 \cdot 5 \cdot 5$

$\dfrac{9}{75} = \dfrac{\cancel{3} \cdot 3}{\cancel{3} \cdot 5 \cdot 5} = \dfrac{3}{25}$

further since 7 and 24 have
no factors in common.

37. The GCF of 26 and 40 is 2.

$\dfrac{26}{40} = \dfrac{\overset{13}{\cancel{26}}}{\underset{20}{\cancel{40}}} = \dfrac{13}{20}$

39. $28 = 2 \cdot 2 \cdot 7$
$60 = 2 \cdot 2 \cdot 3 \cdot 5$

$\dfrac{28}{60} = \dfrac{\cancel{2} \cdot \cancel{2} \cdot 7}{\underset{1 \quad 1}{\cancel{2} \cdot \cancel{2} \cdot 3 \cdot 5}}$

$= \dfrac{7}{3 \cdot 5} = \dfrac{7}{15}$

41. 43 and 79 are both prime
numbers. $\dfrac{43}{79}$ cannot be
simplified.

$\dfrac{28}{60} = \dfrac{\cancel{2} \cdot \cancel{2} \cdot 7}{\underset{1 \quad 1}{\cancel{2} \cdot \cancel{2} \cdot 3 \cdot 5}}$

$= \dfrac{7}{3 \cdot 5} = \dfrac{7}{15}$

43. The GCF of 44 and 80 is 4.

$\dfrac{44}{80} = \dfrac{\overset{11}{\cancel{44}}}{\underset{20}{\cancel{80}}} = \dfrac{11}{20}$

45. The GCF of 40 and 78 is 2.

$\dfrac{40}{78} = \dfrac{\overset{20}{\cancel{40}}}{\underset{39}{\cancel{78}}} = \dfrac{20}{39}$

47. The GCF of 62 and 70 is 2.

$\dfrac{62}{70} = \dfrac{\overset{31}{\cancel{62}}}{\underset{35}{\cancel{70}}} = \dfrac{31}{35}$

49. $65 = 5 \cdot 13$
$79 = 79$

$\dfrac{65}{79}$ cannot be simplified.

## Guide Problems

51. Write $\dfrac{2}{3}$ as an equivalent fraction with a
denominator of 30.

Determine the factor that when multiplied
by the given denominator, 3, gives the
desired denominator, 30. Then multiply
both the numerator and denominator by this
factor. In this case, the factor is 10.

$\dfrac{2}{3} = \dfrac{2 \cdot 10}{3 \cdot 10} = \dfrac{20}{30}$

53. Write $\dfrac{5}{6}$ as an equivalent fraction with a
denominator of 24.

Determine the factor that when multiplied
by the given denominator, 6, gives the
desired denominator, 24. Then multiply
both the numerator and denominator by this
factor. In this case, the factor is 4.

$\dfrac{5}{6} = \dfrac{5 \cdot 4}{6 \cdot 4} = \dfrac{20}{24}$

55. $\dfrac{1}{8} = \dfrac{?}{56}$

Since $8 \cdot 7 = 56$,

$\dfrac{1}{8} = \dfrac{1 \cdot 7}{8 \cdot 7} = \dfrac{7}{56}$

57. $\dfrac{5}{8} = \dfrac{?}{64}$

Since $8 \cdot 8 = 64$,

$\dfrac{5}{8} = \dfrac{5 \cdot 8}{8 \cdot 8} = \dfrac{40}{64}$

59. $\dfrac{5}{8} = \dfrac{?}{48}$

Since $8 \cdot 6 = 48$,

$\dfrac{5}{8} = \dfrac{5 \cdot 6}{8 \cdot 6} = \dfrac{30}{48}$

61. $\dfrac{7}{11} = \dfrac{?}{99}$

Since $11 \cdot 9 = 99$,

$$\dfrac{7}{11} = \dfrac{7 \cdot 9}{11 \cdot 9} = \dfrac{63}{99}$$

63. $\dfrac{1}{3} = \dfrac{?}{27}$

Since $3 \cdot 9 = 27$,

$$\dfrac{1}{3} = \dfrac{1 \cdot 9}{3 \cdot 9} = \dfrac{9}{27}$$

65. $\dfrac{1}{13} = \dfrac{?}{78}$

Since $13 \cdot 6 = 78$,

$$\dfrac{1}{13} = \dfrac{1 \cdot 6}{13 \cdot 6} = \dfrac{6}{78}$$

67. $\dfrac{11}{13} = \dfrac{?}{78}$

Since $13 \cdot 6 = 78$,

$$\dfrac{11}{13} = \dfrac{11 \cdot 6}{13 \cdot 6} = \dfrac{66}{78}$$

69. $\dfrac{5}{13} = \dfrac{?}{52}$

Since $13 \cdot 4 = 52$,

$$\dfrac{5}{13} = \dfrac{5 \cdot 4}{13 \cdot 4} = \dfrac{20}{52}$$

71. $\dfrac{9}{14} = \dfrac{?}{98}$

Since $14 \cdot 7 = 98$,

$$\dfrac{9}{14} = \dfrac{9 \cdot 7}{14 \cdot 7} = \dfrac{63}{98}$$

73. $\dfrac{3}{4} = \dfrac{?}{28}$

Since $4 \cdot 7 = 28$,

$$\dfrac{3}{4} = \dfrac{3 \cdot 7}{4 \cdot 7} = \dfrac{21}{28}$$

75. $\dfrac{5}{8} = \dfrac{?}{40}$

Since $8 \cdot 5 = 40$,

$$\dfrac{5}{8} = \dfrac{5 \cdot 5}{8 \cdot 5} = \dfrac{25}{40}$$

77. $\dfrac{2}{13} = \dfrac{?}{65}$

Since $13 \cdot 5 = 65$,

$$\dfrac{2}{13} = \dfrac{2 \cdot 5}{13 \cdot 5} = \dfrac{10}{65}$$

79. $\dfrac{11}{14} = \dfrac{?}{70}$

Since $14 \cdot 5 = 70$,

$$\dfrac{11}{14} = \dfrac{11 \cdot 5}{14 \cdot 5} = \dfrac{55}{70}$$

81. $\dfrac{7}{16} = \dfrac{?}{48}$

Since $16 \cdot 3 = 48$,

$$\dfrac{7}{16} = \dfrac{7 \cdot 3}{16 \cdot 3} = \dfrac{21}{48}$$

83. $\dfrac{16}{17} = \dfrac{?}{34}$

Since $17 \cdot 2 = 34$,

$$\dfrac{16}{17} = \dfrac{16 \cdot 2}{17 \cdot 2} = \dfrac{32}{34}$$

85. $\dfrac{5}{14} = \dfrac{?}{84}$

Since $14 \cdot 6 = 84$,

$$\dfrac{5}{14} = \dfrac{5 \cdot 6}{14 \cdot 6} = \dfrac{30}{84}$$

## Guide Problems

87. List the fractions $\dfrac{3}{5}$ and $\dfrac{5}{9}$ in ascending order. The strategy when ordering or comparing fractions is to first rewrite the fractions as equivalent fractions with the same denominators, the LCD of the fractions. The fractions can then be compared by comparing their numerators.

a. Find the LCD of the fractions. That is, find the LCM of the denominators 5 and 9.

The LCD of $\dfrac{3}{5}$ and $\dfrac{5}{9}$ is <u>45</u>.

b. Write each fraction as an equivalent fraction with the denominator determined in part a.

$$\frac{3}{5} = \frac{3 \cdot \underline{9}}{5 \cdot \underline{9}} = \frac{27}{\underline{45}}$$

$$\frac{5}{9} = \frac{5 \cdot \underline{5}}{9 \cdot \underline{5}} = \frac{25}{\underline{45}}$$

c. Compare the fractions. Since $25 < 27$,

$$\frac{25}{\underline{45}} < \frac{27}{\underline{45}} \text{ or, in lowest terms, as } \frac{5}{9} < \frac{3}{\underline{5}}$$

89. LCD of $\dfrac{7}{10}$ and $\dfrac{5}{8}$ is 40.

$$\frac{7}{10} = \frac{7 \cdot 4}{10 \cdot 4} = \frac{28}{40}$$

$$\frac{5}{8} = \frac{5 \cdot 5}{8 \cdot 5} = \frac{25}{40}$$

So $\dfrac{25}{40} < \dfrac{28}{40}$ or $\dfrac{5}{8} < \dfrac{7}{10}$.

91. LCD of $\dfrac{1}{14}$ and $\dfrac{3}{8}$ is 56.

$$\frac{1}{14} = \frac{1 \cdot 4}{14 \cdot 4} = \frac{4}{56}$$

$$\frac{3}{8} = \frac{3 \cdot 7}{8 \cdot 7} = \frac{21}{56}$$

So $\dfrac{4}{56} < \dfrac{21}{56}$ or $\dfrac{1}{14} < \dfrac{3}{8}$.

93. LCD of $\dfrac{1}{4}$, $\dfrac{1}{6}$ and $\dfrac{1}{2}$ is 12.

$$\frac{1}{4} = \frac{1 \cdot 3}{4 \cdot 3} = \frac{3}{12}$$

$$\frac{1}{6} = \frac{1 \cdot 2}{6 \cdot 2} = \frac{2}{12}$$

$$\frac{1}{2} = \frac{1 \cdot 6}{2 \cdot 6} = \frac{6}{12}$$

So $\dfrac{2}{12} < \dfrac{3}{12} < \dfrac{6}{12}$ or $\dfrac{1}{6} < \dfrac{1}{4} < \dfrac{1}{2}$.

95. LCD of $\dfrac{5}{6}$, $\dfrac{1}{2}$ and $\dfrac{2}{3}$ is 6.

$$\frac{5}{6} = \frac{5}{6}$$

$$\frac{1}{2} = \frac{1 \cdot 3}{2 \cdot 3} = \frac{3}{6}$$

$$\frac{2}{3} = \frac{2 \cdot 2}{3 \cdot 2} = \frac{4}{6}$$

So $\dfrac{3}{6} < \dfrac{4}{6} < \dfrac{5}{6}$ or $\dfrac{1}{2} < \dfrac{2}{3} < \dfrac{5}{6}$.

97.  LCD of $\dfrac{9}{16}$, $\dfrac{8}{32}$ and $\dfrac{5}{24}$ is 96.

$$\frac{9}{16} = \frac{9 \cdot 6}{16 \cdot 6} = \frac{54}{96}$$

$$\frac{8}{32} = \frac{8 \cdot 3}{32 \cdot 3} = \frac{24}{96}$$

$$\frac{5}{24} = \frac{5 \cdot 4}{24 \cdot 4} = \frac{20}{96}$$

So $\dfrac{20}{96} < \dfrac{24}{96} < \dfrac{54}{96}$ or $\dfrac{5}{24} < \dfrac{8}{32} < \dfrac{9}{16}$.

99.  LCD of $\dfrac{7}{12}$, $\dfrac{1}{18}$, $\dfrac{4}{9}$ and $\dfrac{7}{8}$ is 72.

$$\frac{7}{12} = \frac{7 \cdot 6}{12 \cdot 6} = \frac{42}{72}$$

$$\frac{1}{18} = \frac{1 \cdot 4}{18 \cdot 4} = \frac{4}{72}$$

$$\frac{4}{9} = \frac{4 \cdot 8}{9 \cdot 8} = \frac{32}{72}$$

$$\frac{7}{8} = \frac{7 \cdot 9}{8 \cdot 9} = \frac{63}{72}$$

So $\dfrac{4}{72} < \dfrac{32}{72} < \dfrac{42}{72} < \dfrac{63}{72}$ or $\dfrac{1}{18} < \dfrac{4}{9} < \dfrac{7}{12} < \dfrac{7}{8}$.

101.  If 240 of 380 students are females then there are $380 - 240 = 140$ students that are male. The fraction represented by the males in the class is $\dfrac{140}{380}$. To simplify this fraction, note $140 = 2 \cdot 2 \cdot 5 \cdot 7$ and

$380 = 2 \cdot 2 \cdot 5 \cdot 19$ so $\dfrac{140}{380} = \dfrac{\overset{1}{\cancel{2}} \cdot \overset{1}{\cancel{2}} \cdot \overset{1}{\cancel{5}} \cdot 7}{\underset{1}{\cancel{2}} \cdot \underset{1}{\cancel{2}} \cdot \underset{1}{\cancel{5}} \cdot 19} = \dfrac{7}{19}$.

103.  The down payment, \$6000, as a fraction of the total cost, \$16,000, is given by

$$\frac{\$6000}{\$16,000} = \frac{6 \cdot \overset{1}{\cancel{1000}}}{16 \cdot \underset{1}{\cancel{1000}}} = \frac{6}{16} = \frac{2 \cdot 3}{2 \cdot 8} = \frac{3}{8}.$$

105.  Sergio worked 26 out of his usual 40 hours this week. The fraction representing the hours he worked is

$$\frac{26}{40} = \frac{\overset{13}{\cancel{26}}}{\underset{20}{\cancel{40}}} = \frac{13}{20}.$$

107.  The number of cartons that eventually arrived is $20 + 12 = 32$. Out of 40 ordered, this means 8 cartons never arrived. These cartons are represented by the fraction $\dfrac{8}{40} = \dfrac{\overset{1}{\cancel{8}}}{\underset{5}{\cancel{40}}} = \dfrac{1}{5}$.

109.  The company has a total of $8 + 7 + 4 + 2 + 5 + 1 + 1 = 28$ warehouses. Of these $2 + 1 + 1 = 4$ are in Virginia, Georgia, and Florida. The fraction representing these three states is $\dfrac{4}{28} = \dfrac{\overset{1}{\cancel{4}}}{\underset{7}{\cancel{28}}} = \dfrac{1}{7}$.

111. The 2005 sales were \$122 billion dollars. The 2006 sales were \$136 billion. The 2005 sales as a fraction

of the 2006 sales is $\dfrac{\$122\,\text{billion}}{\$136\,\text{billion}} = \dfrac{122}{136} = \dfrac{\cancel{122}^{61}}{\cancel{136}_{68}} = \dfrac{61}{68}$ .

113. Compare $\dfrac{5}{12}$ and $\dfrac{3}{8}$ by converting to equivalent fractions having 24, the LCD, as denominator.

$$\frac{5}{12} = \frac{5\cdot 2}{12\cdot 2} = \frac{10}{24} \qquad \frac{3}{8} = \frac{3\cdot 3}{8\cdot 3} = \frac{9}{24}$$

The smaller fraction is $\dfrac{9}{24}$ so the smaller washer measures $\dfrac{3}{8}$ inches.

115. First convert the weights of the trucks from mixed numbers to improper fractions. Next, convert the improper fractions to equivalent fractions with the same denominators.

$$1\frac{5}{8} = \frac{1\cdot 8 + 5}{8} = \frac{13}{8} = \frac{13\cdot 15}{8\cdot 15} = \frac{195}{120} \qquad 1\frac{2}{3} = \frac{1\cdot 3 + 2}{3} = \frac{5}{3} = \frac{5\cdot 40}{3\cdot 40} = \frac{200}{120}$$

$$1\frac{7}{10} = \frac{1\cdot 10 + 7}{10} = \frac{17}{10} = \frac{17\cdot 12}{10\cdot 12} = \frac{204}{120}$$

Comparing the resulting numerators, we see $\dfrac{204}{120}$ is the largest fraction, so $1\dfrac{7}{10}$ tons is the largest load.

## Cumulative Skills Review

1. $\begin{array}{r} 4 \\ 12\overline{)55} \\ \underline{48} \\ 7 \end{array}$

$\dfrac{55}{12} = 4\dfrac{7}{12}$

3. $\dfrac{32 + 34 + 16 + 62}{4} = \dfrac{144}{4} = 36$

5. For a square of side 12 feet, the area is $12\cdot 12 = 144$ square feet.

7. $4\dfrac{8}{15} = \dfrac{4\cdot 15 + 8}{15} = \dfrac{68}{15}$

9. The plants will be watered after the least common multiple of 2 and 3 days. The multiples of 2 are 2, 4, 6, 8, 10, 12, … while the multiples of 3 are 3, 6, 9, 12,…The smallest common multiple is 6 so both plants will be watered every sixth day.

## Section 2.4    Multiplying Fractions and Mixed Numbers

### Concept Check

1.  To multiply fractions, multiply the __numerators__ to form the new numerator and multiply the __denominators__ to form the new denominator.

3.  To simplify fractions before multiplying, find a __common__ __factor__ that divides evenly into one of the numerators and one of the denominators.  Then, __divide__ the identified numerator and denominator by this common factor.

### Guide Problems

5.  Multiply $\dfrac{3}{5} \cdot \dfrac{7}{8}$. Simplify, if possible.

When multiplying fractions, multiply the numerators to find the numerator of the product and multiply the denominators to find the denominator of the product.

$$\frac{3}{5} \cdot \frac{7}{8} = \frac{21}{40}$$

7.  Multiply $\dfrac{2}{3} \cdot \dfrac{9}{14}$. Simplify before multiplying.

Common factors in the numerators and denominators may be divided out.

$$\frac{2}{3} \cdot \frac{9}{14} = \frac{\overset{1}{\cancel{2}}}{\underset{1}{\cancel{3}}} \cdot \frac{\overset{3}{\cancel{9}}}{\underset{7}{\cancel{14}}} = \frac{1}{1} \cdot \frac{3}{7} = \frac{3}{7}$$

Here the factor of 2 common to 2 and 14 was divided out as was the 3 common to 3 and 9.

*In problems 9 through 47, simplification (if possible) may be done before or after multiplying.  Both methods are used in the solutions below.  A unique simplified answer must follow from either approach.*

9.  $\dfrac{1}{2} \cdot \dfrac{1}{3} = \dfrac{1 \cdot 1}{2 \cdot 3} = \dfrac{1}{6}$

11.  $\dfrac{1}{5} \cdot \dfrac{4}{7} = \dfrac{1 \cdot 4}{5 \cdot 7} = \dfrac{4}{35}$

13.  $\dfrac{2}{3} \cdot \dfrac{4}{5} = \dfrac{2 \cdot 4}{3 \cdot 5} = \dfrac{8}{15}$

15.  $\dfrac{5}{7} \cdot \dfrac{3}{8} = \dfrac{5 \cdot 3}{7 \cdot 8} = \dfrac{15}{56}$

17.  $\dfrac{4}{7} \cdot \dfrac{6}{11} = \dfrac{4 \cdot 6}{7 \cdot 11} = \dfrac{24}{77}$

19.  $\dfrac{7}{8} \cdot \dfrac{3}{10} = \dfrac{7 \cdot 3}{8 \cdot 10} = \dfrac{21}{80}$

21.  $\dfrac{1}{2} \cdot \dfrac{2}{3} = \dfrac{1 \cdot 2}{2 \cdot 3} = \dfrac{2}{6} = \dfrac{\overset{1}{\cancel{2}}}{\underset{3}{\cancel{6}}} = \dfrac{1}{3}$

23.  $\dfrac{2}{11} \cdot \dfrac{11}{25} = \dfrac{2}{\underset{1}{\cancel{11}}} \cdot \dfrac{\overset{1}{\cancel{11}}}{25} = \dfrac{2 \cdot 1}{1 \cdot 25} = \dfrac{2}{25}$

25.  $\dfrac{3}{6} \cdot \dfrac{4}{11} = \dfrac{\cancel{3}}{\cancel{6}} \cdot \dfrac{\cancel{4}}{11} = \dfrac{1 \cdot 2}{1 \cdot 11} = \dfrac{2}{11}$

27.  $\dfrac{5}{12} \cdot \dfrac{8}{9} = \dfrac{5}{\cancel{12}} \cdot \dfrac{\cancel{8}}{9} = \dfrac{5 \cdot 2}{3 \cdot 9} = \dfrac{10}{27}$

29.  $\dfrac{3}{8} \cdot \dfrac{4}{5} = \dfrac{3}{\cancel{8}} \cdot \dfrac{\cancel{4}}{5} = \dfrac{3 \cdot 1}{2 \cdot 5} = \dfrac{3}{10}$

31.  $\dfrac{4}{9} \cdot \dfrac{5}{12} = \dfrac{\cancel{4}}{9} \cdot \dfrac{5}{\cancel{12}} = \dfrac{1 \cdot 5}{9 \cdot 3} = \dfrac{5}{27}$

33. $\dfrac{4}{3} \cdot \dfrac{3}{8} = \dfrac{\cancel{4}^{\,1}}{\cancel{3}_{\,1}} \cdot \dfrac{\cancel{3}^{\,1}}{\cancel{8}_{\,2}} = \dfrac{1 \cdot 1}{1 \cdot 2} = \dfrac{1}{2}$

35. $\dfrac{5}{14} \cdot \dfrac{14}{25} = \dfrac{\cancel{5}^{\,1}}{\cancel{14}_{\,1}} \cdot \dfrac{\cancel{14}^{\,1}}{\cancel{25}_{\,5}} = \dfrac{1 \cdot 1}{1 \cdot 5} = \dfrac{1}{5}$

37. $\dfrac{5}{12} \cdot \dfrac{9}{10} = \dfrac{\cancel{5}^{\,1}}{\cancel{12}_{\,4}} \cdot \dfrac{\cancel{9}^{\,3}}{\cancel{10}_{\,2}} = \dfrac{1 \cdot 3}{4 \cdot 2} = \dfrac{3}{8}$

39. $\dfrac{5}{21} \cdot \dfrac{14}{25} = \dfrac{\cancel{5}^{\,1}}{\cancel{21}_{\,3}} \cdot \dfrac{\cancel{14}^{\,2}}{\cancel{25}_{\,5}} = \dfrac{1 \cdot 2}{3 \cdot 5} = \dfrac{2}{15}$

41. $\dfrac{5}{9} \cdot \dfrac{9}{16} \cdot \dfrac{7}{5} = \dfrac{\cancel{5}^{\,1}}{\cancel{9}_{\,1}} \cdot \dfrac{\cancel{9}^{\,1}}{16} \cdot \dfrac{7}{\cancel{5}_{\,1}} = \dfrac{1 \cdot 1 \cdot 7}{1 \cdot 16 \cdot 1} = \dfrac{7}{16}$

43. $\dfrac{8}{13} \cdot \dfrac{1}{4} \cdot \dfrac{1}{3} = \dfrac{\cancel{8}^{\,2}}{13} \cdot \dfrac{1}{\cancel{4}_{\,1}} \cdot \dfrac{1}{3} = \dfrac{2 \cdot 1 \cdot 1}{13 \cdot 1 \cdot 3} = \dfrac{2}{39}$

45. $\dfrac{3}{10} \cdot \dfrac{12}{16} \cdot \dfrac{4}{15} = \dfrac{\cancel{3}^{\,1}}{10} \cdot \dfrac{\cancel{12}^{\,3}}{\cancel{16}_{\,4\,1}} \cdot \dfrac{\cancel{4}^{\,1}}{\cancel{15}_{\,5}} = \dfrac{1 \cdot 3 \cdot 1}{10 \cdot 1 \cdot 5} = \dfrac{3}{50}$

47. $\dfrac{5}{8} \cdot \dfrac{2}{3} \cdot \dfrac{4}{9} \cdot \dfrac{7}{20} = \dfrac{\cancel{5}^{\,1}}{\cancel{8}_{\,4\,1}} \cdot \dfrac{\cancel{2}^{\,1}}{3} \cdot \dfrac{\cancel{4}^{\,1}}{9} \cdot \dfrac{7}{\cancel{20}_{\,4}}$

$= \dfrac{7}{3 \cdot 9 \cdot 4} = \dfrac{7}{108}$

## Guide Problems

49. Multiply $6 \cdot \dfrac{2}{3}$.

a. Change each whole number or mixed number to an improper fraction.

$$6 = \dfrac{6}{1}$$

b. Multiply the fractions.  Simplify, if possible. Express your answer as a whole number or mixed number, if possible.

$$6 \cdot \dfrac{2}{3} = \dfrac{6}{1} \cdot \dfrac{2}{3} = \dfrac{\cancel{6}^{\,2}}{1} \cdot \dfrac{2}{\cancel{3}_{\,1}} = \dfrac{2}{1} \cdot \dfrac{2}{1} = \dfrac{4}{1} = 4$$

51. $2\dfrac{2}{3} = \dfrac{2 \cdot 3 + 2}{3} = \dfrac{8}{3}, \quad 2\dfrac{1}{4} = \dfrac{2 \cdot 4 + 1}{4} = \dfrac{9}{4}$

$2\dfrac{2}{3} \cdot 2\dfrac{1}{4} = \dfrac{8}{3} \cdot \dfrac{9}{4} = \dfrac{\cancel{8}^{\,2}}{\cancel{3}_{\,1}} \cdot \dfrac{\cancel{9}^{\,3}}{\cancel{4}_{\,1}} = \dfrac{2 \cdot 3}{1 \cdot 1} = 6$

53. $7\dfrac{1}{2} = \dfrac{7 \cdot 2 + 1}{2} = \dfrac{15}{2}, \quad 3\dfrac{1}{5} = \dfrac{3 \cdot 5 + 1}{5} = \dfrac{16}{5}$

$7\dfrac{1}{2} \cdot 3\dfrac{1}{5} = \dfrac{15}{2} \cdot \dfrac{16}{5} = \dfrac{\cancel{15}^{\,3}}{\cancel{2}_{\,1}} \cdot \dfrac{\cancel{16}^{\,8}}{\cancel{5}_{\,1}} = \dfrac{3 \cdot 8}{1 \cdot 1} = 24$

55. $\dfrac{1}{2}, \quad \dfrac{1}{3}, \quad 6 = \dfrac{6}{1}$

$\dfrac{1}{2} \cdot \dfrac{1}{3} \cdot 6 = \dfrac{1}{\cancel{2}_{\,1}} \cdot \dfrac{1}{\cancel{3}_{\,1}} \cdot \dfrac{\cancel{6}^{\,1}}{1} = 1$

57. $2\dfrac{1}{2} = \dfrac{5}{2}, \quad 16\dfrac{2}{3} = \dfrac{50}{3}$

$2\dfrac{1}{2} \cdot 16\dfrac{2}{3} = \dfrac{5}{2} \cdot \dfrac{50}{3} = \dfrac{5}{\cancel{2}_{\,1}} \cdot \dfrac{\cancel{50}^{\,25}}{3} = \dfrac{125}{3} = 41\dfrac{2}{3}$

59. $\dfrac{3}{4}, \quad 4\dfrac{2}{3} = \dfrac{14}{3}$

$\dfrac{3}{4} \cdot 4\dfrac{2}{3} = \dfrac{3}{4} \cdot \dfrac{14}{3} = \dfrac{\cancel{3}}{\cancel{4}} \cdot \dfrac{\cancel{14}}{\cancel{3}} = \dfrac{7}{2} = 3\dfrac{1}{2}$

61. $\dfrac{1}{2}, \quad 18 = \dfrac{18}{1}$

$\dfrac{1}{2} \cdot 18 = \dfrac{1}{2} \cdot \dfrac{18}{1} = \dfrac{1}{\cancel{2}} \cdot \dfrac{\cancel{18}}{1} = 9$

63. $7 = \dfrac{7}{1}, \quad \dfrac{9}{14}$

$7 \cdot \dfrac{9}{14} = \dfrac{7}{1} \cdot \dfrac{9}{14} = \dfrac{\cancel{7}}{1} \cdot \dfrac{9}{\cancel{14}} = \dfrac{9}{2} = 4\dfrac{1}{2}$

65. $\dfrac{4}{7}, \quad 3\dfrac{1}{4} = \dfrac{13}{4}$

$\dfrac{4}{7} \cdot 3\dfrac{1}{4} = \dfrac{4}{7} \cdot \dfrac{13}{4} = \dfrac{\cancel{4}}{7} \cdot \dfrac{13}{\cancel{4}} = \dfrac{13}{7} = 1\dfrac{6}{7}$

67. $5\dfrac{3}{4} = \dfrac{23}{4}, \quad 7\dfrac{2}{3} = \dfrac{23}{3}$

$5\dfrac{3}{4} \cdot 7\dfrac{2}{3} = \dfrac{23}{4} \cdot \dfrac{23}{3} = \dfrac{529}{12} = 44\dfrac{1}{12}$

69. $2\dfrac{3}{4} = \dfrac{11}{4}, \quad 1\dfrac{1}{3} = \dfrac{4}{3}$

$2\dfrac{3}{4} \cdot 1\dfrac{1}{3} = \dfrac{11}{4} \cdot \dfrac{4}{3} = \dfrac{11}{\cancel{4}} \cdot \dfrac{\cancel{4}}{3} = \dfrac{11}{3} = 3\dfrac{2}{3}$

71. $4\dfrac{1}{3} = \dfrac{13}{3}, \quad 2 = \dfrac{2}{1}, \quad 3\dfrac{1}{2} = \dfrac{7}{2}$

$4\dfrac{1}{3} \cdot 2 \cdot 3\dfrac{1}{2} = \dfrac{13}{3} \cdot \dfrac{2}{1} \cdot \dfrac{7}{2} = \dfrac{13}{3} \cdot \dfrac{\cancel{2}}{1} \cdot \dfrac{7}{\cancel{2}} = \dfrac{91}{3} = 30\dfrac{1}{3}$

73. $7 = \dfrac{7}{1}, \quad \dfrac{1}{4}, \quad \dfrac{8}{21}$

$7 \cdot \dfrac{1}{4} \cdot \dfrac{8}{21} = \dfrac{7}{1} \cdot \dfrac{1}{4} \cdot \dfrac{8}{21} = \dfrac{\cancel{7}}{1} \cdot \dfrac{1}{\cancel{4}} \cdot \dfrac{\cancel{8}}{\cancel{21}} = \dfrac{2}{3}$

75. If $\dfrac{3}{4}$ of 8500 students are from out of the state, the number of students from out of state is

$\dfrac{3}{4} \cdot 8500 = \dfrac{3}{4} \cdot \dfrac{8500}{1} = \dfrac{3}{\cancel{4}} \cdot \dfrac{\overset{2125}{\cancel{8500}}}{1} = 3 \cdot 2125 = 6375$

77. You earned \$12,300 of which $\dfrac{1}{5}$ was withheld. Your total withholdings were

$\dfrac{1}{5} \cdot 12{,}300 = \dfrac{1}{5} \cdot \dfrac{12{,}300}{1} = \dfrac{1}{\cancel{5}} \cdot \dfrac{\overset{2460}{\cancel{12{,}300}}}{1} = 2460 \text{ or } \$2460. \quad \text{You were left with } \$12{,}300 - \$2460 = \$9840.$

79. You will need $6 \cdot 2\dfrac{2}{3}$ cups of flour or $6 \cdot 2\dfrac{2}{3} = \dfrac{6}{1} \cdot \dfrac{8}{3} = \dfrac{\overset{2}{\cancel{6}}}{1} \cdot \dfrac{8}{\cancel{3}} = 2 \cdot 8 = 16$ cups.

81. $\dfrac{3}{16}$ of 6000 is $\dfrac{3}{16} \cdot 6000 = \dfrac{3}{16} \cdot \dfrac{6000}{1} = \dfrac{3}{\cancel{16}} \cdot \dfrac{\overset{375}{\cancel{6000}}}{1} = 3 \cdot 375 = 1125$ so 1125 responded positively.

83. A distance of 1000 miles consists of $\dfrac{1000}{125}$ sections of length 125 miles. Each of these sections requires

$\dfrac{5}{8}$ inches on the map. The total length needed on the map is $\dfrac{1000}{125} \cdot \dfrac{5}{8} = \dfrac{\overset{1}{\cancel{1000}}}{\underset{1}{\cancel{125}}} \cdot \dfrac{5}{\underset{1}{\cancel{8}}} = 5$ inches.

85. The area of Clarissa's flower garden is $5\dfrac{2}{3} \cdot 6\dfrac{3}{4} = \dfrac{17}{3} \cdot \dfrac{27}{4} = \dfrac{17}{\underset{1}{\cancel{3}}} \cdot \dfrac{\overset{9}{\cancel{27}}}{4} = \dfrac{153}{4} = 38\dfrac{1}{4}$ square feet.

87. The new warehouse will have $3\dfrac{5}{16}$ times the current 3600 square feet. Calculating the product gives

$3\dfrac{5}{16} \cdot 3600 = \dfrac{53}{16} \cdot \dfrac{3600}{1} = \dfrac{53}{\underset{1}{\cancel{16}}} \cdot \dfrac{\overset{225}{\cancel{3600}}}{1} = 53 \cdot 225 = 11,925$. The new space occupies 11,925 square feet.

89. The amount of Mario's wages that are dues is $\dfrac{1}{25} \cdot 825 = \dfrac{1}{25} \cdot \dfrac{825}{1} = \dfrac{1}{\underset{1}{\cancel{25}}} \cdot \dfrac{\overset{33}{\cancel{825}}}{1} = 33$ or \$33.

91. The total cost is the cost per pound times the number of pounds. The cost of $2\dfrac{2}{3}$ pounds at \$6.00 per

pound is found by computing $6 \cdot 2\dfrac{2}{3} = \dfrac{6}{1} \cdot \dfrac{8}{3} = \dfrac{\overset{2}{\cancel{6}}}{1} \cdot \dfrac{8}{\underset{1}{\cancel{3}}} = 2 \cdot 8 = 16$. The total cost is \$16.

93. Each person receives $\dfrac{1}{65} \cdot 260,000 = \dfrac{1}{65} \cdot \dfrac{260,000}{1} = \dfrac{1}{\underset{1}{\cancel{65}}} \cdot \dfrac{\overset{4000}{\cancel{260,000}}}{1} = 4000$ or \$4000.

95. In a 10-hour shift, Freddie makes $10 \cdot 6\dfrac{4}{5} = \dfrac{10}{1} \cdot \dfrac{34}{5} = \dfrac{\overset{2}{\cancel{10}}}{1} \cdot \dfrac{34}{\underset{1}{\cancel{5}}} = 68$ chairs. In the same amount of time

Warren makes $10 \cdot 4\dfrac{1}{2} = \dfrac{10}{1} \cdot \dfrac{9}{2} = \dfrac{\overset{5}{\cancel{10}}}{1} \cdot \dfrac{9}{\underset{1}{\cancel{2}}} = 45$. Freddie makes $68 - 45 = 23$ more chairs than Warren.

97. a. You have had $\dfrac{5}{8} \cdot 1300 = \dfrac{5}{\underset{2}{\cancel{8}}} \cdot \dfrac{\overset{325}{\cancel{1300}}}{1} = \dfrac{1625}{2} = 812\dfrac{1}{2}$ mg of calcium.

b. To consume the recommended dosage you need $1300 - 812\dfrac{1}{2} = \dfrac{2600}{2} - \dfrac{1625}{2} = \dfrac{975}{2} = 487\dfrac{1}{2}$ mg more.

c. The fraction representing the calcium you need to consume is $\dfrac{3}{8}$.

## Cumulative Skills Review

1.
$$11\overline{)1716}$$
$$\phantom{11)}156$$
$$\phantom{1)}\underline{11}$$
$$\phantom{1)}61$$
$$\phantom{1)}\underline{55}$$
$$\phantom{1)}66$$
$$\phantom{1)}\underline{66}$$
$$\phantom{1)}0$$

$1716 \div 11 = 156$

3.  There are two whole triangles shaded.  One of two parts is shaded in the third.

The shaded area is represented by $2\dfrac{1}{2}$.

5.
$$1318$$
$$\underline{\phantom{0}612}$$
$$2636$$
$$1318$$
$$\underline{7908\phantom{0}}$$
$$806616$$

$1318 \cdot 612 = 806,616$

7.  $\dfrac{24}{56} = \dfrac{2 \cdot 2 \cdot 2 \cdot 3}{2 \cdot 2 \cdot 2 \cdot 7} = \dfrac{\overset{1}{\cancel{2}} \cdot \overset{1}{\cancel{2}} \cdot \overset{1}{\cancel{2}} \cdot 3}{\underset{1}{\cancel{2}} \cdot \underset{1}{\cancel{2}} \cdot \underset{1}{\cancel{2}} \cdot 7} = \dfrac{3}{7}$

9.  $8\dfrac{2}{5} = \dfrac{8 \cdot 5 + 2}{5} = \dfrac{42}{5}$

## Section 2.5    Dividing Fractions and Mixed Numbers

### Concept Check

1.  In the division problem $\dfrac{a}{b} \div \dfrac{c}{d}$, the fraction $\dfrac{a}{b}$ is called the __dividend__ and the fraction $\dfrac{c}{d}$ is called the __divisor__ .

3.  To divide fractions, __multiply__ the dividend by the __reciprocal__ of the divisor and simplify if possible.

### Guide Problems

5.  Divide $\dfrac{3}{5} \div \dfrac{9}{11}$.

a. Identify the reciprocal of the divisor.
The divisor here is $\dfrac{9}{11}$.  The reciprocal of $\dfrac{9}{11}$ is $\dfrac{11}{9}$.

b. Rewrite the division problem as a multiplication problem.  Replace the operation of division with multiplication and replace the divisor with its reciprocal.

$$\dfrac{3}{5} \div \dfrac{9}{11} = \dfrac{3}{5} \cdot \dfrac{11}{9}$$

c. Multiply the dividend by the reciprocal of the divisor. Multiply using the methods of Section 2.4.

$$\frac{3}{5} \cdot \frac{11}{9} = \frac{\cancel{3}^{1}}{5} \cdot \frac{11}{\cancel{9}_{3}} = \frac{1}{5} \cdot \frac{11}{3} = \frac{11}{15}$$

so $\dfrac{3}{5} \div \dfrac{9}{11} = \dfrac{11}{15}$.

7. The divisor is $\dfrac{3}{4}$ and its reciprocal is $\dfrac{4}{3}$.

Rewrite as a multiplication problem and evaluate.

$$\frac{9}{16} \div \frac{3}{4} = \frac{9}{16} \cdot \frac{4}{3} = \frac{\cancel{9}^{3}}{\cancel{16}_{4}} \cdot \frac{\cancel{4}^{1}}{\cancel{3}^{1}} = \frac{3}{4}$$

9. The divisor is $\dfrac{2}{7}$ and its reciprocal is $\dfrac{7}{2}$.

Rewrite as a multiplication problem and evaluate.

$$\frac{2}{3} \div \frac{2}{7} = \frac{2}{3} \cdot \frac{7}{2} = \frac{\cancel{2}^{1}}{3} \cdot \frac{7}{\cancel{2}_{1}} = \frac{7}{3} = 2\frac{1}{3}$$

11. $\dfrac{7}{14} \div \dfrac{1}{7} = \dfrac{7}{14} \cdot \dfrac{7}{1} = \dfrac{\cancel{7}^{1}}{\cancel{14}_{2}} \cdot \dfrac{7}{1} = \dfrac{7}{2} = 3\dfrac{1}{2}$

13. $\dfrac{1}{3} \div \dfrac{1}{6} = \dfrac{1}{3} \cdot \dfrac{6}{1} = \dfrac{1}{\cancel{3}_{1}} \cdot \dfrac{\cancel{6}^{2}}{1} = \dfrac{2}{1} = 2$

15. $\dfrac{4}{11} \div \dfrac{5}{11} = \dfrac{4}{11} \cdot \dfrac{11}{5} = \dfrac{4}{\cancel{11}_{1}} \cdot \dfrac{\cancel{11}^{1}}{5} = \dfrac{4}{5}$

17. $\dfrac{3}{5} \div \dfrac{2}{3} = \dfrac{3}{5} \cdot \dfrac{3}{2} = \dfrac{9}{10}$

19. $\dfrac{3}{4} \div \dfrac{1}{8} = \dfrac{3}{4} \cdot \dfrac{8}{1} = \dfrac{3}{\cancel{4}_{1}} \cdot \dfrac{\cancel{8}^{2}}{1} = \dfrac{6}{1} = 6$

21. $\dfrac{5}{12} \div \dfrac{25}{36} = \dfrac{5}{12} \cdot \dfrac{36}{25} = \dfrac{\cancel{5}^{1}}{\cancel{12}_{1}} \cdot \dfrac{\cancel{36}^{3}}{\cancel{25}_{5}} = \dfrac{3}{5}$

23. $\dfrac{1}{3} \div \dfrac{1}{9} = \dfrac{1}{3} \cdot \dfrac{9}{1} = \dfrac{1}{\cancel{3}_{1}} \cdot \dfrac{\cancel{9}^{3}}{1} = \dfrac{3}{1} = 3$

25. $\dfrac{1}{10} \div \dfrac{6}{11} = \dfrac{1}{10} \cdot \dfrac{11}{6} = \dfrac{11}{60}$

27. $\dfrac{8}{9} \div \dfrac{4}{5} = \dfrac{8}{9} \cdot \dfrac{5}{4} = \dfrac{\cancel{8}^{2}}{9} \cdot \dfrac{5}{\cancel{4}_{1}} = \dfrac{10}{9} = 1\dfrac{1}{9}$

29. $\dfrac{4}{7} \div \dfrac{4}{7} = \dfrac{4}{7} \cdot \dfrac{7}{4} = \dfrac{\cancel{4}^{1}}{\cancel{7}_{1}} \cdot \dfrac{\cancel{7}^{1}}{\cancel{4}_{1}} = \dfrac{1}{1} = 1$

31. $\dfrac{6}{7} \div \dfrac{3}{5} = \dfrac{6}{7} \cdot \dfrac{5}{3} = \dfrac{\cancel{6}^{2}}{7} \cdot \dfrac{5}{\cancel{3}_{1}} = \dfrac{10}{7} = 1\dfrac{3}{7}$

33. $\dfrac{7}{8} \div \dfrac{3}{4} = \dfrac{7}{8} \cdot \dfrac{4}{3} = \dfrac{7}{\cancel{8}_{2}} \cdot \dfrac{\cancel{4}^{1}}{3} = \dfrac{7}{6} = 1\dfrac{1}{6}$

## Guide Problems

35. Divide $\dfrac{3}{4} \div 9$ .

    a. Change each whole number or mixed number to an improper fraction. Here the whole number 9 appears as the divisor.

$$9 = \frac{9}{1}$$

    b. Identify the reciprocal of the divisor.

The divisor has been rewritten as $\dfrac{9}{1}$ .

The reciprocal of $\dfrac{9}{1}$ is $\dfrac{1}{9}$ .

    c. Rewrite the division problem as a multiplication problem. Replace the operation of division with multiplication and replace the divisor with its reciprocal.

$$\frac{3}{4} \div \frac{9}{1} = \frac{3}{4} \cdot \frac{1}{9}$$

    d. Multiply the dividend by the reciprocal of the divisor. Multiply using the methods of Section 2.4.

$$\frac{3}{4} \cdot \frac{1}{9} = \frac{\overset{1}{\cancel{3}}}{4} \cdot \frac{1}{\underset{3}{\cancel{9}}} = \frac{1}{4} \cdot \frac{1}{3} = \frac{1}{12}$$

so $\dfrac{3}{4} \div 9 = \dfrac{1}{12}$ .

37. Rewrite mixed numbers as improper fractions.

$$5\frac{3}{4} = \frac{5 \cdot 4 + 3}{4} = \frac{23}{4}, \quad 1\frac{1}{2} = \frac{1 \cdot 2 + 1}{2} = \frac{3}{2}$$

$$5\frac{3}{4} \div 1\frac{1}{2} = \frac{23}{4} \div \frac{3}{2}$$

Rewrite the division problem as a multiplication problem using the reciprocal of the divisor.

$$5\frac{3}{4} \div 1\frac{1}{2} = \frac{23}{4} \div \frac{3}{2} = \frac{23}{4} \cdot \frac{2}{3} = \frac{23}{\underset{2}{\cancel{4}}} \cdot \frac{\overset{1}{\cancel{2}}}{3}$$

$$= \frac{23}{2} \cdot \frac{1}{3} = \frac{23}{6} = 3\frac{5}{6}$$

39. $5\dfrac{3}{8} \div 2\dfrac{3}{4} = \dfrac{43}{8} \div \dfrac{11}{4} = \dfrac{43}{8} \cdot \dfrac{4}{11}$

$$= \frac{43}{\underset{2}{\cancel{8}}} \cdot \frac{\overset{1}{\cancel{4}}}{11} = \frac{43}{2} \cdot \frac{1}{11}$$

$$= \frac{43}{22} = 1\frac{21}{22}$$

41. $1\dfrac{3}{4} \div \dfrac{1}{2} = \dfrac{7}{4} \div \dfrac{1}{2} = \dfrac{7}{4} \cdot \dfrac{2}{1}$

$$= \frac{7}{\underset{2}{\cancel{4}}} \cdot \frac{\overset{1}{\cancel{2}}}{1} = \frac{7}{2} \cdot \frac{1}{1}$$

$$= \frac{7}{2} = 3\frac{1}{2}$$

43. $\dfrac{2}{3} \div 2\dfrac{1}{3} = \dfrac{2}{3} \div \dfrac{7}{3} = \dfrac{2}{3} \cdot \dfrac{3}{7}$

$= \dfrac{2}{\cancel{3}_1} \cdot \dfrac{\cancel{3}^1}{7} = \dfrac{2}{1} \cdot \dfrac{1}{7} = \dfrac{2}{7}$

45. $112 \div 2\dfrac{1}{3} = \dfrac{112}{1} \div \dfrac{7}{3} = \dfrac{112}{1} \cdot \dfrac{3}{7}$

$= \dfrac{\cancel{112}^{16}}{1} \cdot \dfrac{3}{\cancel{7}_1} = \dfrac{16}{1} \cdot \dfrac{3}{1}$

$= \dfrac{48}{1} = 48$

47. $22\dfrac{1}{2} \div 2\dfrac{1}{4} = \dfrac{45}{2} \div \dfrac{9}{4} = \dfrac{45}{2} \cdot \dfrac{4}{9}$

$= \dfrac{\cancel{45}^5}{\cancel{2}_1} \cdot \dfrac{\cancel{4}^2}{\cancel{9}_1} = \dfrac{5}{1} \cdot \dfrac{2}{1}$

$= \dfrac{10}{1} = 10$

49. $3\dfrac{3}{5} \div 2 = \dfrac{18}{5} \div \dfrac{2}{1} = \dfrac{18}{5} \cdot \dfrac{1}{2}$

$= \dfrac{\cancel{18}^9}{5} \cdot \dfrac{1}{\cancel{2}_1} = \dfrac{9}{5} \cdot \dfrac{1}{1}$

$= \dfrac{9}{5} = 1\dfrac{4}{5}$

51. $71\dfrac{2}{3} \div 1\dfrac{2}{3} = \dfrac{215}{3} \div \dfrac{5}{3} = \dfrac{215}{3} \cdot \dfrac{3}{5}$

$= \dfrac{\cancel{215}^{43}}{\cancel{3}_1} \cdot \dfrac{\cancel{3}^1}{\cancel{5}_1} = \dfrac{43}{1} \cdot \dfrac{1}{1}$

$= \dfrac{43}{1} = 43$

53. $2\dfrac{2}{3} \div 3\dfrac{1}{2} = \dfrac{8}{3} \div \dfrac{7}{2} = \dfrac{8}{3} \cdot \dfrac{2}{7} = \dfrac{16}{21}$

55. $51 \div 1\dfrac{1}{2} = \dfrac{51}{1} \div \dfrac{3}{2} = \dfrac{51}{1} \cdot \dfrac{2}{3}$

$= \dfrac{\cancel{51}^{17}}{1} \cdot \dfrac{2}{\cancel{3}_1} = \dfrac{17}{1} \cdot \dfrac{2}{1}$

$= \dfrac{34}{1} = 34$

57. $49 \div 1\dfrac{3}{4} = \dfrac{49}{1} \div \dfrac{7}{4} = \dfrac{49}{1} \cdot \dfrac{4}{7}$

$= \dfrac{\cancel{49}^7}{1} \cdot \dfrac{4}{\cancel{7}_1} = \dfrac{7}{1} \cdot \dfrac{1}{1}$

$= \dfrac{28}{1} = 28$

59. $2 \div \dfrac{1}{6} = \dfrac{2}{1} \div \dfrac{1}{6} = \dfrac{2}{1} \cdot \dfrac{6}{1} = \dfrac{12}{1} = 12$

61. a. A vehicle's average miles per gallon (mpg) is determined by dividing the total miles traveled by the total number of gallons of gas required to cover those miles.  In this case, the mpg is

$$730 \div 33\dfrac{1}{3} = \dfrac{730}{1} \div \dfrac{100}{3} = \dfrac{730}{1} \cdot \dfrac{3}{100} = \dfrac{\cancel{730}^{73}}{1} \cdot \dfrac{3}{\cancel{100}_{10}} = \dfrac{219}{10} = 21\dfrac{9}{10} \text{ miles per gallon.}$$

b. From part (a), each gallon of gas will allow you to drive $21\dfrac{9}{10}$ miles.  To travel 1095 miles, you need

$$1095 \div 21\dfrac{9}{10} = \dfrac{1095}{1} \div \dfrac{219}{10} = \dfrac{1095}{1} \cdot \dfrac{10}{219} = \dfrac{\cancel{1095}^5}{1} \cdot \dfrac{10}{\cancel{219}_1} = 50 \text{ gallons.}$$

63. The number of acres in each plot will be $126\frac{7}{8} \div 5 = \frac{1015}{8} \div \frac{5}{1} = \frac{1015}{8} \cdot \frac{1}{5} = \frac{\overset{203}{\cancel{1015}}}{8} \cdot \frac{1}{\cancel{5}} = \frac{203}{8} = 25\frac{3}{8}$.

65. The total number of storage bins possible is

$$19,667 \div 35\frac{1}{2} = \frac{19,667}{1} \div \frac{71}{2} = \frac{19,667}{1} \cdot \frac{2}{71} = \frac{\overset{277}{\cancel{19,667}}}{1} \cdot \frac{2}{\cancel{71}} = \frac{277}{1} \cdot \frac{2}{1} = 554$$

67. Since the individual portions are measured in ounces, first convert $131\frac{1}{4}$ pounds to ounces. There are 16 ounces in a pound so the total amount of sirloin steak on hand is

$16 \cdot 131\frac{1}{4} = \frac{16}{1} \cdot \frac{525}{4} = \frac{\overset{4}{\cancel{16}}}{1} \cdot \frac{525}{\cancel{4}} = \frac{4}{1} \cdot \frac{525}{1} = 2100$ ounces. The number of $10\frac{1}{2}$ ounce portions that can be

made from 2100 ounces is $2100 \div 10\frac{1}{2} = \frac{2100}{1} \div \frac{21}{2} = \frac{\overset{100}{\cancel{2100}}}{1} \cdot \frac{2}{\cancel{21}} = \frac{100}{1} \cdot \frac{2}{1} = 200$. The steakhouse can

serve 200 sirloin steak dinners on Saturday.

69. We need to know the number of $\frac{7}{8}$ mile long segments make up a $38\frac{1}{2}$ mile stretch. This number is

$38\frac{1}{2} \div \frac{7}{8} = \frac{77}{2} \div \frac{7}{8} = \frac{77}{2} \cdot \frac{8}{7} = \frac{\overset{11}{\cancel{77}}}{\underset{1}{\cancel{2}}} \cdot \frac{\overset{4}{\cancel{8}}}{\underset{1}{\cancel{7}}} = 11 \cdot 4 = 44$ so 44 speed limit signs will be needed.

71. Each loaf of bread requires $12\frac{3}{4}$ ounces of flour. Since 7650 ounces of flour were used in total, we must

determine how many measurements of $12\frac{3}{4}$ are in 7650 or $7650 \div 12\frac{3}{4}$.

$$7650 \div 12\frac{3}{4} = \frac{7650}{1} \div \frac{51}{4} = \frac{7650}{1} \cdot \frac{4}{51} = \frac{\overset{150}{\cancel{7650}}}{1} \cdot \frac{4}{\cancel{51}} = 150 \cdot 4 = 600$$

The bakers made 600 loaves of bread.

## Cumulative Skills Review

1.
$$
\begin{array}{r}
192 \\
\times\ 102 \\
\hline
384 \\
1920 \\
\hline
19584
\end{array}
$$

$192 \cdot 102 = 19,584$

3. $\frac{25}{81} \cdot \frac{3}{75} = \frac{\overset{1}{\cancel{25}}}{\underset{27}{\cancel{81}}} \cdot \frac{\overset{1}{\cancel{3}}}{\underset{3}{\cancel{75}}} = \frac{1}{27} \cdot \frac{1}{3} = \frac{1}{81}$

5.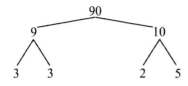

7. $\dfrac{6}{7}\cdot\dfrac{2}{3}\cdot 1\dfrac{1}{2}=\dfrac{6}{7}\cdot\dfrac{2}{3}\cdot\dfrac{3}{2}=\dfrac{6}{7}\cdot\dfrac{\cancel{2}}{\cancel{3}}\cdot\dfrac{\cancel{3}}{\cancel{2}}=\dfrac{6}{7}\cdot\dfrac{1}{1}\cdot\dfrac{1}{1}=\dfrac{6}{7}$

9. $5\cdot 3\dfrac{2}{3}=\dfrac{5}{1}\cdot\dfrac{11}{3}=\dfrac{55}{3}=18\dfrac{1}{3}$

You will need $18\dfrac{1}{3}$ cups of spaghetti.

## Section 2.6    Adding Fractions and Mixed Numbers

### Concept Check

1.  Fractions with the same denominator are known as __like__ fractions.

3.  When adding fractions with different denominators, we must find the __LCD__ of the fractions, write each fraction as an __equivalent__ fraction with this denominator, and then add the resulting like fractions.

5.  When the sum of two mixed numbers contains an improper fraction, we must change the improper fraction to a __mixed number__ and then add the whole number to it.

### Guide Problems

7.  Add $\dfrac{4}{9}+\dfrac{1}{9}$.

The fractions have the same denominators and so are like fractions. To add like fractions, add the numerators and use the denominator 9 as the denominator of the result.

$\dfrac{4}{9}+\dfrac{1}{9}=\dfrac{4+1}{9}=\dfrac{5}{9}$. Note no simplification is possible.

*In each of problems 9–23, the fractions being added are like fractions. The denominator of the sum is the same as the denominator of each fraction. The numerator of the sum is found by adding the numerators of the fraction. The resulting fraction is then simplified if possible.*

9.  $\dfrac{1}{7}+\dfrac{4}{7}=\dfrac{1+4}{7}=\dfrac{5}{7}$

11. $\dfrac{8}{13}+\dfrac{3}{13}=\dfrac{8+3}{13}=\dfrac{11}{13}$

13. $\dfrac{1}{12}+\dfrac{5}{12}=\dfrac{1+5}{12}=\dfrac{6}{12}=\dfrac{6\cdot 1}{6\cdot 2}=\dfrac{1}{2}$

15. $\dfrac{2}{21}+\dfrac{7}{21}=\dfrac{2+7}{21}=\dfrac{9}{21}=\dfrac{3\cdot 3}{3\cdot 7}=\dfrac{3}{7}$

17. $\dfrac{7}{10}+\dfrac{3}{10}=\dfrac{7+3}{10}=\dfrac{10}{10}=1$

19. $\dfrac{8}{9}+\dfrac{5}{9}=\dfrac{8+5}{9}=\dfrac{13}{9}=1\dfrac{4}{9}$

21. $\dfrac{1}{3}+\dfrac{2}{3}+\dfrac{1}{3}=\dfrac{1+2+1}{3}=\dfrac{4}{3}=1\dfrac{1}{3}$       23. $\dfrac{2}{5}+\dfrac{3}{5}+\dfrac{1}{5}=\dfrac{2+3+1}{5}=\dfrac{6}{5}=1\dfrac{1}{5}$

## Guide Problems

25. Add $\dfrac{1}{3}+\dfrac{3}{4}$ .

The fractions being added have different denominators and so are unlike fractions.  The strategy is as follows.

a. Find the LCD of the fractions.

Since $4=2\cdot2$ and 3 is a prime number, the LCD is $2\cdot2\cdot3=12$ .

b. Write each fraction as an equivalent fraction with the LCD found in part a.  The LCD is 12.

For the fraction $\dfrac{1}{3}$, multiply by $\dfrac{4}{4}$ to obtain a denominator of 12.  Similarly multiply $\dfrac{3}{4}$ by $\dfrac{3}{3}$ .

$\dfrac{1}{3}\cdot\dfrac{4}{4}=\dfrac{1\cdot4}{3\cdot4}=\dfrac{4}{12}$          $\dfrac{3}{4}\cdot\dfrac{3}{3}=\dfrac{3\cdot3}{4\cdot3}=\dfrac{9}{12}$

c. Add the fractions.  The fractions have been expressed as equivalent like fractions and can be added using the usual strategy for like fractions.

$\dfrac{1}{3}+\dfrac{3}{4}=\dfrac{4}{12}+\dfrac{9}{12}=\dfrac{13}{12}$ .

The result is an improper fraction and can be written as a mixed number:  $\dfrac{13}{12}=1\dfrac{1}{12}$ .

*In each of problems 27 – 41, the fractions being added are unlike fractions.  In each case, the LCD of the fractions is found, then each fraction is expressed as an equivalent fraction with the LCD as denominator.  The resulting like fractions are added.  The result is then simplified if possible.*

27. $\dfrac{7}{16}+\dfrac{9}{24}$  The LCD of 16 and 24 is 48.

Write as equivalent fractions.
$\dfrac{7}{16}\cdot\dfrac{3}{3}=\dfrac{21}{48}$      $\dfrac{9}{24}\cdot\dfrac{2}{2}=\dfrac{18}{48}$

Add.
$\dfrac{7}{16}+\dfrac{9}{24}=\dfrac{21}{48}+\dfrac{18}{48}=\dfrac{39}{48}=\dfrac{3\cdot13}{3\cdot16}=\dfrac{13}{16}$

29. $\dfrac{1}{6}+\dfrac{1}{8}$  The LCD of 6 and 8 is 24.

Write as equivalent fractions.

$$\frac{1}{6} \cdot \frac{4}{4} = \frac{4}{24} \qquad \frac{1}{8} \cdot \frac{3}{3} = \frac{3}{24}$$

Add.

$$\frac{1}{6} + \frac{1}{8} = \frac{4}{24} + \frac{3}{24} = \frac{7}{24}$$

31. $\dfrac{5}{6} + \dfrac{1}{18}$     LCD $= 18$

$$\frac{5}{6} + \frac{1}{18} = \frac{5}{6} \cdot \frac{3}{3} + \frac{1}{18} = \frac{15}{18} + \frac{1}{18} = \frac{16}{18} = \frac{8}{9}$$

33. $\dfrac{3}{4} + \dfrac{7}{16}$     LCD $= 16$

$$\frac{3}{4} + \frac{7}{16} = \frac{3}{4} \cdot \frac{4}{4} + \frac{7}{16} = \frac{12}{16} + \frac{7}{16} = \frac{19}{16} = 1\frac{3}{16}$$

35. $\dfrac{1}{3} + \dfrac{5}{6} + \dfrac{8}{9}$     LCD $= 18$

$$\frac{1}{3} + \frac{5}{6} + \frac{8}{9} = \frac{1}{3} \cdot \frac{6}{6} + \frac{5}{6} \cdot \frac{3}{3} + \frac{8}{9} \cdot \frac{2}{2} = \frac{6}{18} + \frac{15}{18} + \frac{16}{18} = \frac{37}{18} = 2\frac{1}{18}$$

37. $\dfrac{4}{5} + \dfrac{7}{8} + \dfrac{9}{20}$     LCD $= 40$

$$\frac{4}{5} + \frac{7}{8} + \frac{9}{20} = \frac{4}{5} \cdot \frac{8}{8} + \frac{7}{8} \cdot \frac{5}{5} + \frac{9}{20} \cdot \frac{2}{2} = \frac{32}{40} + \frac{35}{40} + \frac{18}{40} = \frac{85}{40} = \frac{17}{8} = 2\frac{1}{8}$$

39. $\dfrac{1}{5} + \dfrac{1}{9} + \dfrac{13}{15}$     LCD $= 45$

$$\frac{1}{5} + \frac{1}{9} + \frac{13}{15} = \frac{1}{5} \cdot \frac{9}{9} + \frac{1}{9} \cdot \frac{5}{5} + \frac{13}{15} \cdot \frac{3}{3} = \frac{9}{45} + \frac{5}{45} + \frac{39}{45} = \frac{53}{45} = 1\frac{8}{45}$$

41. $\dfrac{5}{6} + \dfrac{2}{9} + \dfrac{4}{15}$     LCD $= 90$

$$\frac{5}{6} + \frac{2}{9} + \frac{4}{15} = \frac{5}{6} \cdot \frac{15}{15} + \frac{2}{9} \cdot \frac{10}{10} + \frac{4}{15} \cdot \frac{6}{6} = \frac{75}{90} + \frac{20}{90} + \frac{24}{90} = \frac{119}{90} = 1\frac{29}{90}$$

## Guide Problems

43. Add $8\dfrac{2}{3} + 7\dfrac{3}{4}$.

The strategy for adding mixed numbers whose fraction parts are unlike fractions is as follows.

a. Find the LCD of the fraction parts. In this case the LCD is the LCM of the denominators 3 and 4. Therefore, the LCD = 12.

b. Write the fraction part of each mixed number as an equivalent fraction with the LCD found in part a. Rewrite each mixed number using the equivalent fraction part.

$$8\frac{2}{3} = 8\frac{2\cdot 4}{3\cdot 4} = 8\frac{8}{12} \qquad\qquad 7\frac{3}{4} = 7\frac{3\cdot 3}{4\cdot 3} = 7\frac{9}{12}$$

c. Add the fraction parts and then add the whole number parts. Simplify, if possible.

$$\begin{array}{r} 8\dfrac{8}{12} \\[1.5em] +\,7\dfrac{9}{12} \\[0.5em] \hline 15\dfrac{17}{12} \end{array} \quad \left(\text{because } \frac{8}{12}+\frac{9}{12}=\frac{17}{12} \text{ and } 8+7=15\right)$$

Note the fraction part of the result is an improper fraction. Rewrite this number so the fraction is proper.

$$15\frac{17}{12} = 15+\frac{17}{12} = 15+1\frac{5}{12} = 16\frac{5}{12}$$

The last step above can be thought of as addition of mixed numbers where the first number has a fraction part of zero.

45. $\dfrac{4}{15}+3\dfrac{9}{25}$  The LCD of $\dfrac{4}{15}$ and $\dfrac{9}{25}$ is 75.

47. $\dfrac{3}{8}+1\dfrac{3}{10}$  The LCD of $\dfrac{3}{8}$ and $\dfrac{3}{10}$ is 40.

Rewrite fractional parts.

$$\frac{4}{15} = \frac{4\cdot 5}{15\cdot 5} = \frac{20}{75}, \qquad 3\frac{9}{25} = 3\frac{9\cdot 3}{25\cdot 3} = 3\frac{27}{75}$$

Rewrite fractional parts.

$$\frac{3}{8} = \frac{3\cdot 5}{8\cdot 5} = \frac{15}{40}, \qquad 1\frac{3}{10} = 1\frac{3\cdot 4}{10\cdot 4} = 1\frac{12}{40}$$

$$\begin{array}{r} \dfrac{20}{75} \\[1.5em] +\,3\dfrac{27}{75} \\[0.5em] \hline 3\dfrac{47}{75} \end{array}$$

$$\begin{array}{r} \dfrac{15}{40} \\[1.5em] +\,1\dfrac{12}{40} \\[0.5em] \hline 1\dfrac{27}{40} \end{array}$$

*Alternate method:*

$$\frac{3}{8}+1\frac{3}{10} = \frac{3}{8}+\frac{13}{10} = \frac{3\cdot 5}{8\cdot 5}+\frac{13\cdot 4}{10\cdot 4}$$
$$= \frac{15}{40}+\frac{52}{40} = \frac{67}{40} = 1\frac{27}{40}$$

49. $\frac{3}{4} + 2\frac{3}{8}$ The LCD of $\frac{3}{4}$ and $\frac{3}{8}$ is 8.

Rewrite fractional parts.

$$\frac{3}{4} = \frac{3 \cdot 2}{4 \cdot 2} = \frac{6}{8}$$

$$\begin{array}{r} \frac{6}{8} \\ + 2\frac{3}{8} \\ \hline 2\frac{9}{8} = 2 + 1\frac{1}{8} = 3\frac{1}{8} \end{array}$$

51. $1\frac{2}{5} = 1\frac{2 \cdot 16}{5 \cdot 16} = 1\frac{32}{80}, \qquad \frac{3}{16} = \frac{3 \cdot 5}{16 \cdot 5} = \frac{15}{80}$

$$\begin{array}{r} 1\frac{32}{80} \\ + \frac{15}{80} \\ \hline 1\frac{47}{80} \end{array}$$

53. $4\frac{7}{24} = 4\frac{7 \cdot 3}{24 \cdot 3} = 4\frac{21}{72}, \quad 8\frac{7}{18} = 8\frac{7 \cdot 4}{18 \cdot 4} = 8\frac{28}{72}$

$$\begin{array}{r} 4\frac{21}{72} \\ + \ 8\frac{28}{72} \\ \hline 12\frac{49}{72} \end{array}$$

55. $7\frac{1}{2} + 7\frac{3}{4} = \frac{15}{2} + \frac{31}{4} = \frac{15 \cdot 2}{2 \cdot 2} + \frac{31}{4}$

$$= \frac{30}{4} + \frac{31}{4} = \frac{61}{4} = 15\frac{1}{4}$$

57. $2\frac{4}{7} + 3\frac{7}{9} = \frac{18}{7} + \frac{34}{9} = \frac{18 \cdot 9}{7 \cdot 9} + \frac{34 \cdot 7}{9 \cdot 7}$

$$= \frac{162}{63} + \frac{238}{63} = \frac{400}{63} = 6\frac{22}{63}$$

59. $3\frac{2}{5} = 3\frac{2 \cdot 8}{5 \cdot 8} = 3\frac{16}{40}, \qquad 3\frac{5}{8} = 3\frac{5 \cdot 5}{8 \cdot 5} = 3\frac{25}{40}$

$$\begin{array}{r} 3\frac{16}{40} \\ + \ 3\frac{25}{40} \\ \hline 6\frac{41}{40} = 6 + 1\frac{1}{40} = 7\frac{1}{40} \end{array}$$

**61.** $1\dfrac{3}{8}+1\dfrac{1}{3}+3\dfrac{1}{4}$

The LCD of $\dfrac{3}{8}$, $\dfrac{1}{3}$ and $\dfrac{1}{4}$ is 24.
Rewrite fractional parts.

$1\dfrac{3}{8}=1\dfrac{3\cdot3}{8\cdot3}=1\dfrac{9}{24}$, $\quad 1\dfrac{1}{3}=1\dfrac{1\cdot8}{3\cdot8}=1\dfrac{8}{24}$

$3\dfrac{1}{4}=3\dfrac{1\cdot6}{4\cdot6}=3\dfrac{6}{24}$

$$\begin{array}{r}1\dfrac{9}{24}\\[6pt]1\dfrac{8}{24}\\[6pt]+\ 3\dfrac{6}{24}\\[2pt]\hline 5\dfrac{23}{24}\end{array}$$

**63.** $\dfrac{5}{30}+\dfrac{3}{40}+2\dfrac{1}{8}$

The LCD of $\dfrac{5}{30}$, $\dfrac{3}{40}$ and $\dfrac{1}{8}$ is 120.
Rewrite fractional parts.

$\dfrac{5}{30}=\dfrac{5\cdot4}{30\cdot4}=\dfrac{20}{120}$, $\quad \dfrac{3}{40}=\dfrac{3\cdot3}{40\cdot3}=\dfrac{9}{120}$

$2\dfrac{1}{8}=2\dfrac{1\cdot15}{8\cdot15}=2\dfrac{15}{120}$

$$\begin{array}{r}\dfrac{20}{120}\\[6pt]\dfrac{9}{120}\\[6pt]+\ 2\dfrac{15}{120}\\[2pt]\hline 2\dfrac{44}{120}=2\dfrac{11}{30}\end{array}$$

**65.** $\dfrac{2}{5}=\dfrac{12}{30}$, $\dfrac{2}{15}=\dfrac{4}{30}$, $2\dfrac{1}{10}=2\dfrac{3}{30}$

$$\begin{array}{r}\dfrac{12}{30}\\[6pt]\dfrac{4}{30}\\[6pt]+\ 2\dfrac{3}{30}\\[2pt]\hline 2\dfrac{19}{30}\end{array}$$

**67.** $2\dfrac{1}{4}=2\dfrac{6}{24}$, $4\dfrac{5}{8}=4\dfrac{15}{24}$, $1\dfrac{1}{6}=1\dfrac{4}{24}$

$$\begin{array}{r}2\dfrac{6}{24}\\[6pt]4\dfrac{15}{24}\\[6pt]+\ 1\dfrac{4}{24}\\[2pt]\hline 7\dfrac{25}{24}=7+1\dfrac{1}{24}=8\dfrac{1}{24}\end{array}$$

**69.** $\dfrac{1}{3}=\dfrac{5}{15}$, $2\dfrac{3}{5}=2\dfrac{9}{15}$, $5\dfrac{1}{5}=5\dfrac{3}{15}$

$$\begin{array}{r}\dfrac{5}{15}\\[6pt]2\dfrac{9}{15}\\[6pt]+\ 5\dfrac{3}{15}\\[2pt]\hline 7\dfrac{17}{15}=7+1\dfrac{2}{15}=8\dfrac{2}{15}\end{array}$$

**71.** $\dfrac{1}{3}=\dfrac{20}{60}$, $\dfrac{5}{12}=\dfrac{25}{60}$, $5\dfrac{4}{5}=5\dfrac{48}{60}$

$$\begin{array}{r}\dfrac{20}{60}\\[6pt]\dfrac{25}{60}\\[6pt]+\ 5\dfrac{48}{60}\\[2pt]\hline 5\dfrac{93}{60}=5+1\dfrac{33}{60}=6\dfrac{33}{60}=6\dfrac{11}{20}\end{array}$$

**73.** James rode a total of $\dfrac{3}{4}+\dfrac{5}{16}+\dfrac{7}{8}=\dfrac{3\cdot4}{4\cdot4}+\dfrac{5}{16}+\dfrac{7\cdot2}{8\cdot2}=\dfrac{12}{16}+\dfrac{5}{16}+\dfrac{14}{16}=\dfrac{31}{16}=1\dfrac{15}{16}$ miles.

75. The total acreage mowed is $16\frac{1}{5}+17\frac{2}{3}+15\frac{1}{12}$ acres. To compute the sum, note the LCD of the fraction parts is 60. Rewrite each fraction part so as to have a denominator of 60 and then add.

$$
\begin{array}{r}
16\dfrac{12}{60} \\[2ex]
17\dfrac{40}{60} \\[2ex]
+\;\;15\dfrac{5}{60} \\[1ex]
\hline
48\dfrac{57}{60}=48\dfrac{19}{20}
\end{array}
$$

Together, Alfred, Larry, and Howard mowed $48\frac{19}{20}$ acres.

77. The length of the beveled edge is the perimeter of the glass or $2\cdot\text{length}+2\cdot\text{width}$. Note

$$2\cdot\text{length}=2\cdot 44\frac{7}{8}=2\cdot\left(44+\frac{7}{8}\right)=2\cdot 44+2\cdot\frac{7}{8}=88+\overset{1}{\cancel{2}}\cdot\frac{7}{\underset{4}{\cancel{8}}}=88\frac{7}{4}=89\frac{3}{4}$$

$$2\cdot\text{width}=2\cdot 29\frac{5}{16}=2\cdot\left(29+\frac{5}{16}\right)=2\cdot 29+2\cdot\frac{5}{16}=58+\overset{1}{\cancel{2}}\cdot\frac{5}{\underset{8}{\cancel{16}}}=58\frac{5}{8}$$

The length is $89\frac{3}{4}+58\frac{5}{8}=89\frac{6}{8}+58\frac{5}{8}=147\frac{11}{8}=148\frac{3}{8}$ inches.

79. The total acreage is $125\frac{1}{3}+65\frac{4}{5}+88\frac{3}{4}$ acres. To compute the sum, note the LCD of the fraction parts is 60. Rewrite each fraction part so as to have a denominator of 60 and then add.

$$
\begin{array}{r}
125\dfrac{20}{60} \\[2ex]
65\dfrac{48}{60} \\[2ex]
+\;\;88\dfrac{45}{60} \\[1ex]
\hline
278\dfrac{113}{60}=278+1\dfrac{53}{60}=279\dfrac{53}{60}
\end{array}
$$

The three crops cover $279\frac{53}{60}$ acres.

81. a. Peter used $5\frac{1}{2}+8\frac{4}{5}+2\frac{5}{6}+4\frac{2}{3}=5\frac{15}{30}+8\frac{24}{30}+2\frac{25}{30}+4\frac{20}{30}=19\frac{84}{30}=19+2\frac{24}{30}=21\frac{24}{30}=21\frac{4}{5}$ gallons.

b. From a. the total amount of paint used was $21\frac{4}{5}$ gallons. At $20 per gallon, the total cost is

$$21\frac{4}{5}\cdot 20 = \frac{109}{5}\cdot\frac{20}{1} = \frac{109}{\cancel{5}}\cdot\frac{\cancel{20}^{\,4}}{1} = 109\cdot 4 = 436 \text{ dollars.}$$

## Cumulative Skills Review

1.  The digit to the right of the tens digit in 12,646, the unit digit, is 6.  To round to the enarest ten, round the tens digit 4 up to get 12,650.

3.  Since the whole number parts of each weight are the same, 25, we only need to find the smallest fraction part to determine the lightest package.  The fraction parts $\frac{2}{5}, \frac{4}{15}$, and $\frac{7}{20}$.  The LCD of these fractions is 60.  Writing each fraction using the LCD gives $\frac{2}{5} = \frac{24}{60}, \frac{4}{15} = \frac{16}{60}$ and $\frac{7}{20} = \frac{21}{60}$.  By comparing numerators, the smallest fraction is $\frac{16}{60}$ or $\frac{4}{15}$.  The lightest package weighs $25\frac{4}{15}$ pounds.

5.  Since $6\cdot 14 = 84$, the fraction $\frac{11}{14}$ is equivalent to $\frac{11}{14} = \frac{11}{14}\cdot\frac{6}{6} = \frac{66}{84}$

7.  The LCD is 48.  Converting each fraction to an equivalent improper fraction gives
$$4\frac{9}{16} = \frac{4\cdot 16 + 9}{16} = \frac{73}{16} = \frac{73}{16}\cdot\frac{3}{3} = \frac{219}{48} \text{ and } 4\frac{7}{12} = \frac{4\cdot 12 + 7}{12} = \frac{55}{12} = \frac{55}{12}\cdot\frac{4}{4} = \frac{220}{48}$$
Since $219 < 220$, the first fraction is the smaller one.  That is $4\frac{9}{16} < 4\frac{7}{12}$ or $4\frac{7}{12} > 4\frac{9}{16}$.

9.  The LCD of $\frac{2}{3}$ and $\frac{1}{7}$ is the least common multiple of the denominators 3 and 7.  The first few multiples of 3 and 7 are 3: 3, 6, 9, 12, 15, 18, 21, 24, … and 7: 7, 14, 21, 28… The least value common to the two lists is 21 so the LCD is 21.

## Section 2.7    Subtracting Fractions and Mixed Numbers

## Concept Check

1.  To subtract like fractions, subtract the second _numerator_ from the first numerator and write this difference over the common _denominator_ .  Simplify, if possible.

3.  When subtracting mixed numbers, we first subtract the _fraction_ parts and then subtract the _whole numbers_ .

5.  When subtracting a mixed number from a whole number, we must borrow 1 from the whole number part and add this in the form of $\dfrac{LCD}{LCD}$ to one less than the whole number part.

## Guide Problems

7.  Subtract $\dfrac{7}{11} - \dfrac{3}{11}$.

$\dfrac{7}{11} - \dfrac{3}{11} = \dfrac{7-3}{11}$  (Keep common denominator)

$\dfrac{7-3}{11} = \dfrac{4}{11}$  (Subtract numerators)

15.  $\dfrac{11}{61} - \dfrac{10}{61} = \dfrac{11-10}{61} = \dfrac{1}{61}$

19.  $\dfrac{17}{48} - \dfrac{7}{48} = \dfrac{17-7}{48} = \dfrac{10}{48} = \dfrac{2\cdot5}{2\cdot24} = \dfrac{5}{24}$

9.  $\dfrac{9}{10} - \dfrac{3}{10} = \dfrac{9-3}{10} = \dfrac{6}{10} = \dfrac{2\cdot3}{2\cdot5} = \dfrac{3}{5}$

11.  $\dfrac{9}{14} - \dfrac{4}{14} = \dfrac{9-4}{14} = \dfrac{5}{14}$

13.  $\dfrac{3}{17} - \dfrac{1}{17} = \dfrac{3-1}{17} = \dfrac{2}{17}$

17.  $\dfrac{16}{25} - \dfrac{11}{25} = \dfrac{16-11}{25} = \dfrac{5}{25} = \dfrac{5\cdot1}{5\cdot5} = \dfrac{1}{5}$

## Guide Problems

21.  Subtract $\dfrac{3}{4} - \dfrac{1}{3}$.

a. Find the LCD of the fraction. The LCD is the least common multiple of the numbers that appear in the denominators of the two fractions. The denominators here are 3 and 4 and their least common multiple is 12, so the LCD is 12.

b. Write each fraction as an equivalent fraction with the LCD found in part a. Each fraction must be multiplied by a fraction equivalent to 1 that yields a fraction with the LCD, 12, as its denominator.

$$\dfrac{3}{4} = \dfrac{3}{4} \cdot \dfrac{3}{3} = \dfrac{3\cdot3}{4\cdot3} = \dfrac{9}{12} \qquad \text{and} \qquad \dfrac{1}{3} = \dfrac{1}{3} \cdot \dfrac{4}{4} = \dfrac{1\cdot4}{3\cdot4} = \dfrac{4}{12}$$

c. Subtract the fractions. The two fractions have been written as equivalent fraction having like denominators. Subtract by keeping the like denominator and subtracting the numerators.

$$\dfrac{3}{4} - \dfrac{1}{3} = \dfrac{9}{12} - \dfrac{4}{12} = \dfrac{9-4}{12} = \dfrac{5}{12}$$

23.  $\dfrac{5}{6} - \dfrac{5}{8}$. The LCD is the LCM of 6 and 8 which is 24.

$$\dfrac{5}{6} - \dfrac{5}{8} = \dfrac{5}{6} \cdot \dfrac{4}{4} - \dfrac{5}{8} \cdot \dfrac{3}{3} = \dfrac{20}{24} - \dfrac{15}{24} = \dfrac{5}{24}$$

25.  $\dfrac{1}{6} - \dfrac{1}{8}$. The LCD is the LCM of 6 and 8 which is 24.

$$\dfrac{1}{6} - \dfrac{1}{8} = \dfrac{1}{6} \cdot \dfrac{4}{4} - \dfrac{1}{8} \cdot \dfrac{3}{3} = \dfrac{4}{24} - \dfrac{3}{24} = \dfrac{1}{24}$$

27. $\dfrac{14}{15} - \dfrac{8}{9}$. The LCD is the LCM of 9 and 15 which is 45.

$$\dfrac{14}{15} - \dfrac{8}{9} = \dfrac{14}{15} \cdot \dfrac{3}{3} - \dfrac{8}{9} \cdot \dfrac{5}{5} = \dfrac{42}{45} - \dfrac{40}{45} = \dfrac{2}{45}$$

29. $\dfrac{4}{5} - \dfrac{3}{10}$. The LCD is the LCM of 5 and 10 which is 10.

$$\dfrac{4}{5} - \dfrac{3}{10} = \dfrac{4}{5} \cdot \dfrac{2}{2} - \dfrac{3}{10} = \dfrac{8}{10} - \dfrac{3}{10} = \dfrac{5}{10} = \dfrac{1}{2}$$

(Remember to simplify after subtracting, if possible.)

31. $\dfrac{19}{25} - \dfrac{11}{50}$. The LCD is the LCM of 25 and 50 which is 50.

$$\dfrac{19}{25} - \dfrac{11}{50} = \dfrac{19}{25} \cdot \dfrac{2}{2} - \dfrac{11}{50} = \dfrac{38}{50} - \dfrac{11}{50} = \dfrac{27}{50}$$

33. $\dfrac{7}{16} - \dfrac{1}{12}$. The LCD is the LCM of 12 and 16 which is 48.

$$\dfrac{7}{16} - \dfrac{1}{12} = \dfrac{7}{16} \cdot \dfrac{3}{3} - \dfrac{1}{12} \cdot \dfrac{4}{4} = \dfrac{21}{48} - \dfrac{4}{48} = \dfrac{17}{48}$$

## Guide Problems

35. Subtract $15\dfrac{3}{4} - 8\dfrac{1}{4}$. The fractional part of the first mixed number, $\dfrac{3}{4}$, is greater than $\dfrac{1}{4}$, the fractional part of the second mixed number. Thus, the subtraction can be formed by subtracting the fractional parts $\left(\dfrac{3}{4} - \dfrac{1}{4}\right)$ and then subtracting the whole number parts $(15 - 8)$.

$$\begin{array}{r} 15\dfrac{3}{4} \\ -8\dfrac{1}{4} \\ \hline 7\dfrac{2}{4} = 7\dfrac{1}{2} \end{array}$$    Note the result could be simplified.

*Note in problems 37 through 43, the fractional part of the first mixed number is greater than or equal to the fractional part of the second mixed number. Further the fractions are like fractions. The subtraction is performed by subtracting fractional parts and subtracting whole number parts.*

37. $\begin{array}{r} 2\dfrac{4}{5} \\ -\ 1\dfrac{1}{5} \\ \hline 1\dfrac{3}{5} \end{array}$          39. $\begin{array}{r} 5\dfrac{11}{12} \\ -\ 3\dfrac{7}{12} \\ \hline 2\dfrac{4}{12} = 2\dfrac{1}{3} \end{array}$

41. $\quad 2\dfrac{5}{6}$

$\quad -\ 1\dfrac{2}{6}$

$\quad\quad 1\dfrac{3}{6} = 1\dfrac{1}{2}$

43. $\quad 12\dfrac{13}{27}$

$\quad -\ 7\dfrac{5}{27}$

$\quad\quad 5\dfrac{8}{27}$

## Guide Problems

45. Subtract $14\dfrac{1}{4} - 3\dfrac{3}{4}$.

    a.  Because the fraction part of the first mixed number is less than the fraction part of the second mixed number, 1 must be borrowed from the whole number of the first mixed number in the form of

$\dfrac{\text{LCD}}{\text{LCD}}$ and added to the fraction part of the first mixed number. The two fraction parts are like fractions with denominators of 4. So 4 is the LCD. The first mixed number is rewritten as

$$14\frac{1}{4} = 13 + 1 + \frac{1}{4} = 13 + \frac{4}{4} + \frac{1}{4} = 13\frac{5}{4}$$

    b.  With the equivalent version of the first mixed number, the subtraction can be performed by first subtracting the fractional parts and then subtracting the whole number parts.

$\quad\quad 13\dfrac{5}{4}$

$\quad -\ \ 3\dfrac{3}{4}$

$\quad\quad 10\dfrac{2}{4} = 10\dfrac{1}{2}$   Simplify the result

*In problems 47 through 53, the approach is the same. The fraction parts have unlike denominators but the fraction part of the minuend is smaller than that of the subtrahend. The first mixed number is rewritten by borrowing 1 from its whole number part before the subtraction is done. Also the subtraction may be written horizontally or vertically but is performed the same, subtract the whole number parts and subtract the fraction parts.*

47. $\quad 4\dfrac{1}{8} = 4 + \dfrac{1}{8} = 3 + \dfrac{8}{8} + \dfrac{1}{8} = 3\dfrac{9}{8}$

$\quad\quad 3\dfrac{9}{8}$

$\quad -\ 2\dfrac{3}{8}$

$\quad\quad 1\dfrac{6}{8} = 1\dfrac{3}{4}$

49. $\quad 6\dfrac{1}{9} = 6 + \dfrac{1}{9} = 5 + \dfrac{9}{9} + \dfrac{1}{9} = 5\dfrac{10}{9}$

$\quad\quad 5\dfrac{10}{9}$

$\quad -\ 2\dfrac{4}{9}$

$\quad\quad 3\dfrac{6}{9} = 3\dfrac{2}{3}$

51. $14\dfrac{1}{7} = 14 + \dfrac{1}{7} = 13 + \dfrac{7}{7} + \dfrac{1}{7} = 13\dfrac{8}{7}$          53. $21\dfrac{3}{8} = 21 + \dfrac{3}{8} = 20 + \dfrac{8}{8} + \dfrac{3}{8} = 20\dfrac{11}{8}$

$13\dfrac{8}{7} - 3\dfrac{5}{7} = 10\dfrac{3}{7}$          $20\dfrac{11}{8} - 16\dfrac{5}{8} = 4\dfrac{6}{8} = 4\dfrac{3}{4}$

## Guide Problems

55. Subtract $17\dfrac{2}{3} - 4\dfrac{1}{4}$.

a. Find the LCD of the fraction parts. Find the least common multiple of 3 and 4.

LCD = $\underline{12}$

b. Write the fractional part of each mixed number as an equivalent fraction with the LCD found in part a.

$$17\dfrac{2}{3} = 17\dfrac{2 \cdot 4}{3 \cdot \underline{4}} = 17\dfrac{8}{\underline{12}}$$

$$4\dfrac{1}{4} = 4\dfrac{1 \cdot 3}{4 \cdot \underline{3}} = 4\dfrac{3}{\underline{12}}$$

c. Subtract the fraction parts and then subtract the whole number parts. Simplify if possible.

$$
\begin{array}{r}
17\dfrac{8}{\underline{12}} \\
- \ \ 4\dfrac{3}{\underline{12}} \\
\hline
13\dfrac{5}{\underline{12}}
\end{array}
$$

*In problems 57 through 75, the approach is the same. The fraction parts have u like denominators. The mixed numbers are rewritten using the LCD as the common denominator. The subtraction is performed by subtracting the whole number parts and subtracting the fraction parts. The result is simplified, if possible. The subtraction may be written vertically or horizontally.*

57. The LCD is 60.

$$1\frac{11}{12} = 1\frac{11 \cdot 5}{12 \cdot 5} = 1\frac{55}{60}$$

$$\frac{3}{20} = \frac{3 \cdot 3}{20 \cdot 3} = \frac{9}{60}$$

$$\begin{array}{r} 1\dfrac{55}{60} \\ -\ \dfrac{9}{60} \\ \hline 1\dfrac{46}{60} = 1\dfrac{23}{30} \end{array}$$

59. The LCD is 30

$$2\frac{5}{6} = 2\frac{5 \cdot 5}{6 \cdot 5} = 2\frac{25}{30}$$

$$\frac{3}{5} = \frac{3 \cdot 6}{5 \cdot 6} = \frac{18}{30}$$

$$\begin{array}{r} 2\dfrac{25}{30} \\ -\ \dfrac{18}{30} \\ \hline 2\dfrac{7}{30} \end{array}$$

61. The LCD is 15.

$$29\frac{3}{5} = 29\frac{3 \cdot 3}{5 \cdot 3} = 29\frac{9}{15}$$

$$20\frac{14}{15} = 20\frac{14}{15}$$

$$\begin{array}{r} 29\dfrac{9}{15} \\ -\ 20\dfrac{14}{15} \\ \hline 8\dfrac{10}{15} = 8\dfrac{2}{3} \end{array}$$

63. The LCD is 6.

$$7\frac{2}{3} = 7\frac{4}{6}, \quad 5\frac{1}{2} = 5\frac{3}{6}$$

$$7\frac{4}{6} - 5\frac{3}{6} = 2\frac{1}{6}$$

65. The LCD is 24.

$$3\frac{5}{6} = 3\frac{20}{24}, \quad \frac{1}{8} = \frac{3}{24}$$

$$3\frac{20}{24} - \frac{3}{24} = 3\frac{17}{24}$$

67. The LCD is 20.

$$30\frac{3}{4} = 30\frac{15}{20},$$

$$10\frac{3}{10} = 10\frac{6}{20}$$

$$30\frac{15}{20} - 10\frac{6}{20} = 20\frac{9}{20}$$

69. The LCD is 30.

$$15\frac{5}{6} = 15\frac{25}{30}, \quad 8\frac{8}{10} = 8\frac{24}{30}$$

$$15\frac{25}{30} - 8\frac{24}{30} = 7\frac{1}{30}$$

71. The LCD is 8.

$$3\frac{1}{4} = 3\frac{2}{8}, \quad 1\frac{1}{8} = 1\frac{1}{8}$$

$$3\frac{2}{8} - 1\frac{1}{8} = 2\frac{1}{8}$$

73. The LCD is 90.

$$9\frac{13}{18} = 9\frac{65}{90}, \quad 1\frac{7}{15} = 1\frac{42}{90}$$

$$9\frac{65}{90} - 1\frac{42}{90} = 8\frac{23}{90}$$

75. The LCD is 6.

$$56\frac{2}{3} = 56\frac{4}{6}, \quad 4\frac{1}{2} = 4\frac{3}{6}$$

$$56\frac{4}{6} - 4\frac{3}{6} = 52\frac{1}{6}$$

## Guide Problems

77. Subtract $18\frac{1}{3} - 5\frac{1}{2}$

a. Find the LCD of the fraction parts. Find the least common multiple of the denominators 2 and 3.

$$LCD = \underline{6}$$

b. Write the fraction part of each mixed number as an equivalent fraction with the LCD found in part a.

$$18\frac{1}{3} = 18\frac{1 \cdot \underline{2}}{3 \cdot \underline{2}} = 18\frac{2}{\underline{6}}$$

$$5\frac{1}{2} = 5\frac{1 \cdot \underline{3}}{2 \cdot \underline{3}} = 5\frac{3}{\underline{6}}$$

c. Because the fraction part of the first mixed number is less than the fraction part of the second mixed number, borrow 1 from the whole number of the first mixed number in the form of $\dfrac{LCD}{LCD}$ and add it to the fraction part of the first mixed number.

$$18\frac{2}{\underline{6}} = 18 + \frac{2}{\underline{6}} = 17 + \frac{6}{\underline{6}} + \frac{2}{\underline{6}} = 17\frac{8}{\underline{6}}$$

d. Subtract the fraction parts and then subtract the whole number parts. Simplify if possible.    .

$$
\begin{array}{r}
17\dfrac{8}{\underline{6}} \\[2ex]
- \ 5\dfrac{3}{\underline{6}} \\[1ex]
\hline
12\dfrac{5}{\underline{6}}
\end{array}
$$

*In problems 79 through 93, the approach is the same. In each case, the fraction parts are unlike fractions. After equivalent mixed numbers are found with a common denominator, it is clear the fraction part of the first number is smaller than that of the second number so 1 must be borrowed from the whole number part of the first mixed number before subtraction can be performed. The subtraction is done either vertically or horizontally by subtracting the fraction parts and then the whole number parts.*

79. The LCD is 15.

$$5\frac{2}{5} = 5\frac{2 \cdot 3}{5 \cdot 3} = 5\frac{6}{15} = 5 + \frac{6}{15}$$
$$= 4 + \frac{15}{15} + \frac{6}{15} = 4\frac{21}{15}$$

$$1\frac{14}{15} = 1\frac{14}{15}$$

$$
\begin{array}{r}
4\dfrac{21}{15} \\[2ex]
- \ 1\dfrac{14}{15} \\[1ex]
\hline
3\dfrac{7}{15}
\end{array}
$$

81. The LCD is 18.

$$13 = 12 + 1 = 12 + \frac{18}{18} = 12\frac{18}{18}$$

$$1\frac{5}{18} = 1\frac{5}{18}$$

$$
\begin{array}{r}
12\dfrac{18}{18} \\[2ex]
- \ 1\dfrac{5}{18} \\[1ex]
\hline
11\dfrac{13}{18}
\end{array}
$$

83. The LCD is 4.

$$69\frac{1}{4} = 69 + \frac{1}{4} = 68 + \frac{4}{4} + \frac{1}{4} = 68\frac{5}{4}$$

$$38\frac{1}{2} = 38\frac{2}{4}$$

$$\begin{array}{r} 68\frac{5}{4} \\ - \ 38\frac{2}{4} \\ \hline 30\frac{3}{4} \end{array}$$

85. The LCD is 8.

$$17\frac{1}{8} = 17\frac{1}{8} = 16\frac{9}{8}, \quad \frac{3}{4} = \frac{6}{8}$$

$$16\frac{9}{8} - \frac{6}{8} = 16\frac{3}{8}$$

87. The LCD is 16.

$$11\frac{5}{16} = 10\frac{21}{16}, \quad 8\frac{3}{4} = 8\frac{12}{16}$$

$$10\frac{21}{16} - 8\frac{12}{16} = 2\frac{9}{16}$$

89. The LCD is 3.

$$4 = 3\frac{3}{3}, \quad \frac{1}{3} = \frac{1}{3}$$

$$3\frac{3}{3} - \frac{1}{3} = 3\frac{2}{3}$$

91. The LCD is 48.

$$10\frac{5}{16} = 10\frac{15}{48} = 9\frac{63}{48}, \quad 4\frac{7}{12} = 4\frac{28}{48}$$

$$9\frac{63}{48} - 4\frac{28}{48} = 5\frac{35}{48}$$

93. The LCD is 16.

$$12 = 11\frac{16}{16}, \quad 9\frac{15}{16} = 9\frac{15}{16}$$

$$11\frac{16}{16} - 9\frac{15}{16} = 2\frac{1}{16}$$

95. The amount remaining is the starting amount minus the amount used so the amount remaining is

$$\frac{7}{8} - \frac{3}{5} = \frac{7 \cdot 5}{8 \cdot 5} - \frac{3 \cdot 8}{5 \cdot 8} = \frac{35}{40} - \frac{24}{40} = \frac{11}{40} \text{ pound.}$$

97. The length remaining is found by subtracting the length of pipe removed from the total length.

$$27\frac{1}{4} - 9\frac{5}{8} = 27\frac{2}{8} - 9\frac{5}{8} = 26\frac{10}{8} - 9\frac{5}{8} = 17\frac{5}{8} \text{ so } 17\frac{5}{8} \text{ inches is the remaining length.}$$

99. The distance remaining is found by subtracting the distance already traveled from the total distance.

$$15\frac{1}{2} - 8\frac{3}{4} = 15\frac{2}{4} - 8\frac{3}{4} = 14\frac{6}{4} - 8\frac{3}{4} = 6\frac{3}{4} \text{ so } 6\frac{3}{4} \text{ miles remain in the hike.}$$

101. The amount of weight lost in cooking was $3\frac{9}{16} - 2\frac{7}{8} = 3\frac{9}{16} - 2\frac{14}{16} = 2\frac{25}{16} - 2\frac{14}{16} = \frac{11}{16}$ pounds.

103. The amount of Kona coffee sold is $22\frac{3}{8} - 6\frac{1}{3} = 22\frac{9}{24} - 6\frac{8}{24} = 16\frac{1}{24}$ pounds.

105. Since the matte and frame is on both sides, they occupy a total of $2\frac{1}{8} + 2\frac{1}{8} = 4\frac{2}{8} = 4\frac{1}{4}$ inches. Subtracting this from the total width gives the width of the photo as $14\frac{3}{4} - 4\frac{1}{4} = 10\frac{2}{4} = 10\frac{1}{2}$ inches.

107. The total amount cut off is $15\frac{3}{8} + 8\frac{4}{5} + 12\frac{1}{6} = 15\frac{45}{120} + 8\frac{96}{120} + 12\frac{20}{120} = 35\frac{161}{120} = 36\frac{41}{120}$ feet. The amount left is then $63 - 36\frac{41}{120} = 62\frac{120}{120} - 36\frac{41}{120} = 26\frac{79}{120}$ feet.

109. The unknown length is computed below.

$$18\frac{3}{16} - 4\frac{3}{8} - 6\frac{1}{2} = 18\frac{3}{16} - 4\frac{6}{16} - 6\frac{8}{16} = 17\frac{19}{16} - 4\frac{6}{16} - 6\frac{8}{16}$$
$$= 13\frac{13}{16} - 6\frac{8}{16} = 7\frac{5}{16}$$

111. At the end of the second week, Susan's weight was
$$134\frac{3}{4} - 2\frac{1}{4} - 1\frac{1}{2} = 134\frac{3}{4} - 2\frac{1}{4} - 1\frac{2}{4} = (134 - 2 - 1) + \left(\frac{3}{4} - \frac{1}{4} - \frac{2}{4}\right) = 131 \text{ pounds.}$$

During the third week, Susan gained $1\frac{1}{8}$ pounds so her final weight was $131 + 1\frac{1}{8} = 132\frac{1}{8}$.

113. a. The total weight of the suitcases is $51\frac{1}{3} + 28\frac{1}{4} = 51\frac{4}{12} + 28\frac{3}{12} = 79\frac{7}{12}$ pounds which is within the limit.

b. The total weight is under the limit by $80 - 79\frac{7}{12} = 79\frac{12}{12} - 79\frac{7}{12} = \frac{5}{12}$ pounds.

## Cumulative Skills Review

1. Since $12 \cdot 5 = 60$, we have $\frac{7}{12} = \frac{7 \cdot 5}{12 \cdot 5} = \frac{35}{60}$.

3. At the end of the day there were $7560 - 2334 = 5226$ gallons left.

5. The LCD of $\frac{3}{7}, \frac{2}{5}$, and $\frac{8}{9}$ is the least common multiple of the denominators 5, 7, and 9. Examining the prime factorizations, $5 = 5$, $7 = 7$, $9 = 3 \cdot 3$, shows the least common multiple has one factor each of 5 and 7 and two factors of 3. The LCD is $3 \cdot 3 \cdot 5 \cdot 7 = 315$.

7. $89 + (4+8)^2 - 12 = 89 + 12^2 - 12$   Simplify within parentheses first
$= 89 + 144 - 12$   Simplify exponents
$= 233 - 12$   Add/subtract from left to right
$= 221$

9. The LCD of $\dfrac{1}{12}$ and $\dfrac{7}{8}$ is the least common multiple of the denominators 8 and 12. Examining the first elements on the list of multiples for each value: 8: 8, 16, 24, 32,.... and 12: 12, 24, 36, ... we see the least multiple common to both lists is 24. The LCD is 24.

## Chapter 2 Numerical Facts of Life

1. The recipe yields 4 dozen cookies. Since you need 12 dozen cookies, the amounts in the recipe must be multiplied by a factor of $\dfrac{12}{4} = 3$.

*The solutions to problems 3-5 are given in the table below. In each case the amount of the ingredient in the original recipe must be multiplied by 3 to arrive at the corresponding amount in the increased recipe.*

| Problem | Ingredient | Original recipe | Increased recipe |
|---------|-----------|-----------------|------------------|
| 3. | brown sugar | $\dfrac{2}{3}$ cup | $3 \cdot \dfrac{2}{3} = \dfrac{3}{1} \cdot \dfrac{2}{3} = \dfrac{6}{3} = 2$ cups |
| 5. | chocolate chips | $11\dfrac{1}{2}$ ounces | $3 \cdot 11\dfrac{1}{2} = \dfrac{3}{1} \cdot \dfrac{23}{2} = \dfrac{69}{2}$ or $34\dfrac{1}{2}$ ounces |

## Chapter 2 Review Exercises

1. The following quotients with 10 as the dividend result in a natural number: $10 \div 1 = 10$, $10 \div 2 = 5$, before factors begin to repeat. The factors of 10 are 1, 2, 5, and 10.

2. The following quotients with 15 as the dividend result in a natural number: $15 \div 1 = 15$, $15 \div 3 = 5$, before factors begin to repeat. The factors of 15 are 1, 3, 5, and 15.

3. The following quotients with 44 as the dividend result in a natural number: $44 \div 1 = 44$, $44 \div 2 = 22$, $44 \div 4 = 11$, before factors begin to repeat. The factors of 44 are 1, 2, 4, 11, 22, and 44.

4. The following quotients with 48 as the dividend result in a natural number: $48 \div 1 = 48$, $48 \div 2 = 24$, $48 \div 3 = 16$, $48 \div 4 = 12$, $48 \div 6 = 8$. before factors begin to repeat. The factors of 44 are 1, 2, 3, 4, 6, 8, 12, 16, 24, and 48.

5. The following quotients with 46 as the dividend result in a natural number: $46 \div 1 = 46$, $46 \div 2 = 23$, before factors begin to repeat. The factors of 46 are 1, 2, 23, and 46.

6. The following quotients with 85 as the dividend result in a natural number: $85 \div 1 = 85$, $85 \div 5 = 17$, before factors begin to repeat. The factors of 85 are 1, 5, 17, and 85.

7. The factors of 89 are 1 and 89 as these are the only natural numbers that divide 89 evenly.

8. The factors of 61 are 1 and 61 as these are the only natural numbers that divide 61 evenly.

9. Find the factors of 24: $24 \div 1 = 24$, $24 \div 2 = 12$, $24 \div 3 = 8$, $24 \div 4 = 6$, $24 \div 6 = 4$ are the quotients that divide evenly. Note the factors begin to repeat with the last division. The factors of 24 are 1, 2, 3, 4, 6, 8, 12, and 24.

10. Find the factors of 63: $63 \div 1 = 63$, $63 \div 3 = 21$, $63 \div 7 = 9$, $63 \div 9 = 7$ are the quotients that divide evenly. Note the factors begin to repeat with the last division. The factors of 63 are 1, 3, 7, 9, 21, and 63.

11. Find the factors of 66: $66 \div 1 = 66$, $66 \div 2 = 33$, $66 \div 3 = 22$, $66 \div 6 = 11$, $66 \div 11 = 6$ are the quotients that divide evenly. Note the factors begin to repeat with the last division. The factors of 66 are 1, 2, 3, 6, 11, 22, 33 and 66.

12. Find the factors of 112: $112 \div 1 = 112$, $112 \div 2 = 56$, $112 \div 4 = 28$, $112 \div 7 = 16$, $112 \div 8 = 14$, $112 \div 14 = 8$ are the quotients that divide evenly. Note the factors begin to repeat with the last division. The factors of 112 are 1, 2, 4, 7, 8, 14, 16, 28, 56. and 112.

13. The only factors of 41 are 1 and 41, so 41 is a prime number.

14. Since 5 divides 50 evenly, $50 \div 5 = 10$, 5 is a factor of 50 different from 1 or 50 so the number 50 is composite.

15. By definition, prime and composite numbers are greater than one. Hence, 1 is neither prime nor composite.

16. Since 7 divides 91 evenly, $91 \div 7 = 13$, 7 is a factor of 91 different from 1 or 91 so the number 91 is composite.

17. The number 5 is a factor of 125 different from 1 or 125 because $125 \div 5 = 25$. Therefore 125 is composite.

18. By definition, prime and composite numbers are greater than one. Hence, 0 is neither prime nor composite.

19. The number 3 divides 57 evenly since $57 \div 3 = 19$. Thus 3 is a factor 57 not equal to 1 or 57 so 57 is composite.

20. The only factors of 97 are 1 and 97, so 97 is a prime number.

21. The number 2 divides 12 evenly since $12 \div 2 = 6$. Thus 2 is a factor 12 not equal to 1 or 12 so 12 is composite.

22. The only factors of 37 are 1 and 37, so 37 is a prime number.

23. Since $81 \div 9 = 9$, the number 9 is a factor of 81 that is not equal to 1 or 81. Thus 81 is composite.

24. The only factors of 13 are 1 and 13, so 13 is a prime number.

25.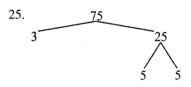

$75 = 3 \cdot 5 \cdot 5 = 3 \cdot 5^2$

26.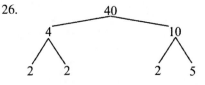

$40 = 2 \cdot 2 \cdot 2 \cdot 5 = 2^3 \cdot 5$

27.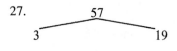

$57 = 3 \cdot 19$

28.

$62 = 2 \cdot 31$

29.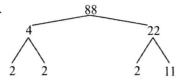

$$88 = 2 \cdot 2 \cdot 2 \cdot 11 = 2^3 \cdot 11$$

30.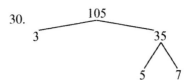

$$105 = 3 \cdot 5 \cdot 7$$

31.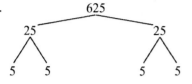

$$625 = 5 \cdot 5 \cdot 5 \cdot 5 = 5^4$$

32.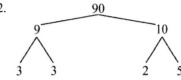

$$90 = 2 \cdot 3 \cdot 3 \cdot 5 = 2 \cdot 3^2 \cdot 5$$

33.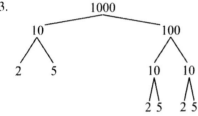

$$1000 = 2 \cdot 2 \cdot 2 \cdot 5 \cdot 5 \cdot 5 = 2^3 \cdot 5^3$$

34.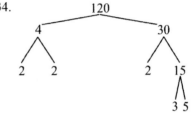

$$120 = 2 \cdot 2 \cdot 2 \cdot 3 \cdot 5 = 2^3 \cdot 3 \cdot 5$$

35.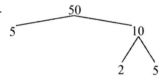

$$50 = 2 \cdot 5 \cdot 5 = 2 \cdot 5^2$$

36.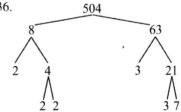

$$504 = 2 \cdot 2 \cdot 2 \cdot 3 \cdot 3 \cdot 7 = 2^3 \cdot 3^2 \cdot 7$$

37. Prime factorizations:
$3 = 3$, $\quad 11 = 11$

The LCM is $3 \cdot 11 = 33$

38. List multiples:
13: 13, 26, 39, 52, ...
26: 26, 52, 78, 104,...
The smallest value common to both
lists is 26. The LCM of 13 and 26 is 26.

39. Prime factorizations:
$2 = 2$, $\quad 43 = 43$

The LCM is $2 \cdot 43 = 86$.

40. Prime factorizations:
$8 = 2 \cdot 2 \cdot 2$, $\quad 20 = 2 \cdot 2 \cdot 5$

Multiply each prime factor the maximum
number of times it appears in any factorization.
The LCM is $2 \cdot 2 \cdot 2 \cdot 5 = 40$.

41. Prime factorizations:
    $6 = 2 \cdot 3, \qquad 8 = 2 \cdot 2 \cdot 2$
    $9 = 3 \cdot 3, \qquad 12 = 2 \cdot 2 \cdot 3$

    Multiply each prime factor the
    maximum number of times it
    appears in any factorization.
    The LCM is $2 \cdot 2 \cdot 2 \cdot 3 \cdot 3 = 72$.

42. Prime factorizations:
    $8 = 2 \cdot 2 \cdot 2, \qquad 12 = 2 \cdot 2 \cdot 3$
    $14 = 2 \cdot 7, \qquad 18 = 2 \cdot 3 \cdot 3$

    Multiply each prime factor the
    maximum number of times it
    appears in any factorization.
    The LCM is $2 \cdot 2 \cdot 2 \cdot 3 \cdot 3 \cdot 7 = 504$.

43. Prime factorizations:
    $6 = 2 \cdot 3, \qquad 7 = 7$
    $12 = 2 \cdot 2 \cdot 3, \qquad 16 = 2 \cdot 2 \cdot 2 \cdot 2$

    The LCM is $2 \cdot 2 \cdot 2 \cdot 2 \cdot 3 \cdot 7 = 336$.

44. Prime factorizations:
    $8 = 2 \cdot 2 \cdot 2, \qquad 12 = 2 \cdot 2 \cdot 3$
    $16 = 2 \cdot 2 \cdot 2 \cdot 2, \qquad 18 = 2 \cdot 3 \cdot 3$

    The LCM is $2 \cdot 2 \cdot 2 \cdot 2 \cdot 3 \cdot 3 = 144$.

45. Prime factorizations:
    $6 = 2 \cdot 3, \qquad 8 = 2 \cdot 2 \cdot 2$
    $12 = 2 \cdot 2 \cdot 3, \qquad 14 = 2 \cdot 7$

    The LCM is $2 \cdot 2 \cdot 2 \cdot 3 \cdot 7 = 168$.

46. Prime factorizations:
    $6 = 2 \cdot 3, \qquad 9 = 3 \cdot 3$
    $12 = 2 \cdot 2 \cdot 3, \qquad 14 = 2 \cdot 7$

    The LCM is $2 \cdot 2 \cdot 3 \cdot 3 \cdot 7 = 252$.

47. Prime factorizations:
    $7 = 7, \qquad 8 = 2 \cdot 2 \cdot 2$
    $9 = 3 \cdot 3, \qquad 16 = 2 \cdot 2 \cdot 2 \cdot 2$

    The LCM is $2 \cdot 2 \cdot 2 \cdot 2 \cdot 3 \cdot 3 \cdot 7 = 1008$.

48. Prime factorizations:
    $3 = 3, \qquad 7 = 7$
    $13 = 13, \qquad 22 = 2 \cdot 11$

    The LCM is $2 \cdot 3 \cdot 7 \cdot 11 \cdot 13 = 6006$.

49. The numerator, 14, is greater than the denominator, 5, so $\dfrac{14}{5}$ is improper.

50. There is both a whole number part, 4, and a fraction part, $\dfrac{7}{9}$, so $4\dfrac{7}{9}$ is mixed.

51. The numerator, 8, is less than the denominator, 15, so $\dfrac{8}{15}$ is proper.

52. There is both a whole number part, 2, and a fraction part, $\dfrac{5}{8}$, so $2\dfrac{5}{8}$ is mixed.

53. The numerator, 9, is less than the denominator, 13, so $\dfrac{9}{13}$ is proper.

54. The numerator, 33, is greater than the denominator, 7, so $\dfrac{33}{7}$ is improper.

55. The rectangle is divided into 24 squares. Of these, 21 are shaded. The fraction represented is $\dfrac{21}{24}$.

56. There are nine disks in all with four shaded. The fraction represented is $\dfrac{4}{9}$.

57. There are three circles shaded wholly while the last circle has three of five sections shaded. The mixed number represented here is $3\frac{3}{5}$.

58. One of the figures is divided into three triangles. One whole figure is shaded while two of the three triangles are shaded in the other. The mixed number represented is $1\frac{2}{3}$.

59. a. There are 8 animals total, 3 of them dogs. The dogs represent $\frac{3}{8}$ of the total.

    b. There are 8 animals total, 5 of them cats. The cats represent $\frac{5}{8}$ of the total.

60. a. There are 7 vehicles total, 2 of them buses. The buses represent $\frac{2}{7}$ of the total.

    b. There are 7 vehicles total, 5 of them cars. The cars represent $\frac{5}{7}$ of the total.

61. 
$$\begin{array}{r} 12 \\ 2\overline{)25} \\ \underline{24} \\ 1 \end{array}$$

$$\frac{25}{2} = 12\frac{1}{2}$$

62. 
$$\begin{array}{r} 4 \\ 8\overline{)37} \\ \underline{32} \\ 5 \end{array}$$

$$\frac{37}{8} = 4\frac{5}{8}$$

63. 
$$\begin{array}{r} 7 \\ 7\overline{)49} \\ \underline{49} \\ 0 \end{array}$$

$$\frac{49}{7} = 7$$

64. 
$$\begin{array}{r} 17 \\ 6\overline{)107} \\ \underline{6} \\ 47 \\ \underline{42} \\ 5 \end{array}$$

$$\frac{107}{6} = 17\frac{5}{6}$$

65. 
$$\begin{array}{r} 11 \\ 5\overline{)55} \\ \underline{5} \\ 5 \\ \underline{5} \\ 0 \end{array}$$

$$\frac{55}{5} = 11$$

66. 
$$\begin{array}{r} 11 \\ 7\overline{)78} \\ \underline{7} \\ 8 \\ \underline{7} \\ 1 \end{array}$$

$$\frac{78}{7} = 11\frac{1}{7}$$

67. $2\frac{3}{5} = \frac{2\cdot 5 + 3}{5} = \frac{13}{5}$

68. $14\frac{8}{9} = \frac{14\cdot 9 + 8}{9} = \frac{134}{9}$

69. $7\frac{3}{8} = \frac{7\cdot 8 + 3}{8} = \frac{59}{8}$

70. $22\frac{1}{5} = \frac{22\cdot 5 + 1}{5} = \frac{111}{5}$

71. $45\frac{2}{15} = \frac{45\cdot 15 + 2}{15} = \frac{677}{15}$

72. $10\frac{6}{17} = \frac{10\cdot 17 + 6}{17} = \frac{176}{17}$

73. The GCF of 9 and 72 is 9.

$$\frac{9}{72} = \frac{\overset{1}{\cancel{9}}}{\underset{8}{\cancel{72}}} = \frac{1}{8}$$

74. The GCF of 25 and 125 is 25.

$$\frac{25}{125} = \frac{\overset{1}{\cancel{25}}}{\underset{5}{\cancel{125}}} = \frac{1}{5}$$

75. $15 = 3\cdot 5$
    $24 = 2\cdot 2\cdot 2\cdot 3$

$$\frac{15}{24} = \frac{\overset{1}{\cancel{3}}\cdot 5}{2\cdot 2\cdot 2\cdot \underset{1}{\cancel{3}}} = \frac{1\cdot 5}{2\cdot 2\cdot 2\cdot 1} = \frac{5}{8}$$

76.  $20 = 2 \cdot 2 \cdot 5$
     $32 = 2 \cdot 2 \cdot 2 \cdot 2 \cdot 2$

$$\frac{20}{32} = \frac{\cancel{2} \cdot \cancel{2} \cdot 5}{\underset{1}{\cancel{2}} \cdot \underset{1}{\cancel{2}} \cdot 2 \cdot 2 \cdot 2}$$
$$= \frac{1 \cdot 1 \cdot 5}{1 \cdot 1 \cdot 2 \cdot 2 \cdot 2} = \frac{5}{8}$$

77.  The GCF of 12 and 36
     is 12.

$$\frac{12}{36} = \frac{\overset{1}{\cancel{12}}}{\underset{3}{\cancel{36}}} = \frac{1}{3}$$

78.  $18 = 2 \cdot 3 \cdot 3$
     $60 = 2 \cdot 2 \cdot 3 \cdot 5$

$$\frac{18}{60} = \frac{\cancel{2} \cdot \cancel{3} \cdot 3}{\underset{1}{\cancel{2}} \cdot 2 \cdot \underset{1}{\cancel{3}} \cdot 5}$$
$$= \frac{1 \cdot 1 \cdot 3}{1 \cdot 2 \cdot 1 \cdot 5} = \frac{3}{10}$$

79.  $25 = 5 \cdot 5$
     $90 = 2 \cdot 3 \cdot 3 \cdot 5$

$$\frac{25}{90} = \frac{\cancel{5} \cdot 5}{2 \cdot 3 \cdot 3 \cdot \underset{1}{\cancel{5}}}$$
$$= \frac{1 \cdot 5}{2 \cdot 3 \cdot 3 \cdot 1} = \frac{5}{18}$$

80.  $54 = 2 \cdot 3 \cdot 3 \cdot 3$
     $90 = 2 \cdot 3 \cdot 3 \cdot 5$

$$\frac{54}{90} = \frac{\cancel{2} \cdot \cancel{3} \cdot \cancel{3} \cdot 3}{\underset{1}{\cancel{2}} \cdot \underset{1}{\cancel{3}} \cdot \underset{1}{\cancel{3}} \cdot 5}$$
$$= \frac{1 \cdot 1 \cdot 1 \cdot 3}{1 \cdot 1 \cdot 1 \cdot 5} = \frac{3}{5}$$

81.  Since $14 \cdot 4 = 56$,

$$\frac{3}{14} = \frac{3 \cdot 4}{14 \cdot 4} = \frac{12}{56}$$

82.  Since $10 \cdot 8 = 80$,

$$\frac{1}{10} = \frac{1 \cdot 8}{10 \cdot 8} = \frac{8}{80}$$

83.  Since $11 \cdot 4 = 44$,

$$\frac{2}{11} = \frac{2 \cdot 4}{11 \cdot 4} = \frac{8}{44}$$

84.  Since $8 \cdot 8 = 64$,

$$\frac{3}{8} = \frac{3 \cdot 8}{8 \cdot 8} = \frac{24}{64}$$

85.  Rewrite using common denominator,
     then compare numerators.

$$\frac{11}{12} = \frac{11}{12}, \quad \frac{5}{6} = \frac{5 \cdot 2}{6 \cdot 2} = \frac{10}{12}$$

Since $10 < 11$, we have

$$\frac{5}{6} < \frac{11}{12}.$$

86.  Rewrite using common denominator,
     then compare numerators.

$$\frac{5}{6} = \frac{5 \cdot 3}{6 \cdot 3} = \frac{15}{18}, \quad \frac{7}{9} = \frac{7 \cdot 2}{9 \cdot 2} = \frac{14}{18}$$

Since $14 < 15$, we have

$$\frac{7}{9} < \frac{5}{6}.$$

87.  Rewrite using common denominator,
     then compare numerators.

$$\frac{5}{8} = \frac{5 \cdot 6}{8 \cdot 6} = \frac{30}{48}, \quad \frac{11}{12} = \frac{11 \cdot 4}{12 \cdot 4} = \frac{44}{48}$$
$$\frac{13}{16} = \frac{13 \cdot 3}{16 \cdot 3} = \frac{39}{48}$$

Since $30 < 39 < 44$, we have

$$\frac{5}{8} < \frac{13}{16} < \frac{11}{12}.$$

88.  Rewrite using common denominator,
     then compare numerators.

$$\frac{7}{9} = \frac{7 \cdot 8}{9 \cdot 8} = \frac{56}{72}, \quad \frac{5}{6} = \frac{5 \cdot 12}{6 \cdot 12} = \frac{60}{72}$$
$$\frac{7}{8} = \frac{7 \cdot 9}{8 \cdot 9} = \frac{63}{72}$$

Since $56 < 60 < 63$, we have

$$\frac{7}{9} < \frac{5}{6} < \frac{7}{8}.$$

89. $\dfrac{7}{9} \cdot \dfrac{3}{9} = \dfrac{7}{9} \cdot \dfrac{\overset{1}{\cancel{3}}}{\underset{3}{\cancel{9}}} = \dfrac{7 \cdot 1}{9 \cdot 3} = \dfrac{7}{27}$

90. $\dfrac{2}{16} \cdot \dfrac{3}{6} = \dfrac{\overset{1}{\cancel{2}}}{\underset{8}{\cancel{16}}} \cdot \dfrac{\overset{1}{\cancel{3}}}{\underset{2}{\cancel{6}}} = \dfrac{1 \cdot 1}{8 \cdot 2} = \dfrac{1}{16}$

91. $\dfrac{2}{3} \cdot \dfrac{10}{11} = \dfrac{2 \cdot 10}{3 \cdot 11} = \dfrac{20}{33}$

92. $\dfrac{2}{5} \cdot \dfrac{1}{3} = \dfrac{2 \cdot 1}{5 \cdot 3} = \dfrac{2}{15}$

93. $\dfrac{4}{5} \cdot \dfrac{1}{4} = \dfrac{\overset{1}{\cancel{4}}}{5} \cdot \dfrac{1}{\underset{1}{\cancel{4}}} = \dfrac{1 \cdot 1}{5 \cdot 1} = \dfrac{1}{5}$

94. $\dfrac{12}{27} \cdot \dfrac{5}{19} = \dfrac{\overset{4}{\cancel{12}}}{\underset{9}{\cancel{27}}} \cdot \dfrac{5}{19} = \dfrac{4 \cdot 5}{9 \cdot 19} = \dfrac{20}{171}$

95. $\dfrac{9}{80} \cdot \dfrac{7}{72} = \dfrac{\overset{1}{\cancel{9}}}{80} \cdot \dfrac{7}{\underset{8}{\cancel{72}}} = \dfrac{1 \cdot 7}{80 \cdot 8} = \dfrac{7}{640}$

96. $\dfrac{7}{8} \cdot \dfrac{5}{7} = \dfrac{\overset{1}{\cancel{7}}}{8} \cdot \dfrac{5}{\underset{1}{\cancel{7}}} = \dfrac{1 \cdot 5}{8 \cdot 1} = \dfrac{5}{8}$

97. $3\dfrac{5}{7} \cdot 2\dfrac{7}{8} = \dfrac{26}{7} \cdot \dfrac{23}{8} = \dfrac{\overset{13}{\cancel{26}}}{7} \cdot \dfrac{23}{\underset{4}{\cancel{8}}}$

   $= \dfrac{13}{7} \cdot \dfrac{23}{4} = \dfrac{299}{28}$

   $= 10\dfrac{19}{28}$

98. $4\dfrac{1}{3} \cdot 7\dfrac{3}{4} = \dfrac{13}{3} \cdot \dfrac{31}{4} = \dfrac{403}{12}$

   $= 33\dfrac{7}{12}$

99. $2\dfrac{1}{4} \cdot 19\dfrac{1}{2} = \dfrac{9}{4} \cdot \dfrac{39}{2} = \dfrac{351}{8}$

   $= 43\dfrac{7}{8}$

100. $6\dfrac{3}{4} \cdot 5\dfrac{1}{2} = \dfrac{27}{4} \cdot \dfrac{11}{2} = \dfrac{297}{8}$

   $= 37\dfrac{1}{8}$

101. $4\dfrac{8}{9} \cdot 1\dfrac{5}{6} = \dfrac{44}{9} \cdot \dfrac{11}{6} = \dfrac{\overset{22}{\cancel{44}}}{9} \cdot \dfrac{11}{\underset{3}{\cancel{6}}}$

   $= \dfrac{22}{9} \cdot \dfrac{11}{3} = \dfrac{242}{27}$

   $= 8\dfrac{26}{27}$

102. $3\dfrac{12}{15} \cdot 4\dfrac{7}{9} = \dfrac{57}{15} \cdot \dfrac{43}{9} = \dfrac{\overset{19}{\cancel{57}}}{15} \cdot \dfrac{43}{\underset{3}{\cancel{9}}}$

   $= \dfrac{19}{15} \cdot \dfrac{43}{3} = \dfrac{817}{45}$

   $= 18\dfrac{7}{45}$

103. $2\dfrac{1}{5} \cdot 12\dfrac{1}{6} = \dfrac{11}{5} \cdot \dfrac{73}{6} = \dfrac{803}{30}$

   $= 26\dfrac{23}{30}$

104. $18\dfrac{2}{3} \cdot 5\dfrac{6}{7} = \dfrac{56}{3} \cdot \dfrac{41}{7} = \dfrac{\overset{8}{\cancel{56}}}{3} \cdot \dfrac{41}{\underset{1}{\cancel{7}}}$

   $= \dfrac{8}{3} \cdot \dfrac{41}{1} = \dfrac{328}{3}$

   $= 109\dfrac{1}{3}$

105. $\dfrac{2}{4} \div \dfrac{1}{8} = \dfrac{2}{4} \cdot \dfrac{8}{1} = \dfrac{2}{\overset{}{\underset{1}{\cancel{4}}}} \cdot \dfrac{\overset{2}{\cancel{8}}}{1} = \dfrac{2 \cdot 2}{1 \cdot 1} = 4$

106. $\dfrac{5}{14} \div \dfrac{1}{2} = \dfrac{5}{14} \cdot \dfrac{2}{1} = \dfrac{5}{\underset{7}{\cancel{14}}} \cdot \dfrac{\overset{1}{\cancel{2}}}{1} = \dfrac{5 \cdot 1}{7 \cdot 1} = \dfrac{5}{7}$

107. $\dfrac{2}{6} \div \dfrac{5}{6} = \dfrac{2}{6} \cdot \dfrac{6}{5} = \dfrac{2}{\underset{1}{\cancel{6}}} \cdot \dfrac{\overset{1}{\cancel{6}}}{5} = \dfrac{2 \cdot 1}{1 \cdot 5} = \dfrac{2}{5}$

108. $\dfrac{1}{3} \div \dfrac{2}{13} = \dfrac{1}{3} \cdot \dfrac{13}{2} = \dfrac{1 \cdot 13}{3 \cdot 2} = \dfrac{13}{6} = 2\dfrac{1}{6}$

109. $\dfrac{6}{10} \div \dfrac{2}{5} = \dfrac{6}{10} \cdot \dfrac{5}{2} = \dfrac{\overset{3}{\cancel{6}}}{\underset{2}{\cancel{10}}} \cdot \dfrac{\overset{1}{\cancel{5}}}{\underset{1}{\cancel{2}}} = \dfrac{3 \cdot 1}{2 \cdot 1} = \dfrac{3}{2} = 1\dfrac{1}{2}$

110. $\dfrac{1}{3} \div \dfrac{7}{8} = \dfrac{1}{3} \cdot \dfrac{8}{7} = \dfrac{1 \cdot 8}{3 \cdot 7} = \dfrac{8}{21}$

111. $\dfrac{2}{3} \div \dfrac{5}{6} = \dfrac{2}{3} \cdot \dfrac{6}{5} = \dfrac{2}{\underset{1}{\cancel{3}}} \cdot \dfrac{\overset{2}{\cancel{6}}}{5} = \dfrac{2 \cdot 2}{1 \cdot 5} = \dfrac{4}{5}$

112. $\dfrac{3}{4} \div \dfrac{1}{2} = \dfrac{3}{4} \cdot \dfrac{2}{1} = \dfrac{3}{\underset{2}{\cancel{4}}} \cdot \dfrac{\overset{1}{\cancel{2}}}{1} = \dfrac{3 \cdot 1}{3 \cdot 1} = \dfrac{3}{2} = 1\dfrac{1}{2}$

113. $1\dfrac{3}{5} \div 2\dfrac{5}{6} = \dfrac{8}{5} \div \dfrac{17}{6} = \dfrac{8}{5} \cdot \dfrac{6}{17}$

$= \dfrac{8}{5} \cdot \dfrac{6}{17} = \dfrac{48}{85}$

114. $29\dfrac{1}{3} \div 2\dfrac{2}{3} = \dfrac{88}{3} \div \dfrac{8}{3} = \dfrac{88}{3} \cdot \dfrac{3}{8}$

$= \dfrac{\overset{11}{\cancel{88}}}{\underset{1}{\cancel{3}}} \cdot \dfrac{\overset{1}{\cancel{3}}}{\underset{1}{\cancel{8}}} = \dfrac{11 \cdot 1}{1 \cdot 1} = 11$

115. $20\dfrac{3}{4} \div 2\dfrac{1}{4} = \dfrac{83}{4} \div \dfrac{9}{4} = \dfrac{83}{4} \cdot \dfrac{4}{9}$

$= \dfrac{83}{\underset{1}{\cancel{4}}} \cdot \dfrac{\overset{1}{\cancel{4}}}{9} = \dfrac{83}{9} = 9\dfrac{2}{9}$

116. $5\dfrac{3}{5} \div 6\dfrac{2}{3} = \dfrac{28}{5} \div \dfrac{20}{3} = \dfrac{28}{5} \cdot \dfrac{3}{20}$

$= \dfrac{\overset{7}{\cancel{28}}}{5} \cdot \dfrac{3}{\underset{5}{\cancel{20}}} = \dfrac{7 \cdot 3}{5 \cdot 5} = \dfrac{21}{25}$

117. $14\dfrac{2}{5} \div 1\dfrac{2}{7} = \dfrac{72}{5} \div \dfrac{9}{7} = \dfrac{72}{5} \cdot \dfrac{7}{9}$

$= \dfrac{\overset{8}{\cancel{72}}}{5} \cdot \dfrac{7}{\underset{1}{\cancel{9}}} = \dfrac{8 \cdot 7}{5 \cdot 1}$

$= \dfrac{56}{5} = 11\dfrac{1}{5}$

118. $10\dfrac{1}{3} \div 3\dfrac{1}{5} = \dfrac{31}{3} \div \dfrac{16}{5} = \dfrac{31}{3} \cdot \dfrac{5}{16}$

$= \dfrac{31 \cdot 5}{3 \cdot 16}$

$= \dfrac{155}{48} = 3\dfrac{11}{48}$

119. $24\dfrac{2}{9} \div 5\dfrac{1}{3} = \dfrac{218}{9} \div \dfrac{16}{3} = \dfrac{218}{9} \cdot \dfrac{3}{16}$

$= \dfrac{\overset{109}{\cancel{218}}}{\underset{3}{\cancel{9}}} \cdot \dfrac{\overset{1}{\cancel{3}}}{\underset{8}{\cancel{16}}} = \dfrac{109 \cdot 1}{3 \cdot 8}$

$= \dfrac{109}{24} = 4\dfrac{13}{24}$

120. $15\dfrac{13}{23} \div 6\dfrac{1}{2} = \dfrac{358}{23} \div \dfrac{13}{2} = \dfrac{358}{23} \cdot \dfrac{2}{13}$

$= \dfrac{358 \cdot 2}{23 \cdot 13} = \dfrac{716}{299}$

$= 2\dfrac{118}{299}$

121. $\dfrac{1}{4} + \dfrac{1}{5} = \dfrac{1 \cdot 5}{4 \cdot 5} + \dfrac{1 \cdot 4}{5 \cdot 4} = \dfrac{5}{20} + \dfrac{4}{20} = \dfrac{9}{20}$

122. $\dfrac{1}{4} + \dfrac{3}{4} = \dfrac{4}{4} = 1$

123. $\dfrac{2}{5}+\dfrac{1}{10}=\dfrac{2\cdot 2}{5\cdot 2}+\dfrac{1}{10}=\dfrac{4}{10}+\dfrac{1}{10}=\dfrac{5}{10}=\dfrac{\overset{1}{\cancel{5}}}{\underset{2}{\cancel{10}}}=\dfrac{1}{2}$  124. $\dfrac{5}{8}+\dfrac{7}{12}=\dfrac{5\cdot 3}{8\cdot 3}+\dfrac{7\cdot 2}{12\cdot 2}=\dfrac{15}{24}+\dfrac{14}{24}=\dfrac{29}{24}=1\dfrac{5}{24}$

125. $\dfrac{6}{7}+\dfrac{1}{5}=\dfrac{6\cdot 5}{7\cdot 5}+\dfrac{1\cdot 7}{5\cdot 7}=\dfrac{30}{35}+\dfrac{7}{35}=\dfrac{37}{35}=1\dfrac{2}{35}$  126. $\dfrac{8}{9}+\dfrac{2}{7}=\dfrac{8\cdot 7}{9\cdot 7}+\dfrac{2\cdot 9}{7\cdot 9}=\dfrac{56}{63}+\dfrac{18}{63}=\dfrac{74}{63}=1\dfrac{11}{63}$

127. $\dfrac{5}{12}+\dfrac{1}{3}=\dfrac{5}{12}+\dfrac{1\cdot 4}{3\cdot 4}=\dfrac{5}{12}+\dfrac{4}{12}=\dfrac{9}{12}=\dfrac{\overset{3}{\cancel{9}}}{\underset{4}{\cancel{12}}}=\dfrac{3}{4}$  128. $\dfrac{7}{15}+\dfrac{5}{6}=\dfrac{7\cdot 2}{15\cdot 2}+\dfrac{5\cdot 5}{6\cdot 5}=\dfrac{14}{30}+\dfrac{25}{30}=\dfrac{39}{30}=\dfrac{\overset{13}{\cancel{39}}}{\underset{10}{\cancel{30}}}=1\dfrac{3}{10}$

129. Convert fraction parts:

$$26\dfrac{1}{2}=26\dfrac{1\cdot 3}{2\cdot 3}=26\dfrac{3}{6},$$

$$28\dfrac{2}{3}=28\dfrac{2\cdot 2}{3\cdot 2}=28\dfrac{4}{6}$$

Add:

$$26\dfrac{3}{6}$$
$$+\ 28\dfrac{4}{6}$$
$$\overline{\qquad\qquad}$$
$$54\dfrac{7}{6}=55\dfrac{1}{6}$$

130. Convert fraction parts:

$$1\dfrac{2}{3}=1\dfrac{2\cdot 4}{3\cdot 4}=1\dfrac{8}{12},$$

$$5\dfrac{5}{12}=5\dfrac{5}{12}$$

Add:

$$1\dfrac{8}{12}$$
$$+\ 5\dfrac{5}{12}$$
$$\overline{\qquad\qquad}$$
$$6\dfrac{13}{12}=7\dfrac{1}{12}$$

131. Convert fraction parts:

$$23\dfrac{1}{4}=23\dfrac{1\cdot 5}{4\cdot 5}=23\dfrac{5}{20},$$

$$20\dfrac{2}{5}=20\dfrac{2\cdot 4}{5\cdot 4}=20\dfrac{8}{20}$$

Add:

$$23\dfrac{5}{20}$$
$$+\ 20\dfrac{8}{20}$$
$$\overline{\qquad\qquad}$$
$$43\dfrac{13}{20}$$

132. Convert fraction parts:

$$37\dfrac{1}{5}=37\dfrac{1\cdot 3}{5\cdot 3}=37\dfrac{3}{15},$$

$$30\dfrac{2}{3}=30\dfrac{2\cdot 5}{3\cdot 5}=30\dfrac{10}{15}$$

Add:

$$37\dfrac{3}{15}$$
$$+\ 30\dfrac{10}{15}$$
$$\overline{\qquad\qquad}$$
$$67\dfrac{13}{15}$$

133. Convert fraction parts:

$$17\dfrac{1}{3}=17\dfrac{1\cdot 4}{3\cdot 4}=17\dfrac{4}{12},$$

$$2\dfrac{11}{12}=2\dfrac{11}{12}$$

134. Convert fraction parts:

$$1\dfrac{5}{8}=1\dfrac{5\cdot 3}{8\cdot 3}=1\dfrac{15}{24},$$

$$9\dfrac{13}{24}=9\dfrac{13}{24}$$

Add:

$$17\frac{4}{12}$$
$$+\quad 2\frac{11}{12}$$
$$\overline{\phantom{+}19\frac{15}{12}=20\frac{3}{12}=20\frac{1}{4}}$$

Add:

$$1\frac{15}{24}$$
$$+\quad 9\frac{13}{24}$$
$$\overline{\phantom{+}10\frac{28}{24}=11\frac{4}{24}=11\frac{1}{6}}$$

**135.** Convert fraction parts:

$$30\frac{1}{3}=30\frac{1\cdot 2}{3\cdot 2}=30\frac{2}{6},$$
$$24\frac{1}{2}=24\frac{1\cdot 3}{2\cdot 3}=24\frac{3}{6}$$

**136.** Convert fraction parts:

$$12\frac{4}{7}=12\frac{4\cdot 6}{7\cdot 6}=12\frac{24}{42},$$
$$15\frac{5}{6}=15\frac{5\cdot 7}{6\cdot 7}=15\frac{35}{42}$$

Add:

$$30\frac{2}{6}$$
$$+\quad 24\frac{3}{6}$$
$$\overline{\phantom{+}54\frac{5}{6}}$$

Add:

$$12\frac{24}{42}$$
$$+\quad 15\frac{35}{42}$$
$$\overline{\phantom{+}27\frac{59}{42}=28\frac{17}{42}}$$

**137.** $\dfrac{4}{5}-\dfrac{3}{5}=\dfrac{4-3}{5}=\dfrac{1}{5}$

**138.** $\dfrac{14}{15}-\dfrac{4}{9}=\dfrac{14\cdot 3}{15\cdot 3}-\dfrac{4\cdot 5}{9\cdot 5}=\dfrac{42}{45}-\dfrac{20}{45}=\dfrac{42-20}{45}=\dfrac{22}{45}$

**139.** $\dfrac{7}{20}-\dfrac{4}{12}=\dfrac{7\cdot 3}{20\cdot 3}-\dfrac{4\cdot 5}{12\cdot 5}=\dfrac{21}{60}-\dfrac{20}{60}=\dfrac{1}{60}$

**140.** $\dfrac{2}{5}-\dfrac{1}{6}=\dfrac{2\cdot 6}{5\cdot 6}-\dfrac{1\cdot 5}{6\cdot 5}=\dfrac{12}{30}-\dfrac{5}{30}=\dfrac{12-5}{30}=\dfrac{7}{30}$

**141.** $\dfrac{5}{9}-\dfrac{7}{18}=\dfrac{5\cdot 2}{9\cdot 2}-\dfrac{7}{18}=\dfrac{10}{18}-\dfrac{7}{18}=\dfrac{3}{18}=\dfrac{\cancel{3}^{1}}{\cancel{18}_{6}}=\dfrac{1}{6}$

**142.** $\dfrac{3}{4}-\dfrac{1}{16}=\dfrac{3\cdot 4}{4\cdot 4}-\dfrac{1}{16}=\dfrac{12}{16}-\dfrac{1}{16}=\dfrac{11}{16}$

**143.** $\dfrac{3}{4}-\dfrac{1}{2}=\dfrac{3}{4}-\dfrac{1\cdot 2}{2\cdot 2}=\dfrac{3}{4}-\dfrac{2}{4}=\dfrac{1}{4}$

**144.** $\dfrac{5}{9}-\dfrac{5}{12}=\dfrac{5\cdot 4}{9\cdot 4}-\dfrac{5\cdot 3}{12\cdot 3}=\dfrac{20}{36}-\dfrac{15}{36}=\dfrac{5}{36}$

**145.** Convert the fraction parts to have a common denominator. If the fraction part of the minuend is smaller than that of the subtrahend, borrow 1 from the whole number part.

$$52\frac{1}{6}=51\frac{7}{6}$$
$$14\frac{2}{3}=14\frac{4}{6}$$

Subtract whole number parts and fraction parts.

**146.** Convert the fraction parts to have a common denominator. If the fraction part of the minuend is smaller than that of the subtrahend, borrow 1 from the whole number part.

$$50\frac{4}{7}=50\frac{12}{21}$$
$$11\frac{5}{21}=11\frac{5}{21}$$

Subtract whole number parts and fraction parts.

$$51\frac{7}{6}$$
$$-\ 14\frac{4}{6}$$
$$37\frac{3}{6} = 37\frac{1}{2}$$

$$50\frac{12}{21}$$
$$-\ 11\frac{5}{21}$$
$$39\frac{7}{21} = 39\frac{1}{3}$$

147. $35\frac{7}{16} = 34\frac{23}{16}$

$13\frac{5}{8} = 13\frac{10}{16}$

$$34\frac{23}{16}$$
$$-\ 13\frac{10}{16}$$
$$21\frac{13}{16}$$

148. $49\frac{1}{3} = 49\frac{2}{6} = 48\frac{8}{6}$

$12\frac{4}{6} = 12\frac{4}{6}$

$$48\frac{8}{6}$$
$$-\ 12\frac{4}{6}$$
$$36\frac{4}{6} = 36\frac{2}{3}$$

149. $32\frac{3}{4} = 32\frac{3}{4}$

$23\frac{1}{2} = 23\frac{2}{4}$

$$32\frac{3}{4}$$
$$-\ 23\frac{2}{4}$$
$$9\frac{1}{4}$$

150. $57\frac{8}{13} = 57\frac{16}{26}$

$21\frac{15}{26} = 21\frac{15}{26}$

$$57\frac{16}{26}$$
$$-\ 21\frac{15}{26}$$
$$36\frac{1}{26}$$

151. $35\frac{2}{3} = 35\frac{8}{12} = 34\frac{20}{12}$

$28\frac{11}{12} = 28\frac{11}{12}$

$$34\frac{20}{12}$$
$$-\ 28\frac{11}{12}$$
$$6\frac{9}{12} = 6\frac{3}{4}$$

152. $56\frac{1}{6} = 55\frac{7}{6}$

$38\frac{2}{3} = 38\frac{4}{6}$

$$55\frac{7}{6}$$
$$-\ 38\frac{4}{6}$$
$$17\frac{3}{6} = 17\frac{1}{2}$$

153. An hour is equal to 60 minutes. The fraction of an hour that is 18 minutes is $\dfrac{18}{60} = \dfrac{\overset{3}{\cancel{18}}}{\underset{10}{\cancel{60}}} = \dfrac{3}{10}$.

154. Since 20 out of 85 stocks went up, the fraction of stocks that went up is $\dfrac{20}{85} = \dfrac{\overset{4}{\cancel{20}}}{\underset{17}{\cancel{85}}} = \dfrac{4}{17}$.

155. Randy's total banana purchase weighs $2\dfrac{3}{4} + 3\dfrac{5}{7} = \dfrac{11}{4} + \dfrac{26}{7} = \dfrac{11 \cdot 7}{4 \cdot 7} + \dfrac{26 \cdot 4}{7 \cdot 4} = \dfrac{77}{28} + \dfrac{104}{28} = \dfrac{181}{28} = 6\dfrac{13}{28}$ pounds.

156. One half of $18\dfrac{3}{4}$ is $\dfrac{1}{2} \cdot 18\dfrac{3}{4} = \dfrac{1}{2} \cdot \dfrac{75}{4} = \dfrac{75}{8} = 9\dfrac{3}{8}$. The halved recipe requires $9\dfrac{3}{8}$ ounces of flour.

157. For 23 sweaters, 23 times the material is needed or $23 \cdot \dfrac{4}{7} = \dfrac{23}{1} \cdot \dfrac{4}{7} = \dfrac{92}{7} = 13\dfrac{1}{7}$ yards.

158. Victor has $3 \cdot 3\dfrac{4}{9} = \dfrac{3}{1} \cdot \dfrac{31}{9} = \dfrac{\overset{1}{\cancel{3}}}{1} \cdot \dfrac{31}{\underset{3}{\cancel{9}}} = \dfrac{31}{3} = 10\dfrac{1}{3}$ feet.

159. The number of jars Granny Nell can fill is given by $50 \div \dfrac{4}{5}$. Dividing gives

$$50 \div \dfrac{4}{5} = \dfrac{50}{1} \cdot \dfrac{5}{4} = \dfrac{250}{4} = 62\dfrac{2}{4} = 62\dfrac{1}{2}$$

so Granny Nell can fill $62\dfrac{1}{2}$ jars holding $\dfrac{4}{5}$ pints each.

160. Since $\dfrac{3}{8}$ of 4800 is $\dfrac{3}{8} \cdot 4800 = \dfrac{3}{8} \cdot \dfrac{4800}{1} = \dfrac{3}{\underset{1}{\cancel{8}}} \cdot \dfrac{\overset{600}{\cancel{4800}}}{1} = \dfrac{3 \cdot 600}{1 \cdot 1} = 1800$, the Lopez family spends $1800 of their monthly income on rent and utilities.

161. Area is equal to the length times the width. Compute $45\dfrac{1}{2} \cdot 15\dfrac{3}{4} = \dfrac{91}{2} \cdot \dfrac{63}{4} = \dfrac{5733}{8} = 716\dfrac{5}{8}$. Since the length and width were given in feet, the area is $716\dfrac{5}{8}$ square feet.

162. Since each gallon allows Guillermo to drive $24\dfrac{1}{6}$ miles, 9 gallons will allow him to travel

$$9 \cdot 24\dfrac{1}{6} = \dfrac{9}{1} \cdot \dfrac{145}{6} = \dfrac{\overset{3}{\cancel{9}}}{1} \cdot \dfrac{145}{\underset{2}{\cancel{6}}} = \dfrac{3 \cdot 145}{1 \cdot 2} = \dfrac{435}{2} = 217\dfrac{1}{2} \text{ miles.}$$

163. The difference in length is given by $5\dfrac{1}{4} - 3\dfrac{5}{16}$ inches. Subtracting gives

$$5\dfrac{1}{4} - 3\dfrac{5}{16} = \dfrac{21}{4} - \dfrac{53}{16} = \dfrac{84}{16} - \dfrac{53}{16} = \dfrac{31}{16} = 1\dfrac{15}{16} \text{ inches.}$$

164. If $12\frac{4}{5}$ inches of rain fell in total, the average per hour over the 7 hour period is

$$12\frac{4}{5} \div 7 = \frac{64}{5} \div 7 = \frac{64}{5} \cdot \frac{1}{7} = \frac{64}{35} = 1\frac{29}{35} \text{ inches.}$$

165. Since $3\frac{1}{4} \cdot 580 = \frac{13}{4} \cdot \frac{580}{1} = \frac{13}{\cancel{4}} \cdot \frac{\overset{145}{\cancel{580}}}{1} = \frac{13 \cdot 145}{1 \cdot 1} = 1885$, the dining room set cost \$1885.

166. First find the total snowfall over the three month period:

$$15\frac{7}{8} + 9\frac{5}{8} + 18 = 24\frac{12}{8} + 18 = 42\frac{12}{8} = 43\frac{4}{8} = 43\frac{1}{2} \text{ inches.  The monthly average is then}$$

$$43\frac{1}{2} \div 3 = \frac{87}{2} \cdot \frac{1}{3} = \frac{\overset{29}{\cancel{87}}}{2} \cdot \frac{1}{\cancel{3}} = \frac{29}{2} = 14\frac{1}{2} \text{ inches per month.}$$

## Chapter 2 Assessment Test

1.  The only whole numbers that divide 31 evenly are 1 and 31 so the factors of 31 are 1 and 31.

2.  Since $64 \div 1 = 64$, $64 \div 2 = 32$, $64 \div 4 = 16$, $64 \div 8 = 8$ are the only quotients that yield whole numbers, before factors repeat, with 64 as the dividend, the factors of 64 are 1, 2, 4, 8, 16, 32, and 64.

3.  The number 5 divides 75 evenly, $75 \div 5 = 15$, and is not equal to 1 or 75, therefore 75 is composite.

4.  By trial and error 43 has no divisors other than itself or 1, so 43 is prime.

5.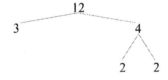

$$12 = 2 \cdot 2 \cdot 3 = 2^2 \cdot 3$$

6.

$$81 = 3 \cdot 3 \cdot 3 \cdot 3 = 3^4$$

7.  List the multiples of each number:
    6:  6, 12, 18, 24, 30, 36,...
    9:  9, 18, 27, 36, 45, ...
    The smallest multiple common to both
    lists is 18.  The LCM of 6 and 9 is 18.

8.  Write the prime factorization of each number:
    $7 = 7$, $\quad 8 = 2 \cdot 2 \cdot 2$, $\quad 14 = 2 \cdot 7$
    The prime numbers involved are 2 and 7.
    The factor 2 appears at most three times in
    any factorization while 7 appears at most once.
    The LCM of 7, 8, and 14 is $2 \cdot 2 \cdot 2 \cdot 7 = 56$.

9.  Since the numerator, 14, is greater than the denominator, 5, the fraction is improper.

10. The number contains both a whole number part and a fraction part, so $4\frac{7}{9}$ is a mixed number.

11. Since the numerator, 8, is less than the denominator, 15, the fraction is proper.

12. Two whole triangles are shaded and one out of two equal portions of another is shaded. The mixed number represented by the figure is $2\frac{1}{2}$.

13. Each circle is divided into eight sections so 8 sections represent a whole. There are a total of 11 sections shaded and so the figure represents the fraction $\frac{11}{8}$. Another way to view the figure is that one whole circle is shaded while 3 out of 8 sections of another are shaded. The figure represents the mixed number $1\frac{3}{8}$.

14. $6\overline{)23}$
    $\underline{18}$
    $5$

    $\frac{23}{6} = 3\frac{5}{6}$

15. $35\overline{)159}$
    $\underline{140}$
    $19$

    $\frac{159}{35} = 4\frac{19}{35}$

16. $27\overline{)81}$
    $\underline{81}$
    $0$

    $\frac{81}{27} = 3$

17. $7\frac{4}{5} = \frac{7 \cdot 5 + 4}{5} = \frac{39}{5}$

18. $2\frac{11}{16} = \frac{2 \cdot 16 + 11}{16} = \frac{43}{16}$

19. $21\frac{1}{3} = \frac{21 \cdot 3 + 1}{3} = \frac{64}{3}$

20. The GCF of 15 and 18 is 3.

    $\frac{15}{18} = \frac{\overset{5}{\cancel{15}}}{\underset{6}{\cancel{18}}} = \frac{5}{6}$

21. $21 = 3 \cdot 7$
    $49 = 7 \cdot 7$

    $\frac{21}{49} = \frac{3 \cdot \overset{1}{\cancel{7}}}{7 \cdot \underset{1}{\cancel{7}}} = \frac{3 \cdot 1}{7 \cdot 1} = \frac{3}{7}$

22. The GCF of 36 and 81 is 9.

    $\frac{36}{81} = \frac{\overset{4}{\cancel{36}}}{\underset{9}{\cancel{81}}} = \frac{4}{9}$

23. $16 = 2 \cdot 2 \cdot 2 \cdot 2$
    $60 = 2 \cdot 2 \cdot 3 \cdot 5$

    $\frac{16}{60} = \frac{\overset{1}{\cancel{2}} \cdot \overset{1}{\cancel{2}} \cdot 2 \cdot 2}{\underset{1}{\cancel{2}} \cdot \underset{1}{\cancel{2}} \cdot 3 \cdot 5} = \frac{2 \cdot 2}{3 \cdot 5} = \frac{4}{15}$

24. Since $8 \cdot 6 = 48$

    $\frac{5}{8} = \frac{5 \cdot 6}{8 \cdot 6} = \frac{30}{48}$

25. Since $9 \cdot 9 = 81$

    $\frac{2}{9} = \frac{2 \cdot 9}{9 \cdot 9} = \frac{18}{81}$

26. Since $6 \cdot 4 = 24$

    $\frac{15}{6} = \frac{15 \cdot 4}{6 \cdot 4} = \frac{60}{24}$

27. Write the fractions with a common denominator.

    $\frac{4}{7} = \frac{4 \cdot 8}{7 \cdot 8} = \frac{32}{56}$

    $\frac{1}{2} = \frac{1 \cdot 28}{2 \cdot 28} = \frac{28}{56}$

28. Write the fractions with a common denominator.

    $\frac{22}{35} = \frac{22 \cdot 2}{35 \cdot 2} = \frac{44}{70}$

    $\frac{37}{70} = \frac{37}{70}$

$$\frac{5}{8} = \frac{5 \cdot 7}{8 \cdot 7} = \frac{35}{56}$$

Since $28 < 32 < 35$,

$$\frac{1}{2} < \frac{4}{7} < \frac{5}{8}$$

$$\frac{6}{10} = \frac{6 \cdot 7}{10 \cdot 7} = \frac{42}{70}$$

Since $37 < 42 < 44$,

$$\frac{37}{70} < \frac{6}{10} < \frac{22}{35}$$

29. $\dfrac{7}{9} \cdot \dfrac{3}{5} = \dfrac{7}{\cancel{9}_{3}} \cdot \dfrac{\cancel{3}^{1}}{5} = \dfrac{7 \cdot 1}{3 \cdot 5} = \dfrac{7}{15}$

30. $\dfrac{1}{5} \cdot \dfrac{5}{10} \cdot 1\dfrac{2}{3} = \dfrac{1}{5} \cdot \dfrac{5}{10} \cdot \dfrac{5}{3} = \dfrac{1}{\cancel{5}_{1}} \cdot \dfrac{\cancel{5}^{1}}{\cancel{10}_{2}} \cdot \dfrac{\cancel{5}^{1}}{3}$

$$= \dfrac{1 \cdot 1 \cdot 1}{1 \cdot 2 \cdot 3} = \dfrac{1}{6}$$

31. $3\dfrac{3}{4} \cdot 8\dfrac{1}{2} \cdot 2 = \dfrac{15}{4} \cdot \dfrac{17}{2} \cdot \dfrac{2}{1} = \dfrac{15}{4} \cdot \dfrac{17}{\cancel{2}_{1}} \cdot \dfrac{\cancel{2}^{1}}{1}$

$$= \dfrac{15 \cdot 17 \cdot 1}{4 \cdot 1 \cdot 1} = \dfrac{255}{4} = 63\dfrac{3}{4}$$

32. $\dfrac{11}{18} \div \dfrac{5}{6} = \dfrac{11}{18} \cdot \dfrac{6}{5} = \dfrac{11}{\cancel{18}_{3}} \cdot \dfrac{\cancel{6}^{1}}{5} = \dfrac{11 \cdot 1}{3 \cdot 5} = \dfrac{11}{15}$

33. $4\dfrac{1}{3} \div 2\dfrac{1}{5} = \dfrac{13}{3} \div \dfrac{11}{5} = \dfrac{13}{3} \cdot \dfrac{5}{11} = \dfrac{65}{33} = 1\dfrac{32}{33}$

34. $120 \div \dfrac{5}{6} = \dfrac{120}{1} \cdot \dfrac{6}{5} = \dfrac{\cancel{120}^{24}}{1} \cdot \dfrac{6}{\cancel{5}_{1}} = \dfrac{24 \cdot 6}{1 \cdot 1} = 144$

35. $\dfrac{5}{9} + \dfrac{1}{2} = \dfrac{5 \cdot 2}{9 \cdot 2} + \dfrac{1 \cdot 9}{2 \cdot 9} = \dfrac{10}{18} + \dfrac{9}{18} = \dfrac{19}{18} = 1\dfrac{1}{18}$

36. $\dfrac{4}{6} + 3\dfrac{5}{8} + \dfrac{7}{12} = \dfrac{4}{6} + \dfrac{29}{8} + \dfrac{7}{12} = \dfrac{4 \cdot 4}{6 \cdot 4} + \dfrac{29 \cdot 3}{8 \cdot 3} + \dfrac{7 \cdot 2}{12 \cdot 2}$

$$= \dfrac{16}{24} + \dfrac{87}{24} + \dfrac{14}{24} = \dfrac{117}{24}$$

$$= 4\dfrac{21}{24} = 4\dfrac{7}{8}$$

37. $\dfrac{1}{10} + 3\dfrac{2}{5} + 1\dfrac{5}{6} + \dfrac{1}{15} = \dfrac{1}{10} + \dfrac{17}{5} + \dfrac{11}{6} + \dfrac{1}{15}$

$$= \dfrac{3}{30} + \dfrac{102}{30} + \dfrac{55}{30} + \dfrac{2}{30}$$

$$= \dfrac{162}{30} = 5\dfrac{12}{30} = 5\dfrac{2}{5}$$

38. $6\dfrac{3}{5} = 6\dfrac{6}{10}, \quad 4\dfrac{1}{2} = 4\dfrac{5}{10}$

$$6\dfrac{6}{10} - 4\dfrac{5}{10} = 2\dfrac{1}{10}$$

39. $12\dfrac{1}{3} = 12\dfrac{4}{12} = 11\dfrac{16}{12}$

$$5\dfrac{3}{4} = 5\dfrac{9}{12}$$

$$12\dfrac{1}{3} - 5\dfrac{3}{4} = 11\dfrac{16}{12} - 5\dfrac{9}{12} = 6\dfrac{7}{12}$$

40. $15\dfrac{18}{25} = 15\dfrac{54}{75}$

$$10\dfrac{7}{15} = 10\dfrac{35}{75}$$

$$15\dfrac{18}{25} - 10\dfrac{7}{15} = 15\dfrac{54}{75} - 10\dfrac{35}{75} = 5\dfrac{19}{75}$$

41. a. The total number of students is $25 + 22 + 8 = 55$. There are 22 of these students from out-of-state.

   The fraction that represents the out-of-state students is $\dfrac{22}{55} = \dfrac{2 \cdot \cancel{11}^{1}}{5 \cdot \cancel{11}_{1}} = \dfrac{2}{5}$.

   b. There are 8 students out of 55 total that are from out of the country. The fraction represented by out-of-country students is $\dfrac{8}{55}$.

   c. The number of in-state and out-of-state students combined is $24 + 22 = 47$. The fraction representing these students is $\dfrac{47}{55}$.

42. The number of $\dfrac{5}{8}$ mile segments in a 75 mile stretch is $75 \div \dfrac{5}{8}$. Performing the division gives

$$75 \div \dfrac{5}{8} = \dfrac{75}{1} \cdot \dfrac{8}{5} = \dfrac{\cancel{75}^{15}}{1} \cdot \dfrac{8}{\cancel{5}_{1}} = \dfrac{15 \cdot 8}{1 \cdot 1} = 120.$$ So 120 signs are on the 75-mile stretch.

43. If each lap is $3\dfrac{3}{5}$ miles long and 54 laps were run, the total number of miles covered is $3\dfrac{3}{5} \cdot 54$.

   Multiplying we find this total is $3\dfrac{3}{5} \cdot 54 = \dfrac{18}{5} \cdot \dfrac{54}{1} = \dfrac{972}{5} = 194\dfrac{2}{5}$ miles.

44. The height of the plant is the sum of its original height and the growth over each of the last two months. Convert the different summands to equivalent mixed numbers with common denominators.

$$12\dfrac{3}{4} = 12\dfrac{9}{12}, \quad 2\dfrac{1}{6} = 2\dfrac{2}{12}, \quad 3\dfrac{1}{2} = 3\dfrac{6}{12}$$

   The total height is $12\dfrac{9}{12} + 2\dfrac{2}{12} + 3\dfrac{6}{12} = 17\dfrac{17}{12} = 18\dfrac{5}{12}$ inches.

45. Two-thirds of the revenue is $\dfrac{2}{3} \cdot 54000 = \dfrac{2}{3} \cdot \dfrac{54000}{1} = \dfrac{2}{\cancel{3}_{1}} \cdot \dfrac{\cancel{54000}^{18000}}{1} = 36,000$. The company's expenses were

   $36,000.

46. The thickness of the copper is $1\dfrac{13}{16} - \dfrac{35}{64}$. Computing this difference gives

$$1\dfrac{13}{16} - \dfrac{35}{64} = \dfrac{29}{16} - \dfrac{35}{64} = \dfrac{29 \cdot 4}{16 \cdot 4} - \dfrac{35}{64} = \dfrac{116}{64} - \dfrac{35}{64} = \dfrac{81}{64} = 1\dfrac{17}{64}$$ centimeters.

# Chapter 3    Decimals

## Section 3.1    Understanding Decimals

### Concept Check

1.   A _decimal_  _fraction_   is a fraction whose denominator is a power of 10.

3.   A number written in decimal notation is called a  _decimal_ .

5.   The names of the place values to the right of the decimal point end in - _ths_ . (For example, ten*ths*, hundred*ths*, thousand*ths*, etc.)

7.   To convert a decimal to a fraction, write the  _digits_  to the right of the decimal point in the numerator, and write the power of 10 that has as many zeros as  _place_    _values_  in the decimal in the denominator.

9.   When rounding, if the digit to the right of the specified place value is  _four_  or less, the digit in the specified place value remains the same, whereas if the digit to the right of the specified place value is  _five_   or more, increase the digit in the specified place value by one.

### Guide Problems

11.  In the decimal 23.5618, identify the digit in each place.

   a. ones place  _3_
   The one places is the first place to the
   *left* of the decimal point.

   c. thousandths place  _1_
   The thousandths place is the third place
   to the *right* of the decimal point.

   b. tenths place  _5_
   The tenths place is the first place to the
   *right* of the decimal point.

   d. ten-thousandths place  _8_
   The ten-thousandths place is the fourth
   place to the *right* of the decimal point.

13.  The digit 5 underlined in 4_5_3.23
   is in the *tens* place, the second
   place to the left of the decimal.

15.  The digit 9 underlined in 3._9_61
   is in the *tenths* place, the first
   place to the right of the decimal.

17.  The digit 4 underlined in 2.876_4_52
   is in the second place of the
   thousandths group and so is in the
   *ten-thousandths* place.

19.  The digit 6 underlined in 13.9014_6_5
   is in the *millionths* place.

### Guide Problems

21.  Write 16.74 in word form.

   a. Write the whole number portion in word
   form. The whole number portion here is
   16 and the word form is "sixteen" .

   c. How many decimal places are in the number?
   Two, since two digits: 7 and 4 follow the
   decimal.

   b. Write the word representing the decimal point.
   A decimal point is always represented by the
   word "and" . Note this is the only time *and* is
   used in a number's word form.

   d. The number of decimal places in part c
   indicates  the _hundredths_ place. From part c.
   there are two places and the second is the
   hundredths place.

e. Write the decimal part of the number in word form. The digits after the decimal form 74 which has word form *seventy-four*. Add the word for the appropriate decimal place found in part d. to get "seventy-four hundredths".

f. Write the entire number in word form. The word form for 16.74 is found by joining the answers to parts a, b, and e and is "sixteen and seventy four hundredths".

23. 0.9 has one decimal place so the word form uses "tenths." The digits after the decimal point form "nine" so 0.9 has word form "nine-tenths." Note that a whole number portion of 0 does not contribute to the word form and the word *and* is not used for the decimal point in this case.

25. 0.0054 has decimal part 54 read "fifty-four." There are four decimal places indicating the ten-thousandths place. The word form is "fifty-four ten-thousandths."

27. For 1.34,
    whole number part of 1 is "one",
    decimal point is "and",
    decimal part of 34 is "thirty-four",
    two decimal places means "hundredths".
    The word form of 1.34 is
    "one and thirty-four hundredths."

29. For 25.3652,
    25 is read "twenty-five"
    decimal point is read "and"
    3652 is read "three thousand six hundred
        fifty-two"
    four decimal places means "ten-thousandths"
    The word form of 25.3652 is "twenty-five and three thousand six hundred fifty-two ten-thousandths."

31. Four decimal places in 0.0062 indicate the ten-thousandths place. The word form is "sixty-two ten-thousandths."

33. One decimal place in 15.7 indicates tenths. Decimal point is read as "and." The word form is "fifteen and seven tenths."

## Guide Problems

35. Write the number sixty-seven and fifteen ten-thousands in decimal notation.

a. Write the whole number part of the number in decimal notation.
    67
   Note the whole number part "sixty-seven" is *before* the "and."

b. Write the decimal part of the number in decimal notation.
    .0015
   Note the decimal part, "fifteen ten-thousands," comes after the "and." Further, "ten-thousands" indicates four decimal places.

c. Write the entire number in decimal notation.
    67.0015
   Simply combine the whole number part from a. and the decimal part from b.

37. "one hundred eighty-three thousandths"
    There is no whole number part so use 0.
    The word "thousandths" indicates three decimal places so the decimal part is .183.
    The decimal notation is 0.183 .

39. "fifteen ten-thousandths"
    There is no whole number part so use 0.
    The phrase "ten-thousandths" indicates four decimal places so the decimal part is .0015. The decimal notation is 0.0015 .

41. "five hundred ninety-eight and eight tenths"
    Whole number part: 598
    Decimal part: .8
    Decimal notation: 598.8

43. "forty-six and three hundredths"
    Whole number part: 46
    Decimal part: .03
    Decimal notation: 46.03

45. "fourteen and thirty-five hundredths"
    Whole number part: 14
    Decimal part: .35
    Decimal notation: 14.35

47. "twenty-nine hundred-thousandths"
    Whole number part: 0
    Decimal part: .00029
    Decimal notation: 0.00029

## Guide Problems

49. Consider the decimal 0.51.

    a. To write 0.51 as a decimal fraction,
       what whole number do we write as
       the numerator?  <u>51</u>
       Use the digits that form the decimal
       part.

    b. What power of 10 do we write as the
       the denominator?  <u>100</u>
       There are two decimal places in the
       given number so the denominator is
       that power of 10 having two zeros.

    c. Write 0.51 as a decimal fraction. Simplify
       if possible. Use the numerator and
       denominator from a. and b. above.

    $$0.51 = \frac{51}{100}$$

51. Consider the decimal 28.37.

    a. To write 28.37 as a mixed number,
       what whole number do we write as
       the numerator of the fraction part?
       <u>37</u>  Use the digits after the decimal.

    b. What power of 10 do we write as the
       the denominator of the fraction part?
       <u>100</u>  There are two decimal places
       in the given number so the denominator
       is that power of 10 having two zeros.

    c. Write 28.37 as a decimal fraction.
       Simplify, if possible.
       Use the whole number part of the
       decimal as the whole number part of
       the mixed number. Form the fraction
       part with the numerator and denominator
       from a. and b. above.

    $$28.37 = 28\frac{37}{100}$$

53. 0.7
    Numerator: 7
    Denominator: 10

    $$0.7 = \frac{7}{10}$$

55. 0.001
    Numerator: 1
    Denominator: 1000

    $$0.001 = \frac{1}{1000}$$

57. 0.64
    Numerator: 64
    Denominator: 100

    $$0.64 = \frac{64}{100} = \frac{16}{25}$$

59. 3.75
    Numerator: 75
    Denominator: 100
    Whole number part: 3

    $$3.75 = 3\frac{75}{100} = 3\frac{3}{4}$$

61. 26.088
    Numerator: 88
    Denominator: 1000
    Whole number part: 26

    $$26.088 = 26\frac{88}{1000} = 26\frac{11}{125}$$

63. 14.5003
    Numerator: 5003
    Denominator: 10,000
    Whole number part: 14

    $$14.5003 = 14\frac{5003}{10,000}$$

## Guide Problems

65. Use the symbol < , >. or = to compare 4.596 and 4.587.  Write 4.596 and 4.587 one above another so that the decimal points are aligned. (Insert zeros, as needed.)

        4.596
        4.587

a. Compare the digits of each number.  Which is the first place value where there is a difference?
<u>hundredths</u> The top number has a 9 in the hundredths place while the bottom number has an 8.

b. Which number is larger?
<u>4.596</u> Comparing the digits in the hundredths places of each number  (9 in 4.596 is more than 8 in 4.587)  we see 4.596 is the larger number.

c. Use the symbol <, >, or = to write a true statement for the numbers.
4.596 > 4.587  or
4.587 < 4.596  since 4.596 is the larger number.

*In problems 67 through 77, the procedure given in the Guide Problems is applied. The numbers are stacked with the decimal places aligned. Zeros are tacked on if necessary. The first place where the numbers disagree is given and the result of comparing the digits in that place is used to determine whether <, >, or = is appropriate.*

67. 0.57 < 0.62

    0.57  Tenths place: 5
    0.62  is less than 6.

69. 4.017 < 4.170

    4.017  Tenths place: 0
    4.170  is less than 1.

71. 0.0023 = 0.00230

    0.00230  Add a zero on top.
    0.00230  Numbers agree in
                every decimal place.

73. 243.33 > 242.33

    243.33  Ones place: 3
    242.33  is more than 2.

75. 133.52 > 133.5

    133.52  Hundredths place: 2
    133.50  is more than 0.

77. 0.730 > 0.7299

    0.7300  Tenths place: 3
    0.7299  is more than 2.

79. Since  0.5 < 0.564 and 0.5 < 0.5654 , the value 0.5 is the smallest. Since 0.564 < 0.5654 , the value 0.5654 is the largest. The numbers in ascending order are 0.5, 0.564, 0.5654.

81. Since  4.6 > 4.576 and 4.6 > 4.57 , the value 4.6 is the largest. Since 4.576 > 4.57 , the value 4.57 is the smallest. The numbers in descending order are 4.6, 4.576, 4.57.

83. Since  1.379 < 1.3856, 1.3856 < 1.3879, and 1.3879 < 1.3898, the numbers in ascending order are 1.379, 1.3856, 1.3879, 1.3898.

## Guide Problems

85. Round 3.07869 to the nearest thousandth.

a. What digit is in the thousandths place?
    <u>8</u>, as shown here: 3.07<u>8</u>69

b. What digit is to the right of the digit in the thousandths place?
    <u>6</u>, as shown here:  3.078<u>6</u>9

c. Explain what to do next.
<u>Increase the specified digit, 8, by one, because the digit in the ten-thousands place is 5 or more. Next, change each digit to the right of the specified place value to 0.</u>

d. Write the rounded number. <u>3.079</u> Note the digits that are changed to zero need not be written since they are at the end of the decimal part.

*In problems 87 through 97, the procedure outlined in the Guide Problems is used. The digit in the specified place is indicated with a single underline. The digit to its right is doubly underlined. If the digit to the right is 5 or greater, increase the specified digit by one, otherwise leave the specified digit the same. In either case, change all digits to the right of the specified digits to zero. If the zeroed digits are on the right of the decimal point they are not written. However, if the digit in the specified place is a zero, as in problems 96 through 98, it is always written.*

87.  14.5734 to the nearest thousandth is 14.573.

89.  8.328 to the nearest tenth is 8.3.

91.  235.88 to the nearest ten is 240.00.

93.  841.9844 to the nearest tenth is 842.0.

95.  0.00394875 to the nearest millionth is 0.003949.

97.  10.3497 to the nearest thousandth is 10.3500.

99.  The two decimal places in 0.34 indicate a place value of hundredths. The length "0.34 millimeters" in word form is "thirty-four hundredths millimeters"

101. In the word form "three and thirty-four hundredths" the word *hundredths* indicates two decimal places while the word *and* states where the decimal point is placed. In decimal notation the number is 3.34 so the length of the wire is 3.34 inches.

103. First, note that 5.4318 > 5.4132 since the first place they disagree is in the hundredths place and 3 is larger than 1. Similarly, 5.4318 > 5.399 by comparing the tenths places. Thus, 5.4318 is the largest number. Further, 5.4132 > 5.399, by comparing the tenths places, so 5.399 is the smallest number. In descending order, the three times are 5.4318 seconds, 5.4132 seconds, 5.399 seconds.

105. The ones place of $349.95 contains a digit of 9. The digit to the right is also 9 so the dollar value is rounded up to $350.

107. The digit in the hundredths place of 0.0441 is 4. The digit to its right, in the thousandths place, is also 4. Leave the digit in the hundredths place alone and zero out the digits after it to obtain a rounded thickness of 0.04 inches.

109. a. When writing a check, the decimal part or cents is written as a decimal fraction. 0.68 has two decimal places and so is $\frac{68}{100}$. Decimal fractions are not reduced when writing a check. The total amount of $17.68 is written as "seventeen and $\frac{68}{100}$ dollars."

   b. The word form "two hundred fifty-one dollars and ten cents" in decimal notation is $251.10.

## Cumulative Review Exercises

1.  The hundreds digit of 34,572 is 5. The digit to its right is 7. Round up to 34,600.

3.  Compute the sum vertically:

$$\begin{array}{r} {\scriptstyle 1\,2} \\ 398 \\ 436 \\ +\ \ 19 \\ \hline 853 \end{array}$$

5. List multiples of 12 and 18.
   12: 12, 24, 36, 48, 60, 72, …
   18: 18, 36, 54, 72, 90, …
   The common multiples are
   36, 72, …The least common
   multiple or LCM is 36.

7. $20 = 2 \cdot 2 \cdot 5 = 2^2 \cdot 5$

9. $5\dfrac{3}{10} = \dfrac{10 \cdot 5 + 3}{10} = \dfrac{53}{10}$

## Section 3.2    Adding and Subtracting Decimals

## Concept Check

1. When adding and subtracting decimals, we write the numbers so that the  decimal   points  are vertically aligned.

3. When adding or subtracting decimals, add or subtract as if working with  whole  numbers.  When adding, carry if necessary.  When subtracting, borrow if necessary.

## Guide Problems

5. Add $4.6 + 2.09 + 15.48$.

a. Write the decimals so that the decimal points are vertically aligned. If necessary, insert extra zeros to the right of the last digit after the decimal point.

$$
\begin{array}{r}
4.60 \\
2.09 \\
+\ 15.48 \\
\end{array}
$$

Note a 0 was inserted in the first addend.

b. Add as with whole numbers. Place the decimal point in the sum.

$$
\begin{array}{r}
\overset{1\,1\ \ 1}{4.60} \\
2.09 \\
+\ 15.48 \\
\hline
22.17 \\
\end{array}
$$

The decimal point in the sum is aligned with the decimal points in the addends.

*In problems 7 through 35, the procedure given in the Guide Problems is followed to add decimal numbers. First write the addends vertically with the decimal points aligned. Insert zeros after the decimal points so all addends have the same number of digits after the decimal. Add the numbers just as whole numbers were added in Chapter 1. Insert a decimal point in the sum so it is aligned with the decimal points in the addends.*

7.
$$
\begin{array}{r}
2.45 \\
+\ 0.24 \\
\hline
2.69 \\
\end{array}
$$

9.
$$
\begin{array}{r}
\overset{1}{30.63} \\
+\ 38.55 \\
\hline
69.18 \\
\end{array}
$$

11.
$$
\begin{array}{r}
\overset{1\,1}{3.189} \\
+\ 0.015 \\
\hline
3.204 \\
\end{array}
$$

13.
$$
\begin{array}{r}
5.134 \\
+\ 0.635 \\
\hline
5.769 \\
\end{array}
$$

15.
$$
\begin{array}{r}
\overset{1\ \ 1}{52.5805} \\
+\ 26.7890 \\
\hline
79.3695 \\
\end{array}
$$

17.
$$
\begin{array}{r}
\overset{1\ 1}{9.64} \\
+\ 6.37 \\
\hline
16.01 \\
\end{array}
$$

19.
$$
\begin{array}{r}
\overset{1}{23.485} \\
11.240 \\
\hline
34.725 \\
\end{array}
$$

21.
$$
\begin{array}{r}
\overset{1\,1\ 1}{49.3500} \\
+\ 0.9928 \\
\hline
50.3428 \\
\end{array}
$$

23.  $\overset{1\,1\,1}{3.3396}$
     $+\ 9.1726$
     ─────────
     $12.5122$

25.  $\overset{1\,1}{1.13}$
     $5.70$
     $+\ 3.85$
     ───────
     $10.68$

27.  $\overset{1\ 1}{21.390}$
     $3.767$
     $+\ 11.001$
     ────────
     $36.158$

29.  $\overset{1\,1}{3.20}$
     $14.82$
     $19.00$
     $+\ 40.27$
     ───────
     $77.29$

31.  $\overset{2\ 2}{1.54}$
     $4.87$
     $0.79$
     $+\ 6.00$
     ───────
     $13.20$
     or  $13.2$

*Problems 33 through 35 are all stated in word form. The keywords and phrases "add" and "the sum of" are indicators of an addition problem. The word forms of each number are first converted to decimal numbers and then the appropriate addition is performed.*

33.  Twelve and two-tenths is 12.2.
     Thirty-two hundredths is 0.32.
     Adding the two gives

          $12.20$
        $+\ 0.32$
          ──────
          $12.52$

35.  The sum of four dollars and fifty-one cents,
     \$4.51, and two dollars and forty cents, \$2.40, is
     \$4.51 + \$2.40 = \$6.91.

## Guide Problems

37.  Subtract $16.4 - 2.91$.

   a. Write the decimals so that the decimal
      points are vertically aligned. If necessary,
      insert extra zeros to the right of the last
      digit after the decimal point.

          $16.40$
        $-\ 2.91$

      Note a 0 was inserted in the minuend.

   b. Subtract as with whole numbers. Place the
      decimal point in the difference.

          $\overset{\quad\;\;13}{\underset{5\;\;\cancel{3}\;10}{1\cancel{6}.\cancel{4}\cancel{0}}}$
        $-\ 2.91$
          ──────
          $13.49$

      The decimal point in the difference is
      aligned with the other decimal points.

*In problems 39 through 67, the procedure given in the Guide Problems is followed to subtract decimal numbers. First write the numbers vertically with the decimal points aligned. Insert zeros after the decimal points so all numbers have the same number of digits after the decimal. Subtract the numbers just as whole numbers were subtracted in Chapter 1. Insert a decimal point in the difference so it is aligned with the decimal points in the other numbers.*

39.  $9.9$
     $-\ 9.5$
     ──────
     $0.4$

41.  $\overset{7\ \ 14}{\cancel{8}.\cancel{4}4}$
     $-\ 2.71$
     ───────
     $5.73$

43.  $\overset{6\ \ \ 10}{6\cancel{7}.\cancel{0}9}$
     $-\ 13.45$
     ────────
     $53.64$

45.  $\overset{\quad\;\;16\ 14}{\underset{8\;\;\cancel{9}\;\cancel{7}\;10}{8\cancel{9}.\cancel{7}\cancel{5}\cancel{0}}}$
     $-\ 87.893$
     ─────────
     $1.857$

47.  $6.4$
     $-\ 1.2$
     ──────
     $5.2$

49.  $\overset{7\ \ 13}{1\cancel{8}.\cancel{3}}$
     $-\ 3.8$
     ──────
     $14.5$

51.  $\overset{8\ \ 12}{\cancel{9}\cancel{2}.58}$
     $-\ 27.21$
     ────────
     $65.37$

53.  $\overset{8\ \ 15}{\cancel{9}\cancel{5}.68}$
     $-\ 6.21$
     ───────
     $89.47$

55. $$\begin{array}{r} \overset{8}{\cancel{5}}\overset{11}{\cancel{1}}\overset{10}{\cancel{2}} \\ 57.\cancel{9}\cancel{2}\cancel{0} \\ -\ 50.823 \\ \hline 7.097 \end{array}$$

57. $$\begin{array}{r} \overset{7}{\cancel{8}}\overset{11}{\cancel{1}}\overset{10}{\cancel{2}} \\ 48\cancel{8}.\cancel{2}\cancel{0} \\ -\ 87.92 \\ \hline 4\ 0\ 0.2\ 8 \end{array}$$

59. $$\begin{array}{r} \overset{5}{\ }\overset{12}{\ } \\ 88\cancel{6}.\cancel{2}85 \\ -202.774 \\ \hline 683.511 \end{array}$$

61. $$\begin{array}{r} \overset{7}{\cancel{8}}\overset{10}{\cancel{1}}\overset{9}{\cancel{1}}\overset{11}{\ } \\ \cancel{8}\cancel{1}.\cancel{0}\cancel{1}6 \\ -\ 37.050 \\ \hline 4\ 3.9\ 6\ 6 \end{array}$$

*Problems 63 through 67 are all stated in word form. The keywords and phrases "minus," "subtracted from," and "difference" are all indicators of a subtraction problem. The word forms of each number are first converted to decimal numbers and then the appropriate subtraction is performed.*

63. Fourteen and two tenths is 14.2.
    Eight and twenty-three hundredths is 8.23.
    The first minus the second is

$$\begin{array}{r} \overset{13}{\cancel{1}}\overset{11}{\cancel{4}}\overset{11}{\cancel{2}}\overset{10}{\ } \\ \cancel{1}\cancel{4}.\cancel{2}\cancel{0} \\ 8.2\ 3 \\ \hline 5.9\ 7 \end{array}$$

65. One and five hundredths is 1.05.
    Subtracting this number from 7 gives
    $7-1.05 = 5.95$.

67. The difference between eighteen , 18, and three and three tenths, 3.3, is $18-3.3 = 14.7$.

## Guide Problems

69. Estimate $199.99 + 19.99$ by rounding each addend to the nearest whole number. The arrow indicates rounding to the specified digit.

$$\begin{array}{rcl} 199.99 & \to & 200 \\ +\ 19.99 & \to & +\ 20 \\ \hline & & 220 \end{array}$$

71. Rounding to the nearest whole number:

$$\begin{array}{rcl} 5.37 & \to & 5 \\ +\ 1.81 & \to & +\ 2 \\ \hline & & 7 \end{array}$$

73. Rounding to the nearest whole number:

$$\begin{array}{rcl} 28.77 & \to & 29 \\ -\ 0.99 & \to & -\ 1 \\ \hline & & 28 \end{array}$$

75. Rounding to the nearest tenth:

$$\begin{array}{rcl} 42.12 & \to & 42.1 \\ +\ 12.88 & \to & +\ 12.9 \\ \hline & & 55.0 \end{array}$$

77. Rounding to the nearest hundredth:

$$\begin{array}{rcl} 14.667 & \to & 14.67 \\ -\ 0.049 & \to & -\ 0.05 \\ \hline & & 14.62 \end{array}$$

79. $$\begin{array}{r} 215.4 \\ 143.7 \\ 190.2 \\ +\ 124.3 \\ \hline 673.6 \end{array}$$

The fence is 673.6 feet long.

81. The perimeter is the sum of the lengths of all the sides or
    $32.46+17.03+12.14+12.14+17.03+32.46 = 123.26$ cm.

83. The savings is the difference in prices

$$\begin{array}{r} \overset{1}{\ }\overset{11}{\cancel{2}}\overset{4}{\cancel{1}}\overset{12}{\cancel{5}} \\ \cancel{2}\cancel{1}\cancel{5}.\cancel{2}5 \\ -\ 1\ 8\ 4.3\ 5 \\ \hline 3\ 0.9\ 0 \end{array}$$   so the saving is $30.90.

85. At night the temperature fell 20.9°F so the temperature that night was $88.7 - 20.9 = 67.8$ degrees Fahrenheit. The temperature then rose 11.3°F to a temperature of $67.8 + 11.3 = 79.1$ degrees Fahrenheit. The temperature at sunrise was 79.1°F.

87. The difference in cost is

$$
\begin{array}{r}
55.78 \\
-\ 52.75 \\
\hline
3.03
\end{array}
$$
so Mandy paid $3.03 less than Leonard.

89. a. Since 0.024 inches were removed from 1.16 inches, the new thickness is $1.16 - 0.024 = 1.136$ inches. The subtraction is shown below.

$$
\begin{array}{r}
1.1\,\overset{5}{\cancel{6}}\,\overset{10}{\cancel{0}} \\
-\ 0.0\,2\,4 \\
\hline
1.1\,3\,6
\end{array}
$$

b. From part a., the plate is now 1.136 inches thick. If a cover that is 0.46 inches thick is attached, the new plate has a thickness of $1.136 + 0.46 = 1.596$ inches. The addition is shown below.

$$
\begin{array}{r}
1.136 \\
+\ 0.460 \\
\hline
1.596
\end{array}
$$

91.

| Number or Code | Date | Transaction Description | Payment, Fee, Withdrawal (–) | | ✓ | Deposit, Credit (+) | | $ Balance | |
|---|---|---|---|---|---|---|---|---|---|
| | 11/1 | | | | | | | 2218 | 90 |
| 078 | 11/12 | Castle Decor | 451 | 25 | | | | 1767 | 65 |
| | 11/19 | deposit | | | | 390 | 55 | 2158 | 20 |
| 079 | 11/27 | Winton Realty | 257 | 80 | | | | 1900 | 40 |

Calculations:

$$
\begin{array}{r}
\overset{1}{\cancel{2}}\,\overset{\overset{11}{\cancel{1}}}{\cancel{2}}\,\overset{11}{\cancel{1}}8.\overset{8}{\cancel{9}}\,\overset{10}{\cancel{0}} \\
-\ 4\,5\,1.2\,5 \\
\hline
1\,7\,6\,7.6\,5
\end{array}
\qquad
\begin{array}{r}
\overset{1}{1}\overset{1}{7}6\overset{1}{7}.\overset{1}{6}5 \\
+\ 390.55 \\
\hline
2158.20
\end{array}
\qquad
\begin{array}{r}
\overset{1}{\cancel{2}}\,\overset{11}{\cancel{1}}5\,\overset{7}{\cancel{8}}.\overset{12}{\cancel{2}}\,0 \\
-\ 2\,5\,7.8\,0 \\
\hline
1\,9\,0\,0.4\,0
\end{array}
$$

## Cumulative Skills Review

1. $\dfrac{3}{28} \div \dfrac{1}{7} = \dfrac{3}{28} \cdot \dfrac{7}{1} = \dfrac{3}{\underset{4}{\cancel{28}}} \cdot \dfrac{\overset{1}{\cancel{7}}}{1} = \dfrac{3}{4}$

3. Multiply vertically:

$$
\begin{array}{r}
88 \\
\times\ 12 \\
\hline
176 \\
880 \\
\hline
1056
\end{array}
$$

5.  $18\frac{1}{9} = \frac{9 \cdot 18 + 1}{9} = \frac{162 + 1}{9} = \frac{163}{9}$

7.  The digit in the tens place of 21,448 is 4 and is followed by an 8 in the ones place. To round to the tens place, increase the 4 to 5 and change the 8 to 0 giving 21,450.

9.  Since $24 \cdot 3 = 72, \frac{13}{24} = \frac{13 \cdot 3}{24 \cdot 3} = \frac{39}{72}.$

## Section 3.3    Multiplying Decimals

## Concept Check

1.  When multiplying decimals, the product has as many decimal places as the total number of decimal places in the two  factors  .

3.  When multiplying a number by a power of 10 greater than one (such as 10, 100, 1000, and so on), move the decimal point to the  right  the same number of places as there are zeros in the power of 10.

## Guide Problems

5.  Multiply $5.26 \cdot 1.4$.

a. How many decimal places are in the factor 5.26.  two

b. How many decimal places are in the factor 1.4.  one

c. How many decimal places will the product have?
Add the answers to a. and b. The product will have $2 + 1 = 3$ places.

d. Determine the product.
Multiply as with whole numbers, ignoring the decimal points. Place the decimal point in the result so the product has the number of decimal places determined in part c.

$$\begin{array}{r} 5.26 \\ \times\ \ 1.4 \\ \hline 2\ 104 \\ 5\ 26 \\ \hline 7.364 \end{array}$$

*In Problems 7 through 33, the followed steps are carried out to perform the indicated multiplication. First, if the problem is written horizontally, rewrite the multiplication in a vertical format. Next, count the number of decimal places in the first factor and the second factor and add the two numbers. This is the number of decimal places in the answer. Multiply the numbers as if they were whole numbers. Lastly, position the decimal point in the answer so that the result has the pre-determined number of decimal places. This may require adding zeros to the left of the multiplication result.*

7.    0.44
   ×  0.8
   ‾‾‾‾‾‾
    0.352

The result must have
2 + 1 = 3 decimal
places.

9.     19
   ×  1.9
   ‾‾‾‾‾‾
    171
    19
   ‾‾‾‾‾‾
    36.1

The result must
have 0 + 1 = 1
decimal place.

11.    5.98
   ×  14.1
   ‾‾‾‾‾‾
    598
    2392
    598
   ‾‾‾‾‾‾
    84.318

13.    0.45
   ×  4.05
   ‾‾‾‾‾‾
    225
    1 800
   ‾‾‾‾‾‾
    1.8225

15.   4.24
   ×    7
   ‾‾‾‾‾‾
    29.68

17.   6.26
   ×  0.06
   ‾‾‾‾‾‾
    0.3756

19.    0.45
   ×  0.22
   ‾‾‾‾‾‾
    90
    90
   ‾‾‾‾‾‾
    0.0990

   or  0.099

21.   200.2
   ×  0.08
   ‾‾‾‾‾‾
    16.016

23.     138
   × 150.25
   ‾‾‾‾‾‾‾‾
    690
    276
    6900
    138
   ‾‾‾‾‾‾‾‾
    20734.50

   or 20,734.5

25.   8050.20
   ×     1.6
   ‾‾‾‾‾‾‾‾
    4830 120
    8050 20
   ‾‾‾‾‾‾‾‾
    12880.320

   or  $12,880.32

27.   21.089
   ×    9.7
   ‾‾‾‾‾‾‾‾
    147623
    189801
   ‾‾‾‾‾‾‾‾
    204.5633

29.   3.000041
   ×      5.02
   ‾‾‾‾‾‾‾‾‾
    6000082
    150002050
   ‾‾‾‾‾‾‾‾‾
    15.06020582

*Problems 31 through 33 are stated in word form. The words "times" and "product" indicate these are multiplication problems. In each case, the factors are converted, if necessary, from word form to decimal notation and then the multiplications are carried out as in the previous problems.*

31.    16.88
   ×   0.75
   ‾‾‾‾‾‾‾
    8440
    11816
   ‾‾‾‾‾‾‾
    12.6600

   or  $12.66

33.  One and three tenths: 1.3
     four and twelve hundredths: 4.12
        1.3
     × 4.12
     ‾‾‾‾‾‾
        26
        13
        5 2
     ‾‾‾‾‾‾
       5.356

## Guide Problems

35. Multiply (3.6)(100).

   Because there are __two__ zeros in 100. move the decimal point __two__ places to the __right__ .
   
   (3.6)(100) = 360

   The decimal point is moved to the *right* since 100 is a power of ten greater than 1.

*Problems 37 through 47 are all multiplication problems where one factor is a power of ten. If the power of ten is greater than 1, count the number of zeros in the power of ten. Move the decimal point in the other factor to the right that number of times to find the product. If necessary, add zeros on the right as you move the decimal point. If the power of ten is smaller than 1, count the number of place values in the power of ten. Move the decimal point in the other factor to the left that number of times to find the product. If necessary, add zeros on the left as you move the decimal point.*

37. $(2.75)(1000) = 2750$
    Three zeros mean
    move the decimal to
    the right three places.

39. $(1.955)(10,000) = 19,550$

41. $(0.75)(10^4) = (0.75)(10,000)$
    $\qquad\qquad = 7500$

43. $(5.4)(0.001) = 0.0054$
    Three place values mean
    move the decimal to
    the left three places.

45. $(0.072)(0.01) = 0.00072$

47. $(32.09)(0.00001) = 0.0003209$

49. 298.99 million is $(298.99)(1,000,000)$
    or 298,990,000.

51. \$8.382 trillion is $(\$8.383)(10^{12})$ or
    $(\$8.383)(1,000,000,000,000)$
    $= \$8,383,000,000,000.$

## Guide Problems

53. Estimate $(10.5)(3.82)$ by rounding each factor to one nonzero digit. The arrow indicates rounding to the specified number of digits.

$$
\begin{array}{rcr}
10.5 & \to & 10 \\
\times\ 3.82 & \to & \times\ 4 \\
\hline
 & & \underline{40}
\end{array}
$$

55. Rounding to one nonzero digit:

$$
\begin{array}{rcr}
3.1 & \to & 3 \\
\times\ 0.49 & \to & \times\ 0.5 \\
\hline
 & & 1.5
\end{array}
$$

57. Rounding to one nonzero digit:

$$
\begin{array}{rcr}
9.4 & \to & 9 \\
\times\ 0.32 & \to & \times\ 0.3 \\
\hline
 & & 2.7
\end{array}
$$

59. Rounding to two nonzero digits:

$$
\begin{array}{rcr}
32.78 & \to & 33 \\
\times\ 2.48 & \to & \times\ 2.5 \\
\hline
 & & 82.5
\end{array}
$$

61. Rounding to two nonzero digits:

$$
\begin{array}{rcr}
51.523 & \to & 52 \\
\times\ 10.49 & \to & \times\ 10 \\
\hline
 & & 520
\end{array}
$$

63. The crew paves 0.42 miles per hour
    for 7 hours gives

$$
\begin{array}{r}
0.42 \\
\times\ \quad 7 \\
\hline
2.94
\end{array}
$$

    a total of 2.94 miles.

65. Thirty slices at 0.375 inches form
    a loaf having thickness

$$
\begin{array}{r}
0.375 \\
\times\ \quad 30 \\
\hline
11.250
\end{array}
$$

    or 11.25 inches.

67. There were 4319 adult tickets at \$18.65 each. The adult ticket revenue was $(\$18.65)(4319) = \$80,549.35$. Similarly, 5322 children's tickets at \$12.40 yielded $(\$12.40)(5322) = \$65,992.80$. The total revenue was $\$80,549.35 + \$65,992.80 = \$146,542.15$.

69. The daily charge is \$35.00 per day for six days totaling $6 \times \$35.00 = \$210.00$. The mileage charge is for 550 miles at \$0.15 per mile for a total of $550 \times \$0.15 = \$82.50$. The total charge is then $\$210.00 + \$82.50 = \$292.50$.

71. a. In standard notation, twenty billion dollars is 20 times 1,000,000,000 or $20,000,000,000.

    b. At $0.075 per bill, the total cost is $0.075 \times 1,000,000,000 = \$75,000,000$.

73. Calder's gross pay total is $(34.5)(\$7.80) = \$269.10$. Deducting \$53.22 leaves a net pay of $\$269.10 - \$53.22 = \$215.88$.

75. Wong's gross pay total is $(42.7)(\$8.25) = \$352.275$ or $\$352.28$. If Wong's net pay was \$261.27 his deductions must have totaled $\$352.28 - \$261.27 = \$91.01$.

77. a. At two pills a day, you will take $(2)(0.05) = 0.10$ mg of copper. Over 30 days, this amounts to $(30)(0.10) = 3$ mg of copper.
    b. At two pills a day, you will take $(2)(0.5) = 1$ mg of manganese. Over 30 days, this amounts to $(30)(1) = 30$ mg of manganese.

79. From the table, the monthly premium for a \$500,000 policy on a 45-year-old female is \$35.67. Over 12 months the annual premium will total

$$
\begin{array}{r}
35.67 \\
\times \quad 12 \\
\hline
7134 \\
3567 \\
\hline
428.04
\end{array}
$$

    or \$428.04.

81. For a \$250,000 policy, the difference in the monthly payment for a 55-year-old male and a 55-year-old female is $\$51.55 - \$36.98 = \$14.57$. Over 12 months this difference amounts to

$$
\begin{array}{r}
14.57 \\
\times \quad 12 \\
\hline
2914 \\
1457 \\
\hline
174.84
\end{array}
$$

    or \$174.84.

83. Toasters total: $(200)(\$69.50) = \$13,900.00$
    Blenders total: $(130)(\$75.80) = \$9854.00$
    Total merchandise: $\$13,900.00 + \$9854.00 = \$23,754.00$
    Total with shipping and insurance:

$$
\begin{array}{r}
\$23,754.00 \\
\$1\,327.08 \\
+ \quad \$644.20 \\
\hline
\$25,725.28
\end{array}
$$

## Cumulative Review Exercises

1.  $0 \div 89 = 0$
    Zero divided by a nonzero number is always 0.

3.  Add vertically:

$$
\begin{array}{r}
\overset{1}{3.6}60 \\
1.299 \\
+ \quad 9.000 \\
\hline
13.959
\end{array}
$$

5.  Three decimal places in 6.852 indicate the thousandths decimal place. Further the decimal point is read as "and" so the word form is "six and eight hundred fifty-two thousandths."

7.  $8\dfrac{5}{6} = \dfrac{6 \cdot 8 + 5}{6} = \dfrac{48 + 5}{6} = \dfrac{53}{6}$

9.   $5\dfrac{3}{5}$ less $2\dfrac{1}{2}$ is

$$5\dfrac{3}{5} - 2\dfrac{1}{2} = \dfrac{28}{5} - \dfrac{5}{2}$$

$$= \dfrac{56}{10} - \dfrac{25}{10}$$

$$= \dfrac{31}{10} = 3\dfrac{1}{10}$$

## Section 3.4    Dividing Decimals

## Concept Check

1.   The number being divided is known as the _dividend_ .

3.   The result of dividing numbers is called the _quotient_ .

5.   When dividing a decimal by a whole number, it is sometimes necessary to write additional zeros to the right of the _last_ digit following the decimal point in the _dividend_ .

7.   When dividing a decimal by a power of 10 greater than one (such as 10, 100, 1000, and so on), move the decimal point in the dividend to the _left_ the same number of places as there are _zeros_ in the power of 10.

## Guide Problems

9.   Divide $6.75 \div 5$.

     Dividing decimals by whole numbers is just like dividing whole numbers. Divide as usual, then place the decimal point in the quotient directly above its position in the dividend.

$$
\begin{array}{r}
1.35 \\
5\overline{)6.75} \\
-5\phantom{.75} \\
\hline
17\phantom{5} \\
-15\phantom{5} \\
\hline
25 \\
-25 \\
\hline
0
\end{array}
$$

$$
\begin{array}{r}
2.7 \\
11.\ 4\overline{\smash{)}10.8} \\
-\underline{8} \\
28 \\
-\underline{28} \\
0
\end{array}
$$

$$
\begin{array}{r}
5.6 \\
13.\ 6\overline{\smash{)}33.6} \\
-\underline{30} \\
36 \\
-\underline{36} \\
0
\end{array}
$$

$$
\begin{array}{r}
1.74 \\
15.\ 25\overline{\smash{)}43.50} \\
-\underline{25} \\
185 \\
-\underline{175} \\
100 \\
-\underline{100} \\
0
\end{array}
$$

$$
\begin{array}{r}
0.4 \\
17.\ 93\overline{\smash{)}37.2} \\
-\underline{37\ 2} \\
0
\end{array}
$$

$$
\begin{array}{r}
0.075 \\
19.\ 40\overline{\smash{)}3.000} \\
-\underline{280} \\
200 \\
-\underline{200} \\
0
\end{array}
$$

$$
\begin{array}{r}
6.27 \\
21.\ 11\overline{\smash{)}68.97} \\
-\underline{66} \\
29 \\
-\underline{22} \\
77 \\
-\underline{77} \\
0
\end{array}
$$

$$
\begin{array}{r}
0.044 \\
23.\ 50\overline{\smash{)}2.200} \\
-\underline{2\,00} \\
200 \\
-\underline{200} \\
0
\end{array}
$$

$$
\begin{array}{r}
0.07 \\
25.\ 67\overline{\smash{)}4.69} \\
-\underline{4\ 69} \\
0
\end{array}
$$

## Guide Problems

27.  Divide $4.7 \div 100$.

Because there are the __two__ zeros in 100, move the decimal point __two__ places to the __left__ .

$$4.7 \div 100 = \underline{0.047}$$

*The rules for dividing by powers of 10 are applied in problems 29 through 35. If dividing by a power of 10 that is greater than one, such as 10, 100, etc., count the number of zeros in the divisor and move the decimal place in the dividend that many places to the left. If dividing by a power of 10 less than one, such as 0.1, 0.01, etc. ,count the total number of decimal places in the divisor and move the decimal point in the dividend that many places to the right.*

29.  $\dfrac{24.78}{100}$

Divisor is greater than one.
Zeros in divisor: two.
Move decimal point to the left.

$$\dfrac{24.78}{100} = 0.2478$$

31.  $\dfrac{67}{0.001}$

Divisor is less than one.
Decimal places in divisor: three.
Move decimal point to the right.

$$\dfrac{67}{0.001} = 67,000$$

33. $\dfrac{56.003}{0.00001}$

Divisor is less than one.
Decimal places in divisor: five.
Move decimal point to the right.

$\dfrac{56.003}{0.00001} = 5,600,300$

35. $\dfrac{76}{100,000}$

Divisor is greater than one.
Zeros in divisor: five.
Move decimal point to the left.

$\dfrac{76}{100,000} = 0.00076$

## Guide Problems

37. Consider $3.5\overline{)1.96}$

a. Write an equivalent division problem in which the divisor is a whole number. Since the divisor has one decimal place, move the decimal one place to the right in both the dividend and divisor.

$3.5\overline{)1.96}$ is equivalent to $\underline{35\overline{)19.6}}$

b. Divide.

$$\begin{array}{r} 0.56 \\ 35\overline{)19.60} \\ -17\,5 \\ \hline 2\,1\,0 \\ -2\,1\,0 \\ \hline 0 \end{array}$$

*In problems 39 through 49, if the divisor is a decimal number, the division is rewritten with a whole number divisor by moving the decimal point in both the divisor and dividend the same number of places to the right until the divisor is a whole number. Once the problem is rewritten the division is carried as usual when dividing a decimal number by a whole number.*

39. $0.8\overline{)64} \rightarrow 8\overline{)640}$

$$\begin{array}{r} 80 \\ 8\overline{)640} \\ -64 \\ \hline 0 \end{array}$$

41. $1.5\overline{)40.2} \rightarrow 15\overline{)402}$

$$\begin{array}{r} 26.8 \\ 15\overline{)402.0} \\ -30 \\ \hline 102 \\ -90 \\ \hline 120 \\ -120 \\ \hline 0 \end{array}$$

43. $2.55\overline{)7.905} \rightarrow 255\overline{)7905}$

$$\begin{array}{r} 3.1 \\ 255\overline{)790.5} \\ -765 \\ \hline 255 \\ -255 \\ \hline 0 \end{array}$$

45. $0.81\overline{)3.564} \rightarrow 81\overline{)356.4}$

$$\begin{array}{r} 4.4 \\ 81\overline{)356.4} \\ -324 \\ \hline 324 \\ -324 \\ \hline 0 \end{array}$$

47. $6.586 \div 7.4 \rightarrow 65.86 \div 74$

$$\begin{array}{r} 0.89 \\ 74\overline{)65.86} \\ -592 \\ \hline 666 \\ -666 \\ \hline 0 \end{array}$$

49. The divisor is not a decimal number so no rewriting is required.

$$\begin{array}{r} 0.45 \\ 84\overline{)37.80} \\ -33\,6 \\ \hline 4\,20 \\ -4\,20 \\ \hline 0 \end{array}$$

*Problems 51 through 61 require the answer to be rounded to a specified place. Division is carried out as usual, however the division is stopped one place past the specified place value since that is all that is needed for rounding. For example, in 51 through 54 the answer is rounded to the nearest whole number so division stops when the value in the tenths place of the quotient is determined.*

51.  $49 \div 2.6 \rightarrow 490 \div 26$

$$
\begin{array}{r}
18.8 \\
26\overline{)490.0} \\
-26 \\ \hline
230 \\
-208 \\ \hline
220 \\
-208 \\ \hline
12
\end{array}
$$

Rounded to the nearest whole number the quotient is 19.

53.  $12.42 \div 0.39 \rightarrow 1242 \div 39$

$$
\begin{array}{r}
31.8 \\
39\overline{)1242.0} \\
-117 \\ \hline
72 \\
-39 \\ \hline
330 \\
-312 \\ \hline
18
\end{array}
$$

Rounded to the nearest whole number the quotient is 32.

55.  $34.5 \div 12.8 \rightarrow 345 \div 128$

$$
\begin{array}{r}
2.69 \\
128\overline{)345.00} \\
-256 \\ \hline
890 \\
-768 \\ \hline
1220 \\
-1152 \\ \hline
68
\end{array}
$$

Rounded to the nearest tenth the quotient is 2.7.

57.  $2.3\overline{)1.6} \rightarrow 23\overline{)16}$

$$
\begin{array}{r}
0.69 \\
23\overline{)16.00} \\
-138 \\ \hline
220 \\
-207 \\ \hline
13
\end{array}
$$

Rounded to the nearest tenth the quotient is 0.7.

59.
$$
\begin{array}{r}
0.076 \\
13\overline{)1.000} \\
-91 \\ \hline
90 \\
-78 \\ \hline
12
\end{array}
$$

Rounded to the nearest hundredth the quotient is 0.08.

61.  $239 \div 1.1 \rightarrow 2390 \div 11$

$$
\begin{array}{r}
217.272 \\
11\overline{)2390.000} \\
-22 \\ \hline
19 \\
-11 \\ \hline
80 \\
-77 \\ \hline
30 \\
-22 \\ \hline
80 \\
-77 \\ \hline
30 \\
-22 \\ \hline
8
\end{array}
$$

## Guide Problems

63.  Estimate $39.1 \div 7.6$ by rounding the dividend and divisor to one nonzero digit.

a. Round the dividend and divisor to one nonzero digit.

$39.1 \rightarrow \underline{40}$

$7.6 \rightarrow 8$

b. Divide the rounded numbers.

$$
\begin{array}{r}
5 \\
8\overline{)40}
\end{array}
$$

An estimate of $39.1 \div 7.6$ is 5.

65. $29.5 \to 30$

       $5.9 \to 6$

       $6\overline{)30}$ with quotient $5$

     $29.5 \div 5.9$ is about 5.

67. $0.0212 \to 0.02$

       $0.39 \to 0.4$

     $0.4\overline{)0.02} \to 4\overline{)0.20}$ with quotient $0.05$

     $0.0212 \div 0.39$ is about 0.05.

69. If 24 bottles of water cost $5.33, the cost per bottle is $5.33 \div 24 = 0.22208...$ or about 22 cents per bottle.

71. If hair grows 4.89 inches per year then the length grown in one day is $4.89 \div 365 = 0.013397...$ or, rounding to the nearest ten-thousandth, 0.0134 inches per day.

73. Bernard's hourly rate is found by dividing his total wages by the total hours worked. This amounts to $\$411.25 \div 32.9 = \$4112.5 \div 329 = \$12.50$. Bernard earned $12.50 an hour.

75. a. Computing minutes per mile means total minutes divided by total miles., or in this case, $52.7 \div 6.2$. the division is equivalent to $527 \div 62 = 8.5$. Terry's average time was 8.5 minutes per mile.

     b. Since a distance of 10K is equivalent to a distance of 6.2 miles, 5K is equivalent to $6.2 \div 2 = 3.1$ miles. At 8.5 minutes per mile, a 5K race would take Terry $3.1 \cdot 8.5 = 26.35$ minutes.

77. If three cantaloupes cost $4.89 then one cantaloupe costs $\$4.89 \div 3 = \$1.63$. At that individual price, two cantaloupes will cost $2 \times \$1.63 = \$3.26$.

79. Since 4 dozen cans is $4 \times 12 = 48$ cans, the cost per can was $\$103.68 \div 48 = \$2.16$.

81. The average weight loss per month is found by dividing the total pounds lost by the number of months. Dividing $57.6 \div 9$ gives 6.4 pounds per month.

83. a. The total in rent was $341,000 and rent is $3.10 per square foot so the total number of square feet must be $3.1\overline{)341,000} = 31\overline{)3,410,000}$ or 110,000 square feet.

     b. From a. there are 110,000 square feet with a cost of $3960 to maintain. The cost per square foot is $\$3960 \div 110,000 = \$0.036$ or about $0.04 per square foot.

## Cumulative Skills Review

1.
$$\begin{array}{r} 10.45 \\ \times\ 0.65 \\ \hline 5225 \\ 6270 \\ \hline 6.7925 \end{array}$$

Each factor has two decimal places so the result has $2 + 2 = 4$ places.

3. $\dfrac{18}{30} = \dfrac{6 \cdot 3}{6 \cdot 5} = \dfrac{\cancel{6} \cdot 3}{\cancel{6} \cdot 5} = \dfrac{3}{5}$

5.
$$\begin{array}{r} 2\overset{3}{\cancel{4}}.\overset{11}{\cancel{1}}1 \\ -\ \ 3.60 \\ \hline 20.51 \end{array}$$

7. First compute the remaining acreage. Since 178 of 532 acres are reserved, what remains is $532 - 178 = 354$ acres. The number of 3-acre home sites that this acreage can be divided into is $354 \div 3 = 118$.

9.  Marcus needs $4\frac{1}{3}$ times the required amount of cheese or $4\frac{1}{3} \cdot 24 = \frac{13}{3} \cdot \frac{24}{1} = \frac{13}{\cancel{3}} \cdot \frac{\overset{8}{\cancel{24}}}{1} = 13 \cdot 8 = 104$ ounces.

## Section 3.5    Working with Fractions and Decimals

## Concept Check

1.  To convert a fraction to a decimal,  __divide__  the numerator by the denominator.

3.  The repeating decimal 0.14141414… is written as  $\overline{0.14}$  .

5.  Convert $\frac{5}{16}$ to a decimal.

To convert a fraction to a decimal, divide the denominator into the numerator.  Continue the division until a zero remainder is obtained or until the digits in the quotient repeat in an obvious pattern.

$$
\begin{array}{r}
0.3125 \\
16\overline{)5.0000} \\
-48\phantom{000} \\
\hline
20\phantom{00} \\
-16\phantom{00} \\
\hline
40\phantom{0} \\
-32\phantom{0} \\
\hline
80 \\
-80 \\
\hline
0
\end{array}
\qquad \frac{5}{16} = \underline{0.3125}
$$

*In problems 7 through 25, fractions and mixed numbers are converted to decimal numbers as in the Guide problems. Divide the denominator into the numerator using long division. Continue the division until the decimal quotient terminates or begins to repeat. If the decimal form repeats, indicate the portion that repeats with a bar over the repeating set of digits. When converting a mixed number, there are two approaches. One way is to convert the fraction part and then place the integer part on the left side of the decimal point. The other is convert the mixed number to an improper fraction and then convert to a decimal in the usual fashion.*

7.
$$
\begin{array}{r}
0.15 \\
20\overline{)3.00} \\
-20\phantom{0} \\
\hline
100 \\
-100 \\
\hline
0
\end{array}
\qquad \frac{3}{20} = 0.15
$$

9.
$$
\begin{array}{r}
0.5625 \\
16\overline{)9.0000} \\
-80\phantom{000} \\
\hline
100\phantom{0} \\
-96\phantom{0} \\
\hline
40 \\
-32 \\
\hline
80 \\
-80 \\
\hline
0
\end{array}
\qquad \frac{9}{16} = 0.5625
$$

11.  $50\overline{)19.00}$    $\dfrac{0.38}{}$      $\dfrac{19}{50} = 0.38$

$\quad\quad -15\,0$
$\quad\quad\overline{\phantom{-}4\,00}$
$\quad\quad\quad -4\,00$
$\quad\quad\quad\overline{\phantom{-4\,0}0}$

13.  $11\overline{)5.000000}$    $\dfrac{0.4545.....}{}$      $\dfrac{5}{11} = 0.\overline{45}$

$\quad\quad -44$
$\quad\quad\overline{\phantom{-}60}$
$\quad\quad\quad -55$
$\quad\quad\quad\overline{\phantom{-}50}$
$\quad\quad\quad\quad -44$
$\quad\quad\quad\quad\overline{\phantom{-}60}$
$\quad\quad\quad\quad\quad -55$
$\quad\quad\quad\quad\quad\overline{\phantom{-5}5}$

15.  $1\dfrac{3}{5} = \dfrac{8}{5}$

$\quad\quad 5\overline{)8.0}$    $\dfrac{1.6}{}$     $1\dfrac{3}{5} = 1.6$

$\quad\quad -5$
$\quad\quad\overline{\phantom{-}30}$
$\quad\quad\quad -30$
$\quad\quad\quad\overline{\phantom{-3}0}$

17.  $20\overline{)13.00}$    $\dfrac{0.65}{}$      $\dfrac{13}{20} = 0.65$

$\quad\quad -120$
$\quad\quad\overline{\phantom{-}100}$
$\quad\quad\quad -100$
$\quad\quad\quad\overline{\phantom{-10}0}$

19.  $6\overline{)11.000}$    $\dfrac{1.833...}{}$     $\dfrac{11}{6} = 1.8\overline{3}$

$\quad\quad -6$
$\quad\quad\overline{\phantom{-}50}$
$\quad\quad\quad -48$
$\quad\quad\quad\overline{\phantom{-}20}$
$\quad\quad\quad\quad -18$
$\quad\quad\quad\quad\overline{\phantom{-}20}$
$\quad\quad\quad\quad\quad -18$
$\quad\quad\quad\quad\quad\overline{\phantom{-1}2}$

21.  $16\overline{)3.0000}$    $\dfrac{0.1875}{}$     $\dfrac{3}{16} = 0.1875$

$\quad\quad -16$
$\quad\quad\overline{\phantom{-}140}$
$\quad\quad\quad -128$
$\quad\quad\quad\overline{\phantom{-}120}$
$\quad\quad\quad\quad -112$
$\quad\quad\quad\quad\overline{\phantom{-}80}$
$\quad\quad\quad\quad\quad -80$
$\quad\quad\quad\quad\quad\overline{\phantom{-8}0}$

23.  $4\dfrac{7}{18} = 4 + \dfrac{7}{18}$

$\quad\quad 18\overline{)7.000}$    $\dfrac{0.388...}{}$     $4\dfrac{7}{18} = 4.38\overline{8}$

$\quad\quad -54$
$\quad\quad\overline{\phantom{-}160}$
$\quad\quad\quad -144$
$\quad\quad\quad\overline{\phantom{-}160}$
$\quad\quad\quad\quad -144$
$\quad\quad\quad\quad\overline{\phantom{-}16}$

25.  $5\dfrac{1}{8} = 5 + \dfrac{1}{8}$

$\quad\quad 8\overline{)1.000}$    $\dfrac{0.125}{}$     $5\dfrac{1}{8} = 5.125$

$\quad\quad -8$
$\quad\quad\overline{\phantom{-}20}$
$\quad\quad\quad -16$
$\quad\quad\quad\overline{\phantom{-}40}$
$\quad\quad\quad\quad -40$
$\quad\quad\quad\quad\overline{\phantom{-4}0}$

*In Problems 27 through 33, fractions are converted to decimals rounded to a specified place. Division only needs to be carried out to one place beyond the specified place in order to round. In Problems 27 through 29, the division stops once the hundredths digit is determined so we can round to tenths. Similarly, in Problems 31 through 33, where rounding to hundredths is required, division stops at the thousandths place.*

27. $6\overline{)5.00}$ gives $0.83...$ so $\dfrac{5}{6} = 0.83...$ or 0.8 to

    the nearest tenth.

29. $10\dfrac{3}{8} = \dfrac{83}{8}$ and $8\overline{)83.00}$ gives $10.37...$ so $10\dfrac{3}{8}$

    is 10.37... or 10.4 to the nearest tenth.

31. $23\overline{)4.000}$ gives $0.173...$ so $\dfrac{4}{23} = 0.173...$ or 0.17 to

    the nearest hundredth.

33. $34\overline{)39.000}$ gives $1.147...$ so $\dfrac{39}{34} = 1.147...$ or 1.15 to

    the nearest hundredth.

## Guide Problems

35. Calculate $6.27 + 5\dfrac{1}{10} \cdot 8.08$.

    a. Convert the decimal to a fraction.

$$6.27 + 5\dfrac{1}{10} \cdot 8.08 = 6.27 + 5.\underline{1} \cdot 8.08$$

    b. Simplify the expression.

$$6.27 + 5.\underline{1} \cdot 8.08$$
$$6.27 + \underline{41.208}$$
$$\underline{47.478}$$

*In Problems 37 through 53, the following steps are applied. First convert any fraction or mixed numbers to decimals. Second, simplify the expression by applying the order of operations: 1. Simplify within parentheses and other grouping symbols including numerators and denominators of fractions, 2. Evaluate exponents, 3. Evaluate multiplications and divisions from left to right, and 4. Evaluate additions and subtractions from left to right.*

37. $76.3 - 4 \cdot 11$

    $76.3 - 44$

    $32.3$

39. $10.3 + 4\dfrac{1}{5} \cdot 2\dfrac{1}{10}$

    $10.3 + 4.2 \cdot 2.1$

    $10.3 + 8.82$

    $19.12$

41. $\dfrac{1}{5} + \dfrac{0.25 + 2.77}{10}$

    $0.2 + \dfrac{3.02}{10}$

    $0.2 + 0.302$

    $0.502$

43. $12.4 - 2\left(3\dfrac{3}{5} + 1.2\right)$

    $12.4 - 2(3.6 + 1.2)$

    $12.4 - 2 \cdot 4.8$

    $12.4 - 9.6$

    $2.8$

45. $4\left(\dfrac{6}{25} + 0.56\right) - 1\dfrac{1}{10} \cdot 2$

    $4(0.24 + 0.56) - 1.1 \cdot 2$

    $4 \cdot 0.8 - 2.2$

    $3.2 - 2.2$

    $1$

47. $(5.2 \cdot 4.5) - 1\dfrac{1}{2} + 0.2$

    $(5.2 \cdot 4.5) - 1.5 + 0.2$

    $23.4 - 1.5 + 0.2$

    $21.9 + 0.2$

    $22.1$

49. $2\dfrac{5}{8}+3\dfrac{1}{2}(4.9-4.1)$

    $2.625+3.5(4.9-4.1)$

    $2.625+3.5\cdot0.8$

    $2.625+2.8$

    $5.425$

51. $20.73-3\left(4\dfrac{1}{10}-3\right)^2$

    $20.73-3(4.1-3)^2$

    $20.73-3\cdot1.1^2$

    $20.73-3\cdot1.21$

    $20.73-3.63$

    $17.1$

53. $4\dfrac{1}{2}+0.08^2+\dfrac{33.5-16}{\dfrac{1}{4}}$

    $4.5+0.08^2+\dfrac{33.5-16}{0.25}$

    $4.5+0.08^2+\dfrac{17.5}{0.25}$

    $4.5+0.0064+\dfrac{17.5}{0.25}$

    $4.5+0.0064+70$

    $4.5064+70$

    $74.5064$

55. Half of $\$265.12$ is $\$265.12\div2$.

$$\begin{array}{r}132.56\\2\overline{)265.12}\\-\underline{2}\phantom{65.12}\\6\phantom{5.12}\\-\underline{6}\phantom{5.12}\\5\phantom{.12}\\-\underline{4}\phantom{.12}\\11\phantom{2}\\-\underline{10}\phantom{2}\\12\\\underline{12}\\0\end{array}$$

The insurance company pays $\$132.56$.

57. $A=\dfrac{1}{2}bh=\dfrac{1}{2}\cdot48.8\cdot60.7$

   $=0.5\cdot48.8\cdot60.7$

   $=24.4\cdot60.7$

   $=1481.08$

or 1481.1 square inches rounded.

59. a. Almonds: $2\dfrac{2}{5}\cdot\$5.89=2.4\cdot\$5.89=\$14.136$ or $\$14.14$. Pears: $3\dfrac{7}{8}\cdot\$3.30=3.875\cdot\$3.30=\$12.7875$

    or $\$12.79$. Grapes: $1\dfrac{1}{4}\cdot\$2.17=1.25\cdot\$2.17=\$2.7125$ or $\$2.71$.

    b. The total cost to Fran was $\$14.14+\$12.79+\$2.71=\$29.64$.

61. If 22.8 feet is cut from $68\dfrac{3}{5}$ feet, the piece remaining is $68\dfrac{3}{5}-22.8=68.6-22.8=45.8$ feet. Rewriting as

    a fraction gives $45.8=45\dfrac{8}{10}=45\dfrac{4}{5}$ feet.

## Cumulative Skills Review

1. The digit in the thousandths place in 0.58<u>5</u>60 is 5. The digit to its right is a 6. so round the 5 up to 6 and zero the digits to the right to get 0.586 rounded.

3. $\dfrac{3}{4}\cdot\dfrac{2}{15}=\dfrac{\cancel{3}^{\,1}}{\cancel{4}_{\,2}}\cdot\dfrac{\cancel{2}^{\,1}}{\cancel{15}_{\,5}}=\dfrac{1\cdot1}{2\cdot5}=\dfrac{1}{10}$

5. First move the decimal one place in both the divisor and dividend so the divisor is a whole number. Then divide

$$
\begin{array}{r}
2.855 \\
12\overline{)34.260} \\
-24 \phantom{....} \\
\hline
102 \phantom{..} \\
-96 \phantom{..} \\
\hline
66 \phantom{.} \\
-60 \phantom{.} \\
\hline
60 \\
-60 \\
\hline
0
\end{array}
$$

7. First multiply 14 by 22.

$$
\begin{array}{r}
14 \\
\times\ 22 \\
\hline
28 \\
28 \phantom{.} \\
\hline
308
\end{array}
$$

Now multiply by 10,000 by adding four zeros to the result.
$$14 \cdot 22 \cdot 10,000 = 3,080,000$$

9. $\quad 4 \cdot \dfrac{1}{3} \cdot 2\dfrac{5}{6} = \dfrac{\cancel{4}}{1} \cdot \dfrac{1}{3} \cdot \dfrac{17}{\cancel{6}} = \dfrac{2 \cdot 1 \cdot 17}{1 \cdot 3 \cdot 3} = \dfrac{34}{9} = 3\dfrac{7}{9}$

## Chapter 3 Numerical Facts Of Life

1. For each figure in the 2006 Payroll column, the millions digit has been underlined and the digit to its right is doubly underlined. If the double underlined digit is 5 or greater, round the millions up. Otherwise, leave the millions digit alone. The rounded figure is given in the third column.

| Team | 2006 Payroll | Rounded 2006 Payroll |
|---|---|---|
| New York Yankees | $19<u>4</u>,<u>6</u>63,079 | $195,000,000 |
| Boston Red Sox | $12<u>0</u>,<u>0</u>99,824 | $120,000,000 |
| Los Angeles Angels | $10<u>3</u>,<u>4</u>72,000 | $103,000,000 |
| Colorado Rockies | $4<u>1</u>,<u>2</u>33,000 | $41,000,000 |
| Tampa Bay Devil Rays | $3<u>5</u>,<u>4</u>17,967 | $35,000,000 |
| Florida Marlins | $1<u>4</u>,<u>9</u>98,500 | $15,000,000 |

3. For each figure in the Rounded 2006 Payroll column, the corresponding average payroll per player is found by diving the rounded payroll by 30, the number of players on a baseball roster. In each case, the division is carried to one decimal place since the answer has to be rounded to the nearest dollar. For example,

$$\frac{\$41,000,000}{30} = \$1,366,666.6... \text{ is } \$1,366,667 \text{ to the nearest dollar.}$$

| Team | Rounded 2006 Payroll | Average Payroll Per Game |
|---|---|---|
| New York Yankees | $195,000,000 | $6,500,000 |
| Boston Red Sox | $120,000,000 | $4,000,000 |
| Los Angeles Angels | $103,000,000 | $3,433,333 |
| Colorado Rockies | $41,000,000 | $1,366,667 |
| Tampa Bay Devil Rays | $35,000,000 | $1,166,667 |
| Florida Marlins | $15,000,000 | $500,000 |

## Chapter 3 Review Exercises

1. The indicated digit in 13.$\underline{3}$512 is in the first place to the right of the decimal point and so is in the *tenths* place.

2. The indicated digit in 0.145$\underline{7}$919 is in the fourth place to the right of the decimal point and so is in the *ten-thousandths* place.

3. The indicated digit in 314.09$\underline{2}$45 is in the third place to the right of the decimal point and so is in the *thousandths* place.

4. The indicated digit in 89.2$\underline{5}$901 is in the second place to the right of the decimal point and so is in the *hundredths* place.

5. The indicated digit in 0.35021$\underline{8}$ is in the sixth place to the right of the decimal point and so is in the *millionths* place.

6. The indicated digit in 1476.00215$\underline{9}$62 is in the seventh place to the right of the decimal point and so is in the *ten-millionths* place.

7. Given 28.355 the three decimal places indicate thousandths. The decimal point is written as *and*. The word form is "twenty-eight and three hundred fifty-five thousandths."

8. Given 0.00211 the five decimal places indicate hundred-thousandths. The word form is "two hundred eleven hundred-thousandths."

9. Given 0.158 the three decimal places indicate thousandths. The word form is "one hundred fifty-eight thousandths."

10. Given 142.12 the two decimal places indicate hundredths. The decimal point is written as *and*. The word form is "one hundred forty-two and twelve hundredths."

11. Given 59.625 the three decimal places indicate thousandths. The decimal point is written as *and*. The word form is "fifty-nine and six hundred twenty-five thousandths."

12. Given 0.39 the two decimal places indicate hundredths. The word form is "thirty-nine hundredths."

13. The phrase *ten-thousandths* tells us there are four decimal places while the absence of and tells us there is a zero to the left of the decimal point. The number is 0.0298.

14. The word *thousandths* indicates three decimal places, while *and* indicates the decimal point. The number is 22.324.

15. *Hundredths* indicates two decimal places. The number is 178.13.

16. *Ten-thousandths* indicates four decimal places. The number is 0.0735.

17. *Hundredths* indicates two decimal places. The number is 912.25.

18. *Hundred-thousandths* indicates five decimal places. The number is 0.00016.

19. $9.57 = 9\dfrac{57}{100}$
Two decimal places indicate a denominator of 100.

20. $0.315 = \dfrac{315}{1000} = \dfrac{63}{200}$
Three decimal places indicate a denominator of 1000.

21. $5.006 = 5\dfrac{6}{1000} = 5\dfrac{3}{500}$

22. $1.19 = 1\dfrac{19}{100}$

*In problems 23 through 28, the first place where the numbers disagree is underlined. Compare these digits to determine if a less than symbol, greater than symbol, or equal sign should be inserted. If that digit is smaller in the number on the left, then a less than symbol, <, is inserted. If it is greater on the left then a greater than symbol, >, is inserted. If there is no place where the digits disagree the numbers are equal.*

23. $23.51\underline{2} < 23.51\underline{9}$

24. $0.81\underline{2}4 < 0.81\underline{3}3$

25. $3.458\underline{8}7 > 3.458\underline{7}7$

26. $12\underline{5}.6127 > 12\underline{4}.78$

27. $0.02324 = 0.02324$

28. $55.3\underline{9}8 > 55.3\underline{8}9$

*In problems 29 through 34, the specified place to which to round the number is underlined. The digit to its right determines whether we round up or down in the usual fashion.*

29. $1.8\underline{5}3$ rounded to the nearest hundredth is 1.85.

30. $2.14\underline{8}7$ rounded to the nearest thousandth is 2.149.

31. $3.\underline{3}96$ rounded to the nearest tenth is 3.4.

32. $4.114\underline{5}8$ rounded to the nearest ten-thousandth is 4.1146.

33. $1.5885\underline{5}6$ rounded to the nearest hundred-thousandth is 1.58856.

34. $7.45\underline{1}2$ rounded to the nearest thousandth is 7.451.

35.
$$
\begin{array}{r}
\overset{1}{2.13}5 \\
+\ 3.447 \\
\hline
5.582
\end{array}
$$

36.
$$
\begin{array}{r}
\overset{1}{6.0}98 \\
+\ 1.211 \\
\hline
7.309
\end{array}
$$

37.
$$
\begin{array}{r}
5.5\overset{2}{1}73 \\
0.0991 \\
+\ 6.0070 \\
\hline
11.6234
\end{array}
$$

38.
$$
\begin{array}{r}
\overset{1\ 1}{1.2}34 \\
0.022 \\
+\ 8.455 \\
\hline
9.711
\end{array}
$$

39.
$$
\begin{array}{r}
15.4\overset{1}{4}50 \\
+\ 0.3369 \\
\hline
15.7819
\end{array}
$$

40.
$$
\begin{array}{r}
\overset{1\ 1\ 1}{12.6}45 \\
+\ 0.856 \\
\hline
13.501
\end{array}
$$

41.
$$
\begin{array}{r}
\overset{1\ 1}{0.08}9 \\
+\ 9.652 \\
\hline
9.741
\end{array}
$$

42.
$$
\begin{array}{r}
6.2\overset{1}{4}4 \\
+\ 0.048 \\
\hline
6.292
\end{array}
$$

43.
$$
\begin{array}{r}
22.1\overset{1}{2}3 \\
9.003 \\
+\ 0.450 \\
\hline
31.576
\end{array}
$$

44.
$$
\begin{array}{r}
\overset{1\ 1}{0.0}33 \\
11.920 \\
+\ 18.200 \\
\hline
30.153
\end{array}
$$

45.
$$
\begin{array}{r}
24.\overset{5}{6}\overset{15}{6}5 \\
-\ 2.362 \\
\hline
22.293
\end{array}
$$

46.
$$
\begin{array}{r}
18.\overset{2}{3}\overset{12}{2}9 \\
-\ 6.154 \\
\hline
12.175
\end{array}
$$

47.
$$
\begin{array}{r}
12.127 \\
-\ 6.015 \\
\hline
6.112
\end{array}
$$

48.
$$
\begin{array}{r}
10.5\overset{1}{2}\overset{17}{7} \\
-\ 8.519 \\
\hline
2.008
\end{array}
$$

49.
$$
\begin{array}{r}
0.0\overset{6}{7}\overset{13}{4}\overset{10}{1}\overset{10}{0} \\
-\ 0.00562 \\
\hline
0.06848
\end{array}
$$

50.
$$
\begin{array}{r}
11.\overset{0}{1}\overset{14}{5}\overset{14}{5}\overset{10}{0} \\
-\ 0.0877 \\
\hline
11.0673
\end{array}
$$

51.
$$
\begin{array}{r}
0.0\overset{7}{8}\overset{15}{6}\overset{14}{4} \\
-\ 0.0596 \\
\hline
0.0268
\end{array}
$$

52.
$$
\begin{array}{r}
6.3\overset{3}{4}\overset{14}{5}\overset{10}{0} \\
-\ 2.0089 \\
\hline
4.3361
\end{array}
$$

53.      15.629
      −  0.609
      ‾‾‾‾‾‾‾‾
         15.020
      or 15.02

54.       0.988
       −  0.036
       ‾‾‾‾‾‾‾‾
          0.952

55.      5.025
      ×  1.25
      ‾‾‾‾‾‾‾‾
         25125
       1 0050
       5 025
      ‾‾‾‾‾‾‾‾
         6.28125

56.      3.972
      ×  0.035
      ‾‾‾‾‾‾‾‾
         19860
         11916
      ‾‾‾‾‾‾‾‾
         0.139020
      or 0.13902

57.      9.041
      ×  1.44
      ‾‾‾‾‾‾‾‾
         36164
       3 6164
       9 041
      ‾‾‾‾‾‾‾‾
         13.01904

58.          ¹
         7.221
      ×  0.009
      ‾‾‾‾‾‾‾‾
         0.064989

59. $(0.0945)(100) = 9.45$
    Move the decimal point two
    places to the right.

60. $(11.33)(10) = 113.3$
    Move the decimal point one
    place to the right.

61.      1.46
      ×  8.1
      ‾‾‾‾‾‾‾‾
         146
        1168
      ‾‾‾‾‾‾‾‾
         11.826

62.      15.66
      ×  1.75
      ‾‾‾‾‾‾‾‾
         7830
        10962
        1566
      ‾‾‾‾‾‾‾‾
         27.4050
      or 27.405

63.      19.02
      ×  0.92
      ‾‾‾‾‾‾‾‾
         3804
       17 118
      ‾‾‾‾‾‾‾‾
         17.4984

64.      21.14
      ×  0.005
      ‾‾‾‾‾‾‾‾
         0.10570
      or 0.1057

65. 145.9 million is 145.9 times 1,000,000.
    Move the decimal six places to the right
    to get the standard notation 145,900,000.

66. 1.25 trillion is 1.25 times 1,000,000,000,000.
    Move the decimal twelve places to the right
    to get the standard notation 1,250,000,000,000.

67. 455.2 billion is 455.2 times 1,000,000,000.
    Move the decimal nine places to the right
    to get the standard notation 455,200,000,000.

68. $16.78 million is $16.78 times 1,000,000.
    Move the decimal six places to the right
    to get the standard notation $16,780,000.

69. $100\overline{)8.9}$

    100 has two zeros so move the
    decimal in the dividend two
    places to the left.
    $$100\overline{)8.9} \quad 0.089$$

70.
$$
\begin{array}{r}
1.409 \\
13\overline{)18.317} \\
-13 \\
\hline
53 \\
-52 \\
\hline
117 \\
-117 \\
\hline
0
\end{array}
$$

71.
$$
\begin{array}{r}
1.79 \\
20\overline{)35.80} \\
-20 \\
\hline
158 \\
-140 \\
\hline
180 \\
-180 \\
\hline
0
\end{array}
$$

72. $\dfrac{15.95}{10} = 1.595$

10 has one zeros so move the decimal in the dividend one place to the left.

73.
$$
\begin{array}{r}
8.275 \\
12\overline{)99.300} \\
-96 \\
\hline
33 \\
-24 \\
\hline
90 \\
-84 \\
\hline
60 \\
\underline{60} \\
0
\end{array}
$$

74.
$$
\begin{array}{r}
36.352 \\
15\overline{)545.280} \\
-45 \\
\hline
95 \\
-90 \\
\hline
52 \\
-45 \\
\hline
78 \\
75 \\
\hline
30 \\
\underline{30} \\
0
\end{array}
$$

75. $0.1\overline{)49.88} \rightarrow 1\overline{)498.8}$

$$
\begin{array}{r}
498.8 \\
\rightarrow 1\overline{)498.8}
\end{array}
$$

76. $2.6\overline{)32.76} \rightarrow 26\overline{)327.6}$

$$
\begin{array}{r}
12.6 \\
26\overline{)327.6} \\
-26 \\
\hline
67 \\
-52 \\
\hline
156 \\
-156 \\
\hline
0
\end{array}
$$

77. $\dfrac{21.8}{4.5} \rightarrow \dfrac{218}{45}$

$$
\begin{array}{r}
4.844... \\
45\overline{)218.000} \\
-180 \\
\hline
380 \\
-360 \\
\hline
200 \\
-180 \\
\hline
200 \\
\underline{180} \\
20
\end{array}
$$

or 4.8 rounded.

78. $\dfrac{92.6}{2.3} \rightarrow \dfrac{926}{23}$

$$
\begin{array}{r}
40.260... \\
23\overline{)926.000} \\
-92 \\
\hline
60 \\
-46 \\
\hline
140 \\
-138 \\
\hline
20
\end{array}
$$

or 40.3 rounded.

79. $5 \div 0.82 \rightarrow 500 \div 82$

$$
\begin{array}{r}
6.097... \\
82\overline{)500.000} \\
-492 \\
\hline
800 \\
-738 \\
\hline
620 \\
-574 \\
\hline
46
\end{array}
$$

or 6.1 rounded.

80. $55 \div 1.6 \rightarrow 550 \div 16$

$$
\begin{array}{r}
34.375 \\
16\overline{)550.000} \\
-48 \\
\hline
70 \\
-64 \\
\hline
60 \\
-48 \\
\hline
120 \\
\underline{112} \\
80 \\
\underline{80} \\
0
\end{array}
$$

or 34.4 rounded.

81.
$$
\begin{array}{r}
0.2 \\
60\overline{)12.0} \\
-12\ 0 \\
\hline
0
\end{array}
$$

$$\frac{12}{60} = 0.2$$

82.
$$
\begin{array}{r}
0.04 \\
25\overline{)1.00} \\
-1\ 00 \\
\hline
0
\end{array}
$$

$$\frac{1}{25} = 0.04$$

83.
$$
\begin{array}{r}
0.9166... \\
12\overline{)11.0000} \\
-108 \\
\hline
20 \\
-12 \\
\hline
80 \\
-72 \\
\hline
80 \\
72 \\
\hline
8
\end{array}
$$

$$\frac{11}{12} = 0.9166... = 0.91\overline{6}$$

84. $3\dfrac{7}{11} = \dfrac{40}{11}$

$$
\begin{array}{r}
3.6363... \\
11\overline{)40.0000} \\
-33 \\
\hline
70 \\
-66 \\
\hline
40 \\
-33 \\
\hline
70 \\
66 \\
\hline
40 \\
33 \\
\hline
7
\end{array}
$$

$$3\frac{7}{11} = 3.6363... = 3.\overline{63}$$

85.
$$
\begin{array}{r}
0.856... \\
7\overline{)6.000} \\
-5\ 6 \\
\hline
40 \\
-35 \\
\hline
50 \\
-42 \\
\hline
8
\end{array}
$$

$\dfrac{6}{7}$ rounded to hundredths is 0.86.

86. $4\dfrac{3}{16} = \dfrac{67}{16}$

$$
\begin{array}{r}
4.187... \\
16\overline{)67.000} \\
-64 \\
\hline
30 \\
-16 \\
\hline
140 \\
-128 \\
\hline
120 \\
112 \\
\hline
8
\end{array}
$$

$4\dfrac{3}{16}$ rounded to hundredths is 4.19.

87. $25 + (130.99 - 5.3^2)$
$25 + (130.99 - 28.09)$
$25 + 102.9$
$127.9$

88. $1000 \div 125 + 9.2^2$
$1000 \div 125 + 84.64$
$8 + 84.64$
$92.64$

89. $\dfrac{1.5^3}{5}(96.6 \div 12) \cdot 10^2$

$\dfrac{3.375}{5}(96.6 \div 12) \cdot 10^2$

$\dfrac{3.375}{5} \cdot 8.05 \cdot 10^2$

$\dfrac{3.375}{5} \cdot 8.05 \cdot 100$

$0.675 \cdot 8.05 \cdot 100$

$5.43375 \cdot 100$

$543.375$

90. $85.3 - 4^3 - \left(1\dfrac{1}{5} \cdot 5\right)$
$85.3 - 4^3 - (1.2 \cdot 5)$
$85.3 - 4^3 - 6$
$85.3 - 64 - 6$
$21.3 - 6$
$15.3$

91. $\dfrac{(45.3 \div 9.06)^2}{10} + 12.1$

$\dfrac{5^2}{10} + 12.1$

$\dfrac{25}{10} + 12.1$

$2.5 + 12.1$

$14.6$

92. $30 \div 0.1 \cdot \dfrac{2.6 + 7^2}{10}$

$30 \div 0.1 \cdot \dfrac{2.6 + 49}{10}$

$30 \div 0.1 \cdot \dfrac{51.6}{10}$

$30 \div 0.1 \cdot 5.16$

$300 \cdot 5.16$

$1548$

93. The numbers 12.65, 12.18, and 12.27 agree until the tenths place where the digits are 6, 1, and 2 respectively. Order the original values as you would order these digits: 12.18 feet, 12.27 feet, 12.65 feet.

94. The four weights agree in the hundreds place but all disagree in the tens place where the digits are 2, 5, 1, 8. Order the weights in descending order by ordering these digits in descending order: 189.44 pounds, 155.65 pounds, 126.32 pounds, 114.18 pounds.

95. Mike and Morley climbed a total of $1265.38 + 1389.12 = 2654.50$ or 2654.5 feet.

96. The new bill is the old bill plus the additional fee or $\$46.95 + \$5.75 = \$52.70$.

97. First, find the total of Trish's bills. Rent, food, and the car payment total $\$1975.12 + \$322.45 + \$655.24 = \$2952.81$. Subtract this amount from her paycheck amount $\$4789.25 - \$2952.81 = \$1836.44$. Trish has \$1836.44 remaining.

98. The first tree was $145.54 - 103.92 = 41.62$ feet taller.

99. Twelve reams at \$5.81 cost a total of $12 \cdot \$5.81 = \$69.72$.

100. If Greg earns \$825.45 dollars per week, then in one year Greg earns $52 \cdot \$825.45 = \$42{,}923.40$.

101. The number 161.27 million is 161.27 times 1,000,000. To find standard notation for 161.27 million move the decimal point six places to the right. In standard notation the distance is 161,270,000 kilometers.

102. The number 778.1 million is 778.1 times 1,000,000. To find standard notation for 778.1 million move the decimal point six places to the right. In standard notation the distance is 778,100,000 kilometers.

103. With 14.5 gallons of gasoline at 55 miles per gallon, the car will travel $14.5 \cdot 55 = 797.5$ miles.

104. First, determine the cost Kool-Beanz paid per cap. There were 12 caps at a total of $134.40 so each cap cost
$134.40 \div 12 = \$11.20$. For a profit of $8.75 on each cap, they should charge $\$11.20 + \$8.75 = \$19.95$.

105. a. In currency, you have $54 \cdot \$1 + 18 \cdot \$5 + 11 \cdot \$10 + 27 \cdot \$20 + 2 \cdot \$50 = \$54 + \$90 + \$110 + \$540 + \$100$ or $894.00.

b. In coins, you have $28 \cdot \$0.01 + 24 \cdot \$0.05 + 16 \cdot \$0.10 + 13 \cdot \$0.25 = \$0.28 + \$1.20 + \$1.60 + \$3.25 = \$6.33$.

c. The total deposit is $\$894.00 + \$6.33 = \$900.33$.

d. In words, you are depositing nine hundred dollars and thirty-three cents.

106. a. In currency, you have $26 \cdot \$1 + 15 \cdot \$5 + 19 \cdot \$10 + 32 \cdot \$20 + 4 \cdot \$50 = \$26 + \$75 + \$190 + \$640 + \$200$ or $1131.00.

b. In coins, you have
$16 \cdot \$0.01 + 42 \cdot \$0.05 + 36 \cdot \$0.10 + 28 \cdot \$0.25 = \$0.16 + \$2.10 + \$3.60 + \$7.00 = \$12.86$.

c. The total deposit is $\$1131.00 + \$12.86 = \$1143.86$.

d. In words, you are depositing one thousand, one hundred forty-three dollars and eighty-six cents.

107.

| Number or Code | Date | Transaction Description | Payment, Fee, Withdrawal (−) | | ✓ | Deposit, Credit (+) | | $ Balance | |
|---|---|---|---|---|---|---|---|---|---|
| | 7/1 | | | | | | | 1694 | 20 |
| 228 | 7/12 | Wal-Mart | 183 | 40 | | | | 1510 | 80 |
| | 7/16 | deposit | | | | 325.50 | | 1836 | 30 |
| | 7/24 | ATM withdrawal | 200 | 00 | | | | 1636 | 30 |

Calculations:

$$1694.20 - 183.40 = 1510.80 \qquad 1510.80 + 325.50 = 1836.30 \qquad 1836.30 - 200.00 = 1636.30$$

108.

| Number or Code | Date | Transaction Description | Payment, Fee, Withdrawal (−) | | ✓ | Deposit, Credit (+) | | $ Balance | |
|---|---|---|---|---|---|---|---|---|---|
| | 3/1 | | | | | | | 2336 | 40 |
| | 3/11 | deposit | | | | 1550 | 35 | 3886 | 75 |
| 357 | 3/19 | Visa | 253 | 70 | | | | 3633 | 05 |
| 358 | 3/23 | FedEx | 45 | 10 | | | | 3587 | 95 |

Calculations:

$$2336.40 + 1550.35 = 3886.75 \qquad 3886.75 - 253.70 = 3633.05 \qquad 3633.05 - 45.10 = 3587.95$$

109. a. Three shirts at $56.75 each will cost $3 \cdot \$56.75 = \$170.25$.

b. If a $75.00 shirt is on sale for $56.75, you save $\$75.00 - \$56.75 = \$18.25$ for each shirt you buy. If you buy three shirts, your total savings will be $3 \cdot \$18.25 = \$54.75$.

110. a. Vickie's new job will pay her $\$60,000 - \$56,000 = \$4000$ more per year. This is $\$4000 \div 52 = \$76.9230...$ more per week or about $\$76.92$.

   b. From part a., Vickie's extra income is $\$76.92$ per week. If she saves this amount every week, it will take her $\$2307.60 \div \$76.92 = 30$ weeks to save the down payment.

111. First, add the three temperatures: $52.6 + 42.8 + 40.9 = 136.3$ degrees. Next, divide by three: $136.3 \div 3 = 45.4333...$ degrees. To the nearest tenth, the average temperature was $45.4°F$.

112. Find the total of the four GPAs: $3.56 + 3.48 + 3.72 + 3.88 = 14.64$. The average is $14.64 \div 4 = 3.66$.

113. a. First, find the total cost of the pizza. With tip and delivery the total cost is $\$18.80 + \$1.14 + \$3.50$ or $\$23.44$. George's share of the cost is three-fourths of $\$23.44$ or $\frac{3}{4} \cdot \$23.44 = 0.75 \cdot \$23.44 = \$17.58$.

   b. Clarissa's share can be found by subtracting George's share from the total. Clarissa' share is $\$23.44 - \$17.58 = \$5.86$.

114. a. The charity received $\frac{7}{8}$ of $\$158,700$ or $\frac{7}{8} \cdot \$158,700 = 0.875 \cdot \$158,700 = \$138,862.50$.

   b. Using part a., the balance was $\$158,700 - \$138,862.50 = \$19,837.50$ so $\$19,837.50$ went to printing costs and administrative expenses.

115. The total weight of the purchase is $6.7 + 3.9 + 4\frac{1}{2} = 6.7 + 3.9 + 4.5 = 15.1$ pounds. In fraction notation, this is $15.1 = 15\frac{1}{10}$ pounds.

116. The total weight of the purchase is $2.1 + 1\frac{4}{5} + 3\frac{3}{4} = 2.1 + 1.8 + 3.75 = 7.65$ pounds. In fraction notation, this is $7.65 = 7\frac{65}{100} = 7\frac{13}{20}$ pounds.

117. First, round the cost of the amplifier to two nonzero digits: $\$567.68$ rounds to $\$570.00$. Next, estimate the required monthly savings by dividing this number by twelve: $\$570 \div 12 = \$47.50$. Toby should save roughly $\$47.50$ each month.

118. First, round the yearly average to two nonzero digits: $653.6$ rounds to $650$. To estimate the weekly average, divide this amount by the number of weeks in a year. Since $650 \div 52 = 12.5$, each household recycles roughly $12.5$ pounds of glass and aluminum per week.

## Chapter 3 Assessment Test

1. The 7 in 23.0719 is in the second place to the right of the decimal point which is the *hundredths* place.

2. The 9 in 0.360914 is in the fourth place to the right of the decimal point which is the *ten-thousandths* place.

3.  The three decimal places in 42.949 indicate *thousandths* while the decimal point is read as *and*. The word form is "forty-two and nine hundred forty-nine thousandths."

4.  The four decimal places in 0.0365 indicate *ten-thousandths*. The word form is "three hundred sixty-five ten-thousandths."

5.  In "twenty-one hundred-thousandths" the phrase *hundred-thousandths* tells us there are five decimal places. The decimal notation is 0.00021.

6.  In "sixty-one and two hundred eleven thousandths" the word *thousandths* tells us there are three decimal places. The word *and* locates the decimal point and sixty-one is to the left of and so the decimal notation is 61.211.

7.  $8.85 = 8\dfrac{85}{100} = 8\dfrac{17}{20}$

8.  $0.125 = \dfrac{125}{1000} = \dfrac{1}{8}$

9.  The thousandths place is the first place where 0.66$\underline{4}$3 and 0.66$\underline{3}$49 disagree. Since 4 is larger than 3, then $0.6643 > 0.66349$.

10.  The hundredths place is the first place where 12.1$\underline{1}$8 and 12.1$\underline{8}$1 disagree. Since 1 is smaller than 8, we have $12.118 < 12.181$.

11.  $2.14530 = 2.145300$ since adding a zero to 2.14530 does not change its value.

12.  The digit in the tenths place in 1.$\underline{5}$97 is 5. The digit to its right is 9 which is more than five so change the 5 to 6 and delete the digits to the right to get 1.6.

13.  The digit in the hundredths place in 4.1$\underline{1}$089 is 1. The digit to its right is 0 which is less than five so leave the 1 unchanged but delete the digits to the right to get 4.11.

14.
$$\begin{array}{r} \overset{1}{\phantom{0}}3.490 \\ 0.006 \\ +\ 5.800 \\ \hline 9.296 \end{array}$$

15. Add vertically, tacking on zeros so all values have the same number of decimal places.

$$\begin{array}{r} \overset{1}{\phantom{0}}13.4400 \\ 10.9370 \\ +\ 0.1009 \\ \hline 24.4779 \end{array}$$

16.
$$\begin{array}{r} \overset{2\ \overset{13}{\cancel{3}}\ 10}{\cancel{3}\,\cancel{4}.\cancel{0}\,29} \\ -\ 6.5\,12 \\ \hline 2\,7.5\,17 \end{array}$$

17. Subtract vertically.
$$\begin{array}{r} 0.0\,\overset{8}{\cancel{9}}\,\overset{13}{\cancel{3}}\,8 \\ -\ 0.0\,0\,4\,5 \\ \hline 0.0\,8\,9\,3 \end{array}$$

18.
$$\begin{array}{r} 7.228 \quad \leftarrow 3 \text{ decimal places} \\ \times\ \ 1.3 \quad \leftarrow 1 \text{ decimal place} \\ \hline 2\,1684 \\ 7\,228\phantom{0} \\ \hline 9.3964 \quad \leftarrow 3+1 = 4 \text{ decimal places} \end{array}$$

19. Reverse the order for easier multiplication.

$$\begin{array}{r} 15.42 \quad \leftarrow 2 \text{ decimal places} \\ \times\ \ 0.008 \quad \leftarrow 3 \text{ decimal place} \\ \hline 0.12336 \quad \leftarrow 2+3 = 5 \text{ decimal places} \end{array}$$

20. Standard notation for 218.6 million is found by moving the decimal six places to the right giving 218,600,000.

21. Standard notation for 3.37 billion is found by moving the decimal nine places to the right giving 3,370,000,000.

22. To divide by 1000 move the decimal point three places to the left.

$$\frac{92.8}{1000} = 0.0928$$

23. To divide $1.6\overline{)40.96}$, first move the decimal point one place in the divisor and dividend, then use long division.

$$
\begin{array}{r}
25.6 \\
16\overline{)409.6} \\
\underline{32\phantom{0.0}} \\
89\phantom{.0} \\
\underline{80\phantom{.0}} \\
96 \\
\underline{96}
\end{array}
$$

24. $\dfrac{4.2^2}{8} \cdot (12.5 - 3.6) - 10.6045 =$

$\dfrac{4.2^2}{8} \cdot 8.9 - 10.6045 =$

$\dfrac{17.64}{8} \cdot 8.9 - 10.6045 =$

$2.205 \cdot 8.9 - 10.6045 =$

$19.6245 - 10.6045 = 9.02$

25. $36.3 \div 6.6 + (3.34 - 2.64)^3 =$

$36.3 \div 6.6 + 0.7^3 =$

$36.3 \div 6.6 + 0.343 =$

$5.5 + 0.343 = 5.843$

26. a. Emerson earned $\$14 \times 19.25$ or $\$269.50$.

b. A total of $\$20.62 + \$64.20 = \$84.82$ was deducted from her check. Using the answer in a., her take-home pay was $\$269.50 - \$84.82 = \$184.68$.

27. a. Sam spent a total of $3 \cdot \$12.69 + 2 \cdot \$16.50 + \$3.70 = \$38.07 + \$33.00 + \$3.70 = \$74.77$.

b. Sam pays with four twenties or $\$80$. From part a., his change is $\$80 - \$74.77 = \$5.23$.

28. First write $465\dfrac{6}{10}$ as the decimal number 465.6. The average miles per gallon are given by $23.1\overline{)465.6} = 231\overline{)4656}$. Dividing out to two decimal places gives 20.15…. After rounding to the nearest tenth, the average miles per gallon was 20.2 mpg.

29. The average temperature was $(62.3 + 39.6 + 43.4) \div 3 = 145.3 \div 3 = 48.433...$ or 48.4 degrees.

30. Total cost of the Power Play Advance ($\$99.99$), the carrying case ($\$7.99$), and two games ($\$26.99$ each) is $\$99.99 + \$7.99 + 2 \cdot \$26.99 = \$99.99 + \$7.99 + \$53.98 = \$161.96$.

31. Rounded to the nearest dollar, the magnifier costs $\$8.00$ while each game costs $\$27.00$, so the magnifier and four games cost approximately $\$8.00 + 4 \cdot \$27.00 = \$8.00 + \$108.00 = \$116.00$.

32. The total cost of the Power Play Advance and three games would be $\$99.99 + 3 \cdot \$26.99 = \$99.99 + \$80.97 = \$180.96$. If twelve equal payments are made, with no other charges, each payment will be $\$180.96 \div 12 = \$15.08$.

# Chapter 4   Ratio and Proportion

## Section 4.1    Understanding Ratios

## Concept Check

1.  A _ratio_ is a comparison of two quantities by division.

3.  Ratios may be written as two numbers separated by the word _to_ , as two numbers separated by a _colon_ , or as a _fraction_ .

5.  A ratio may express a comparison of a part to a _whole_ or a part to _another part_ .

7.  Explain the procedure to simplify a ratio that contains decimals.

> 1. Write the ratio in fraction notation
> 2. Rewrite as a ratio of whole numbers.
>    Multiply the ratio by 1 in the form
>    _n/n_ where _n_ is a power of 10 large
>    enough to remove any decimals in both
>    the numerator and the denominator.
> 3. Simplify, if possible.

## Guide Problems

9.  Write the ratio of 7 to 12.

a. Use the word *to*.  _7 to 12_
   Insert the word *to* between
   the values being compared.

b. Use a colon.  _7 : 12_
   Insert a colon between the
   values being compared.

c. Write in fraction notation
   $$\frac{7}{12}$$
   Create a fraction using the first
   term of the ratio as the
   numerator and the second term
   as denominator.

11. The ratio of 5 to 17 may be written

> using the word *to*,  5 to 17
> using a colon,  5 : 17
> as a fraction,  $\frac{5}{17}$ .

13. The ratio of 3 to $8\frac{1}{4}$ may be written

> using the word *to*,  3 to $8\frac{1}{4}$
> using a colon,  3 : $8\frac{1}{4}$
> as a fraction,  $\dfrac{3}{8\frac{1}{4}}$ .

15. The ratio of 2.7 to 9 may be written

> 2.7 to 9    2.7 : 9    $\dfrac{2.7}{9}$

17. The ratio of 5 to 2 may be written

> 5 to 2    5 : 2    $\dfrac{5}{2}$

19. The ratio of 8 to 15 may be written

    8 to 15    8 : 15    $\dfrac{8}{15}$

21. The ratio of 44 to 1.2 may be written

    44 to 1.2    44 : 1.2    $\dfrac{44}{1.2}$

## Guide Problems

23. Consider the ratio 18 to 27.

    a. Write the ratio in fraction notation.

        $\dfrac{18}{27}$

    Create a fraction using the first
    term of the ratio as the numerator
    and the second term as denominator.

    b. Simplify.
    Divide out the common factor of 9.

    $\dfrac{18}{27} = \dfrac{\overset{2}{\cancel{18}}}{\underset{3}{\cancel{27}}} = \dfrac{2}{3}$

25. Consider the ratio 2.6 to 50.

    a. Write the ratio in fraction notation.

        $\dfrac{2.6}{50}$

    Create a fraction using the first
    term of the ratio as the numerator
    and the second term as denominator.

    b. Identify 1 in the form where n is a power
    of 10 large enough to remove any decimals
    in the numerator and the denominator.

    The numerator has one digit after the
    decimal place while the denominator has
    none. Thus a factor of 10 will clear
    the decimals so we use 1 in the form

        $\dfrac{10}{10}$

    c. Multiply the ratio by the fraction in part b.

    $\dfrac{2.6}{50} = \dfrac{2.6}{50} \cdot \dfrac{10}{10} = \dfrac{2.6(10)}{50(10)} = \dfrac{26}{500}$

    d. Simplify.
    Divide out the common factor of 2.

    $\dfrac{26}{500} = \dfrac{\overset{13}{\cancel{26}}}{\underset{250}{\cancel{500}}} = \dfrac{13}{250}$

*In problems 27-45, a ratio must be simplified. The following strategy is used. The ratio is first written as a
fraction. If decimals appear in the numerator or denominator, we multiply by the appropriate form of 1 to clear
all decimals. If mixed numbers are involved, each mixed number is converted to an improper fraction and then
division is performed. In all cases, the resulting fraction is simplified, if possible.*

27. 20 to 6

    Convert to fraction: $\dfrac{20}{6}$
    Simplify by dividing out 6.

    $\dfrac{20}{6} = \dfrac{\overset{10}{\cancel{20}}}{\underset{3}{\cancel{6}}} = \dfrac{10}{3}$

29. 144 to 12

    $\dfrac{144}{12} = \dfrac{\overset{12}{\cancel{144}}}{\underset{1}{\cancel{12}}} = \dfrac{12}{1}$

**31.** 16 to 64

$$\frac{16}{64} = \frac{\overset{1}{\cancel{16}}}{\underset{4}{\cancel{64}}} = \frac{1}{4}$$

**33.** 500 to 1000

$$\frac{500}{1000} = \frac{\overset{1}{\cancel{500}}}{\underset{2}{\cancel{1000}}} = \frac{1}{2}$$

**35.** 0.3 to 1.1

Convert to fraction: $\dfrac{0.3}{1.1}$

Multiplying each term by 10 would clear the decimal so multiply the ratio by 1 in the form 10/10.

$$\frac{0.3}{1.1} = \frac{0.3}{1.1} \cdot \frac{10}{10} = \frac{0.3(10)}{1.1(10)} = \frac{3}{11}$$

No further simplification is possible.

**37.** 3.2 to 6

$$\frac{3.2}{6} = \frac{3.2}{6} \cdot \frac{10}{10} = \frac{32}{60}$$

$$\frac{32}{60} = \frac{\overset{8}{\cancel{32}}}{\underset{15}{\cancel{60}}} = \frac{8}{15}$$

**39.** 1.25 to 1

$$\frac{1.25}{1} = \frac{1.25}{1} \cdot \frac{100}{100} = \frac{125}{100}$$

$$\frac{125}{100} = \frac{\overset{5}{\cancel{125}}}{\underset{4}{\cancel{100}}} = \frac{5}{4}$$

**41.** 9 to $2\frac{1}{4}$

Convert to fraction: $\dfrac{9}{2\frac{1}{4}}$

Write any mixed numbers as improper fraction and then divide.

$$\frac{9}{\frac{9}{4}} = 9 \div \frac{9}{4} = 9 \cdot \frac{4}{9} = \frac{\overset{1}{\cancel{9}}}{1} \cdot \frac{4}{\underset{1}{\cancel{9}}} = \frac{4}{1}$$

No further simplification is possible.

**43.** $1\frac{1}{3}$ to $1\frac{2}{3}$

$$\frac{1\frac{1}{3}}{1\frac{2}{3}} = \frac{\frac{4}{3}}{\frac{5}{3}} = \frac{4}{3} \div \frac{5}{3} = \frac{4}{3} \cdot \frac{3}{5} = \frac{4}{\cancel{3}} \cdot \frac{\overset{1}{\cancel{3}}}{5} = \frac{4}{5}$$

**45.** 9 to $1\frac{1}{3}$

$$\frac{9}{1\frac{1}{3}} = \frac{9}{\frac{4}{3}} = 9 \div \frac{4}{3} = \frac{9}{1} \cdot \frac{3}{4} = \frac{27}{4}$$

## Guide Problems

**47.** Consider the ratio 6 quarts to 2 gallons.

a. Identify the smaller units.
quarts

b. Convert gallons to quarts.
Express the quantity with the larger unit in terms of the smaller unit.

Since 1 gallon = 4 quarts,
2 gallons = 2 · 4 quarts = 8 quarts

c. Write the ratio.

$$\frac{6}{8} \left( \frac{6 \text{ quarts}}{2 \text{ gallons}} = \frac{6 \text{ quarts}}{8 \text{ quarts}} = \frac{6}{8} \right)$$

d. Simplify.

$$\frac{6}{8} = \frac{3}{4}$$

*In problems 49-59, the steps outlined above are applied. First, identify the smaller of the two units involved. Second, write the quantity having the larger unit in terms of the smaller unit. Write the desired ratio in terms of the smaller unit. The unit will cancel out. Lastly, simplify the resulting fraction, if possible. Remember, when a fraction represents a ratio and the denominator simplifies to one, you always write the one in the denominator.*

49. 3 feet to 4 yards
Smaller unit: feet
4 yards = 4 · 3 feet = 12 feet

$$\frac{3 \text{ feet}}{4 \text{ yards}} = \frac{3}{12} = \frac{1}{4}$$

51. 8 cups to 10 pints
Smaller unit: cups
10 pints = 10 · 2 cups
= 20 cups

$$\frac{8 \text{ cups}}{10 \text{ pints}} = \frac{8}{20} = \frac{2}{5}$$

53. 144 inches to 3.5 feet
Smaller unit: inches
3.5 feet = 3.5 · 12 inches
= 42 inches

$$\frac{144 \text{ inches}}{3.5 \text{ feet}} = \frac{144}{42} = \frac{24}{7}$$

55. 10 pounds to 150 ounces
Smaller unit: ounces
10 pounds = 10 · 16 ounces
= 160 ounces

$$\frac{10 \text{ pounds}}{150 \text{ ounces}} = \frac{160}{150} = \frac{16}{15}$$

57. 12 yards to 2 feet
Smaller unit: feet
12 yards = 12 · 3 feet
= 36 feet

$$\frac{12 \text{ yards}}{2 \text{ feet}} = \frac{36}{2} = \frac{18}{1}$$

59. 5 ounces to 20 drams
Smaller unit: drams
5 ounces = 5 · 16 drams
= 80 drams

$$\frac{5 \text{ ounces}}{20 \text{ drams}} = \frac{80}{20} = \frac{4}{1}$$

61. a. The ratio of 9 goals in 12 games can be written as $\frac{9}{12}$ which simplifies to $\frac{3}{4}$. The three ways to write this ratio are

$$\frac{3}{4}, \text{ 3 to 4, 3 : 4}$$

b. The ratio of 152 goals in 284 games can be written as $\frac{152}{284}$ which simplifies to $\frac{38}{71}$. The three ways to write this ratio are

$$\frac{38}{71}, \text{ 38 to 71, 38 : 71}$$

63. We have 3 cups of milk, 7 cups of flour and 1 cup of butter. Note all three quantities are in terms of the same unit, cups, so no conversion is necessary when finding ratios.

a. The ratio of milk to flour is $\frac{3}{7}$.

b. The ratio of flour to butter is $\frac{7}{1}$.

c. The total amount of ingredients is $3 + 7 + 1 = 11$ cups. The ratio of butter to the total is $\frac{1}{11}$.

65. Using ounces, the smaller measurement unit, the ratio of Arabica beans to Special Blend coffee is

$$\frac{12 \text{ ounces}}{3 \text{ pounds}} = \frac{12 \text{ ounces}}{3 \cdot 16 \text{ ounces}} = \frac{12}{48} = \frac{1}{4}$$

67. Using inches, the smaller measurement unit, the ratio of window's height to its width is

$$\frac{3 \text{ feet}}{16 \text{ inches}} = \frac{3 \cdot 12 \text{ inches}}{16 \text{ inches}} = \frac{36}{16} = \frac{9}{4}$$

69. The ratio of the longest side, length 10, to the shortest side, length 6, of the triangle is
$$\frac{10}{6} = \frac{5}{3}.$$

71. The ratio of the longest side, length 12, to the shortest side, length 7, of the rectangle is
$$\frac{12}{7}.$$

73. From the chart, there are 741 restaurants in Canada and 1156 in Japan. The ratio of Canadian to Japanese restaurants can be expressed as
$$741 \text{ to } 1156, \ 741 : 1156, \ \frac{741}{1156}.$$

75. There are 578 restaurants in the UK and 1000 in China/Hong Kong. The ratio is $\frac{578}{1000} = \frac{289}{500}$.

77. From the table there are 31,100 McDonald's to 6700 Wendy's. As a fraction, this ratio is
$$\frac{31,100}{6700} = \frac{311}{67}.$$

79. There are $31,100 + 6700 = 37,800$ combined McDonald's and Wendy's. The ratio of Burger King restaurants to this total is $\frac{11,200}{37,800} = \frac{8}{27}$.

81. The online purchase of flowers totaled $3.7 billion. The online purchase of tools totaled $7.0 billion. The corresponding ratio can be expressed in three ways. First, determine and simplify the fraction form to use whole numbers. Then write the other two forms. The three forms are
$$\frac{3.7}{7.0} = \frac{37}{70}, \ 37 \text{ to } 70, \ 37 : 70$$

83. The total online purchase were approximately $17.4 + $14.1 + $7.0 + $3.7 = $42.2 billion. The ratio of online tool purchases, $7.0 billion, to the total can be written as
$$\frac{7.0}{42.2} = \frac{70}{422} = \frac{35}{211}, \ 35 \text{ to } 211, \ 35 : 211$$

## Cumulative Skills Review

1. $400 \div 5^2 - 4(2 + 1)$
   $400 \div 5^2 - 4 \cdot 3$
   $400 \div 25 - 4 \cdot 3$
   $16 - 4 \cdot 3$
   $16 - 12$
   $4$

3. First rewrite the fraction as a decimal. Then convert the problem to a division with a whole number divisor.
$$65.155 \div 3\frac{2}{5} = 65.155 \div 3.4 = 651.55 \div 34$$
   Next use long division.
$$34 \overline{)651.5500} \quad 19.1632...$$
   To the nearest thousandth, the result is 19.163.

5. Add vertically.
$$\begin{array}{r} {}^{1}\ {}^{1\,1} \\ 127 \\ 5\,652 \\ 78 \\ + \ 10,322 \\ \hline 16,179 \end{array}$$

7. The LCD is $47 \cdot 5 \cdot 3 = 705$.
$$\frac{23}{47} = \frac{345}{705}, \ \frac{4}{5} = \frac{564}{705}, \ \frac{8}{15} = \frac{376}{705}$$
   When written over a common denominator, the numerators in descending order are 564, 376, 345 so the fractions in descending order are
$$\frac{4}{5}, \ \frac{8}{15}, \ \frac{23}{47}.$$

9. The hundredths place is underlined in 8975.4$\underline{5}$5.
   The digit to its right is 5 which is five or larger so the hundredths digit is rounded up to 6 yielding 8975.46 as the desired rounded value.

## Section 4.2    Working with Rates and Units

### Concept Check

1.  A _ rate _ is a ratio that compares two quantities that have different units.

3.  In rates, we include the units because they are *different* and therefore do not _divide_ out.

5.  A unit rate is written in fraction notation.  We then _divide_ the numerator by the denominator.

7.  To write a unit price, we write the rate in fraction notation with the _price_ as the numerator and the quantity (number of items or units)_ as the denominator.

### Guide Problems

9.  a. Write the rate 8 pages in 12 minutes
       in fraction notation.

$$\frac{8 \text{ pages}}{12 \text{ minutes}}$$

   b. Simplify.
      Divide out the common factor of 4.

$$\frac{8 \text{ pages}}{12 \text{ minutes}} = \frac{\overset{2}{\cancel{8}} \text{ pages}}{\underset{3}{\cancel{12}} \text{ minutes}} = \frac{2 \text{ pages}}{3 \text{ minutes}}$$

   c. Write the rate in word form.

   _ 2 pages for every 3 minutes

*Problems 11 through 27 are solved exactly as above. The given rate is first written in fraction form. The numerical fraction is simplified in the usual manner. Lastly, the simplified rate is written in word form.*

11. 85 fence panels for 1350 feet
$$\frac{85 \text{ fence panels}}{1350 \text{ feet}} = \frac{17 \text{ fence panels}}{270 \text{ feet}}$$
Word form: 17 fence panels for every 270 feet

13. 9 vans for 78 people
$$\frac{9 \text{ vans}}{78 \text{ people}} = \frac{3 \text{ vans}}{26 \text{ people}}$$
Word form: 3 vans for every 26 people

15. 32 bags for 24 passengers
$$\frac{32 \text{ bags}}{24 \text{ passengers}} = \frac{4 \text{ bags}}{3 \text{ passengers}}$$
Word form: 4 bags for every 3 passengers

17. 2500 revolutions for 8 minutes
$$\frac{2500 \text{ revolutions}}{8 \text{ minutes}} = \frac{625 \text{ revolutions}}{2 \text{ minutes}}$$
Word form: 625 revolutions for every 2 minutes

19. 75 patients for 9 doctors
$$\frac{75 \text{ patients}}{9 \text{ doctors}} = \frac{25 \text{ patients}}{3 \text{ doctors}}$$
Word form: 25 patients for every 3 doctors

21. 182 gallons of milk for 34 cows
$$\frac{182 \text{ gallons}}{34 \text{ cows}} = \frac{91 \text{ gallons}}{17 \text{ cows}}$$
Word form: 91 gallons of milk for every 17 cows

23. 562 students for 28 teachers
$$\frac{562 \text{ students}}{28 \text{ teachers}} = \frac{281 \text{ students}}{14 \text{ teachers}}$$
Word form: 281 students for every 14 teachers

25. 6284 square feet for 14 gallons of paint
$$\frac{6284 \text{ square feet}}{14 \text{ gallons}}$$
$$= \frac{3142 \text{ square feet}}{7 \text{ gallons}}$$
Word form: 3142 square feet for every 7 gallons of paint

27. 55 hits for 180 at bats
$$\frac{55 \text{ hits}}{180 \text{ at bats}} = \frac{11 \text{ hits}}{36 \text{ at bats}}$$
Word form: 11 hits for every 36 at bats

## Guide Problems

29.  a. Write the rate 280 miles on 20 gallons of fuel
in fraction notation.

$$\frac{280 \text{ miles}}{20 \text{ gallons}}$$

b. Divide the numerator by the denominator and
write the unit rate as a fraction.

$$\frac{280 \text{ miles}}{20 \text{ gallons}} = \frac{280 \div 20 \text{ miles}}{20 \div 20 \text{ gallons}} = \frac{14 \text{ miles}}{1 \text{ gallon}}$$

c. Write the unit rate in word form.
14 miles per gallon or 14 miles/gallon
or 14 mpg

*The solutions to Problems 31 through 47 all follow the strategy as above. The given rate is written in fraction form by placing the first quantity over the second. Next divide both the numerator and denominator by the value in the denominator so the denominator is now 1 and the fraction represents a unit rate. Lastly, write the unit rate in word form. This typically involves writing the numerator and the unit of the denominator separated by the word per. If the division of thenumerator resulted in a decimal number, round the numerator to the nearest tenth.*

31.  $3600 in 12 months

$$\frac{\$3600}{12 \text{ months}} = \frac{\$3600 \div 12}{12 \div 12 \text{ months}} = \frac{\$300}{1 \text{ month}}$$

$300 per month

33.  16 touchdowns in 3 games

$$\frac{16 \text{ touchdowns}}{3 \text{ games}} = \frac{16 \div 3 \text{ touchdowns}}{3 \div 3 \text{ games}}$$
$$= \frac{5.33... \text{ touchdowns}}{1 \text{ game}}$$

5.3 touchdowns per game

35.  30 parking spaces for 30 apartments

$$\frac{30 \text{ parking spaces}}{30 \text{ apartments}} = \frac{30 \div 30 \text{ parking spaces}}{30 \div 30 \text{ apartments}}$$
$$= \frac{1 \text{ parking space}}{1 \text{ apartment}}$$

1 parking space per apartment

37.  325 yards of material for 85 shirts

$$\frac{325 \text{ yards of material}}{85 \text{ shirts}} = \frac{325 \div 85 \text{ yards of material}}{85 \div 85 \text{ shirts}}$$
$$= \frac{3.823... \text{ yards of material}}{1 \text{ shirt}}$$

3.8 yards of material per shirt

39.  28 servings for 7 pizzas

$$\frac{28 \text{ servings}}{7 \text{ pizzas}} = \frac{28 \div 7 \text{ servings}}{7 \div 7 \text{ pizzas}} = \frac{4 \text{ servings}}{1 \text{ pizza}}$$

4 servings per pizza

41.  176 roses in 8 vases

$$\frac{176 \text{ roses}}{8 \text{ vases}} = \frac{176 \div 8 \text{ roses}}{8 \div 8 \text{ vases}} = \frac{22 \text{ roses}}{1 \text{ vase}}$$

22 roses per vase

43.  19 kilowatts in 6 hours

$$\frac{19 \text{ kilowatts}}{6 \text{ hours}} = \frac{19 \div 6 \text{ kilowatts}}{6 \div 6 \text{ hours}}$$
$$= \frac{3.166... \text{ kilowatts}}{1 \text{ hour}}$$

3.2 kilowatts per hour

45.  8835 branches per 95 trees

$$\frac{8835 \text{ branches}}{95 \text{ tree}} = \frac{8835 \div 95 \text{ branches}}{95 \div 95 \text{ tree}}$$
$$= \frac{93 \text{ branches}}{1 \text{ tree}}$$

93 branches per tree

47.  5040 words in 12 pages

$$\frac{5040 \text{ words}}{12 \text{ pages}} = \frac{5040 \div 12 \text{ words}}{12 \div 12 \text{ pages}} = \frac{420 \text{ words}}{1 \text{ page}}$$

420 words per page

## Guide Problems

49. a. Set up the rate, $22.50 for 12 golf balls, in fraction notation with price as the numerator and the quantity (number of items or units) as the denominator.

$$\frac{\$22.50}{12 \text{ golf balls}}$$

b. Divide the numerator by the denominator and write the unit price in fraction notation.

$$\frac{\$22.50 \div 12}{12 \div 12 \text{ golf balls}} = \frac{\$1.875}{1 \text{ golf ball}}$$

c. Write the unit price in word form. Round to the nearest cent.
   $1.88 per golf ball
   ($1.875 rounds to $1.88)

*In Problems 51 through 65 a unit price is found using the steps outlined above. Write the given rate in fraction notation. Always put the price or cost in the numerator and the quantity in the denominator. Next divide both the numerator and denominator by the value in the denominator so the denominator is now 1 and the fraction represents a unit price. Lastly, write the unit price in word form. If necessary, round the dollar amount in the numerator to the nearest cent.*

51. $24 for 300 minutes of long distance
$$\frac{\$24}{300 \text{ minutes}} = \frac{\$24 \div 300}{300 \div 300 \text{ minute}} = \frac{\$0.08}{1 \text{ minute}}$$
The unit price is $0.08 per minute of long distance.

53. 55 ounces of detergent for $7.15
$$\frac{\$7.15}{55 \text{ ounces}} = \frac{\$7.15 \div 55}{55 \div 55 \text{ ounces}} = \frac{\$0.13}{1 \text{ ounce}}$$
The unit price is $0.13 per ounce.

55. 45 oranges for $12.60
$$\frac{\$12.60}{45 \text{ oranges}} = \frac{\$12.60 \div 45}{45 \div 45 \text{ oranges}} = \frac{\$0.28}{1 \text{ orange}}$$
The unit price is $0.28 per orange.

57. $9.30 for 6 milkshakes
$$\frac{\$9.30}{6 \text{ milkshakes}} = \frac{\$9.30 \div 6}{6 \div 6 \text{ milkshakes}} = \frac{\$1.55}{1 \text{ milkshake}}$$
The unit price is $1.55 per milkshake.

59. 4 batteries for $4.64
$$\frac{\$4.64}{4 \text{ batteries}} = \frac{\$4.64 \div 4}{4 \div 4 \text{ batteries}} = \frac{\$1.16}{1 \text{ battery}}$$
The unit price is $1.16 per battery.

61. $23.20 for a 16-pound turkey
$$\frac{\$23.20}{16 \text{ pounds}} = \frac{\$23.20 \div 16}{16 \div 16 \text{ pounds}} = \frac{\$1.45}{1 \text{ pound}}$$
The unit price is $1.45 per pound of turkey.

63. 16 pies for $143.20
$$\frac{\$143.20}{16 \text{ pies}} = \frac{\$143.20 \div 16}{16 \div 16 \text{ pies}} = \frac{\$8.95}{1 \text{ pie}}$$
The unit price is $8.95 per pie.

65. $5.50 for 24 bottles of spring water
$$\frac{\$5.50}{24 \text{ bottles}} = \frac{\$5.50 \div 24}{24 \div 24 \text{ bottles}} = \frac{\$0.229...}{1 \text{ bottle}}$$
The unit price is $0.23 per bottle of spring water.

67. The rate 14 patients for 6 nurses can be written as $\dfrac{14 \text{ patients}}{6 \text{ nurses}} = \dfrac{7 \text{ patients}}{3 \text{ nurses}}$ or in word form as

7 patients for every 3 nurses.

69. The fraction form for 28 pounds of cheese in 21 orders is $\dfrac{28 \text{ pounds}}{21 \text{ orders}} = \dfrac{4 \text{ pounds}}{3 \text{ orders}}$. The word form is

4 pounds of cheese for every 3 orders.

71. a. Todd's unit rate:

$$\frac{29 \text{ miles}}{4.3 \text{ hours}} = \frac{29 \div 4.3 \text{ miles}}{4.3 \div 4.3 \text{ hours}} = \frac{6.744... \text{ miles}}{1 \text{ hour}}$$

or 6.7 miles per hour to the nearest tenth.

b. Kimberly's unit rate:

$$\frac{25 \text{ miles}}{4 \text{ hours}} = \frac{25 \div 4 \text{ miles}}{4 \div 4 \text{ hours}} = \frac{6.25 \text{ miles}}{1 \text{ hour}}$$

or 6.3 miles per hour to the nearest tenth.

c. Todd has the higher unit rate so Todd ran faster.

75. From the table, there are 612,274 total pilots and 211,446 general aviation aircraft. The simplified fraction for the ratio of pilots to aircraft is

$$\frac{612,274 \text{ pilots}}{211,446 \text{ aircraft}} = \frac{306,137 \text{ pilots}}{105,723 \text{ aircraft}}$$

Note, 2 is the only common factor.
In word form, this ratio is
306,137 pilots to 105,723 general aviation aircraft.

79. $610 for 12 purses gives a unit price of

$$\frac{\$610}{12 \text{ purses}} = \frac{\$610 \div 12}{12 \div 12 \text{ purses}} = \frac{\$50.833...}{1 \text{ purse}}$$

or $50.83 per purse.
$916 for 20 purses gives a unit price of

$$\frac{\$916}{20 \text{ purses}} = \frac{\$916 \div 20}{20 \div 20 \text{ purses}} = \frac{\$45.80}{1 \text{ purse}}$$

or $45.80 per purse.
The best buy is 20 purses at $45.80 per purse.

83. The unit prices for each size are summarized in the table below.

| Size | Price | Unit price |
|------|-------|------------|
| 25 pounds | $41.25 | $41.25 ÷ 25 = $1.65 |
| 30 pounds | $51.90 | $51.90 ÷ 30 = $1.73 |
| 45 pounds | $67.50 | $67.50 ÷ 45 = $1.50 |

45 pounds at $1.50 per pound is the best buy.

73. a. The X400's unit rate:

$$\frac{15 \text{ inches}}{9 \text{ seconds}} = \frac{15 \div 9 \text{ inches}}{9 \div 9 \text{ seconds}} = \frac{1.666... \text{ inches}}{1 \text{ seconds}}$$

or 1.7 inches per second to the nearest tenth.

b. The T300's unit rate:

$$\frac{13 \text{ inches}}{8 \text{ seconds}} = \frac{13 \div 8 \text{ inches}}{8 \div 8 \text{ seconds}} = \frac{1.625 \text{ inches}}{1 \text{ second}}$$

or 1.6 inches per second to the nearest tenth.

c. The X400 has the higher unit rate so it is the faster machine.

77. From the table, there are 675,243,265 passengers and 5026 public airports. The corresponding unit rate is

$$\frac{675,243,265 \text{ passengers}}{5026 \text{ airports}}$$

$$= \frac{675,243,265 \div 5026 \text{ passengers}}{5026 \div 5026 \text{ airports}}$$

$$= \frac{134,350.03... \text{ passengers}}{1 \text{ airport}}$$

In word form, this unit rate is
134,350 passengers for every airport.

81. $3.65 for 12 ounces gives a unit price of

$$\frac{\$3.65}{12 \text{ ounces}} = \frac{\$3.65 \div 12}{12 \div 12 \text{ ounces}} = \frac{\$0.3041...}{1 \text{ ounce}}$$

or $0.30 per ounce.
$5.28 for 16 ounces gives a unit price of

$$\frac{\$5.28}{16 \text{ ounces}} = \frac{\$5.28 \div 16}{16 \div 16 \text{ ounces}} = \frac{\$0.33}{1 \text{ ounce}}$$

or $0.33 per ounce.
The 12 ounce tube at $0.30 an ounce is the best buy.

85. The unit prices for each size are summarized in the table below.

| Size | Price | Unit price |
|------|-------|------------|
| 4 gallons | $65.00 | $65.00 ÷ 4 = $16.25 |
| 5 gallons | $79.10 | $79.10 ÷ 5 = $15.82 |
| 6 gallons | $100.50 | $100.50 ÷ 6 = $16.75 |

5 gallons at $15.82 per gallon is the best buy.

87. a. The unit price for the 14.7-ounce size is $0.99 \div 14.7 = \$0.06734...$ per ounce or, after rounding, \$0.07 per ounce.

    b. The unit price for the 50-ounce size is $2.99 \div 50 = \$0.0598$ per ounce or, after rounding, \$0.06 per ounce.

    c. The unit price for the 45-ounce gel would be $2.25 \div 45 = \$0.05$ per ounce.
    This is less than the unit price for the 50-ounce size of 0.06 found in part b. so the 45-ounce gel is the better buy.

## Cumulative Skills Review

1. $41 \div 0$ is undefined. Division by 0 is always undefined.

3. $15\dfrac{5}{12} - 6\dfrac{1}{6} = 15\dfrac{5}{12} - 6\dfrac{2}{12}$

    $= 9\dfrac{3}{12} = 9\dfrac{1}{4}$

5. $$\overset{\overset{\scriptstyle 14}{\phantom{4}}\,\overset{\scriptstyle 9}{\phantom{5}}\,\overset{\scriptstyle 9}{\phantom{0}}\,\overset{\scriptstyle 9}{\phantom{0}}\,\overset{\scriptstyle 9}{\phantom{0}}\,10}{4\,5\,.\,0\,0\,0\,0}$$

    $$\begin{array}{r} -\ 5\,.\,9\,9\,9\,9 \\ \hline 3\,9\,.\,0\,0\,0\,1 \end{array}$$

7. a. The smaller unit is feet so

    $\dfrac{8 \text{ yards}}{14 \text{ feet}} = \dfrac{8 \cdot 3 \text{ feet}}{14 \text{ feet}} = \dfrac{24 \text{ feet}}{14 \text{ feet}} = \dfrac{12}{7}$

    b. The smaller unit is feet so

    $\dfrac{3 \text{ yards}}{7 \text{ feet}} = \dfrac{3 \cdot 3 \text{ feet}}{7 \text{ feet}} = \dfrac{9 \text{ feet}}{7 \text{ feet}} = \dfrac{9}{7}$

9. $6\dfrac{1}{2} + 5\dfrac{3}{8} + 4\dfrac{3}{4} = 6\dfrac{4}{8} + 5\dfrac{3}{8} + 4\dfrac{6}{8}$

    $= 15\dfrac{13}{8} = 15 + 1\dfrac{5}{8}$

    $= 16\dfrac{5}{8}$

    Nick purchased $16\dfrac{5}{8}$ pounds in all.

## Section 4.3    Understanding and Solving Proportions

## Concept Check

1. A proportion is a mathematical statement that two ratios are __equal__ .

3. If $\dfrac{a}{b}$ and $\dfrac{c}{d}$ are equal ratios, their proportion is written as $\underline{\dfrac{a}{b} = \dfrac{c}{d}}$ .

5. To verify whether two ratios are proportional, use __cross__ __multiplication__ to be sure that the cross products are __equal__ .

7. When one of the terms of a proportion is unknown, we can __solve__ the proportion for that term.

9.  To verify the answer after solving a proportion, we replace the unknown in the original proportion with the answer, and check to see that the __cross__   __products__  are equal.

## Guide Problems

11. Write 5 is to 10 as 9 is to 18 as a proportion.

a. Write 5 to 10 as a ratio.

$$\frac{5}{10}$$

b. Write 9 to 18 as a ratio.

$$\frac{9}{18}$$

c. Write a proportion by separating the two ratios with an equal sign.

$$\frac{5}{10}=\frac{9}{18}$$

*Problems 13 through 23 are solved in the same manner. A proportion is stated as a sentence and consists of two ratios separated by the word "as." Write each ratio as a fraction, then separate the two fractions with an equal sign to convert the sentence to a proportion.*

13. 22 is to 44 as 7 is to 14.

22 to 44: $\dfrac{22}{44}$

7 to 14: $\dfrac{7}{14}$

$$\frac{22}{44}=\frac{7}{14}$$

15. 8 suits is to 3 weeks as 32 suits is to 12 weeks.

8 suits to 3 weeks: $\dfrac{8 \text{ suits}}{3 \text{ weeks}}$

32 suits to 12 weeks: $\dfrac{32 \text{ suits}}{12 \text{ weeks}}$

$$\frac{8 \text{ suits}}{3 \text{ weeks}}=\frac{32 \text{ suits}}{12 \text{ weeks}}$$

17. 3.6 is to 5.8 as 14.4 is to 23.2.

$$\frac{3.6}{5.8}=\frac{14.4}{23.2}$$

19. 5 cans is to $8 as 15 cans is to $24.

$$\frac{5 \text{ cans}}{\$8}=\frac{15 \text{ cans}}{\$24}$$

21. 12 is to 7 as 3.6 is to 2.1.

$$\frac{12}{7}=\frac{3.6}{2.1}$$

23. 150 calories is to 7 ounces as 300 calories is to 14 ounces.

$$\frac{150 \text{ calories}}{7 \text{ ounces}}=\frac{300 \text{ calories}}{14 \text{ ounces}}$$

*Problems 25 through 35 reverse the process of Problems 13 through 23, turning a proportion into sentence form. Given a proportion, write the fraction on the left in word form, do the same with the fraction on the right. Separate the two with "as" to write the proportion in word form. Remember that a ratio such as "5 to 2" is written as "5 is to 2" when part of a proportion.*

25. $\dfrac{6}{3}=\dfrac{30}{15}$

left hand side: 6 to 3
right hand side: 30 to 15
Sentence form:
6 is to 3 as 30 is to 15.

27. $\dfrac{15 \text{ pages}}{2 \text{ minutes}}=\dfrac{75 \text{ pages}}{10 \text{ minutes}}$

left hand side:
   15 pages to 2 minutes
right hand side:
   75 pages to 10 minutes
Sentence form:
15 pages is to 2 minutes as 75 pages is to 10 minutes.

29. $\dfrac{3 \text{ strikeouts}}{2 \text{ hits}}=\dfrac{27 \text{ strikeouts}}{18 \text{ hits}}$

Sentence form:
3 strikeouts is to 2 hits as 27 strikeouts is to 18 hits.

31. $\dfrac{16}{5} = \dfrac{80}{25}$

Sentence form:
16 is to 5 as 80 is to 25.

33. $\dfrac{22.2}{65.3} = \dfrac{44.4}{130.6}$

Sentence form:
22.2 is to 65.3 as 44.4 is to 130.6.

35. $\dfrac{25 \text{ songs}}{2 \text{ CDs}} = \dfrac{125 \text{ songs}}{10 \text{ CDs}}$

Sentence form:
25 songs is to 2 CDs as 125 songs is to 10 CDs.

## Guide Problems

37. Determine whether the ratios $\dfrac{55}{11}$ and $\dfrac{80}{16}$ are proportional.

a. Multiply the denominator of the first ratio by the numerator of the second ratio.

$11 \cdot 80 = 880$

b. Multiply the numerator of the first ratio by the denominator of the second ratio.

$55 \cdot 16 = 880$

c. Are the cross products equal?

Yes. They both equal 880.

d. Are the ratios proportional?

Yes. The cross products are equal.

e. If the ratios are proportional, write a proportion.

Since these ratios are proportional, write a proportion by separating the ratios with an equal sign.

$\dfrac{55}{11} = \dfrac{80}{16}$

*In Problems 39 through 51, determine if two ratios are proportional by computing the cross products. If the cross products are equal, the ratios are proportional and are written as a proportion. If the cross products are not equal, the rations are not proportional.*

39. $\dfrac{45}{15} \overset{?}{=} \dfrac{33}{12}$

$15 \cdot 33 = 495, \quad 45 \cdot 12 = 540$

The cross products are not equal. The ratios are not proportional.

41. $\dfrac{16}{23} \overset{?}{=} \dfrac{48}{79}$

$23 \cdot 48 = 1104, \quad 16 \cdot 79 = 1264$

The cross products are not equal. The ratios are not proportional.

43. $\dfrac{80}{5} \overset{?}{=} \dfrac{400}{25}$

$5 \cdot 400 = 2000, \quad 80 \cdot 25 = 2000$

The cross products are equal. The ratios are proportional.

$\dfrac{80}{5} = \dfrac{400}{25}$

45. $\dfrac{5}{17} \overset{?}{=} \dfrac{35}{119}$

$17 \cdot 35 = 595, \quad 5 \cdot 119 = 595$

The cross products are equal. The ratios are proportional.

$\dfrac{5}{17} = \dfrac{35}{119}$

47. $\dfrac{63}{7} \overset{?}{=} \dfrac{26}{3}$

$7 \cdot 26 = 182, \quad 63 \cdot 3 = 189$

The cross products are not equal. The ratios are not proportional.

49. $\dfrac{25}{39} \overset{?}{=} \dfrac{75}{117}$

$39 \cdot 75 = 2925, \quad 25 \cdot 117 = 2925$

The cross products are equal. The ratios are proportional.

$\dfrac{25}{39} = \dfrac{75}{117}$

51. $\dfrac{54}{13} \overset{?}{=} \dfrac{108}{26}$

$13 \cdot 108 = 1404, \quad 54 \cdot 26 = 1404$

The cross products are equal. The ratios are proportional.

$\dfrac{54}{13} = \dfrac{108}{26}$

## Guide Problems

53. Solve for the unknown quantity in the proportion

$$\frac{c}{20} = \frac{14}{56}.$$

a. Cross multiply to find the cross products.
$$20 \cdot 14 = 280$$
$$c \cdot 56 = 56c$$

b. Separate the cross products by an equal sign to form an equation.
$$280 = 56c$$

c. Divide both sides of the equation by the number on the side with the unknown. In this case, the number on the side with the unknown is 56.
$$\frac{280}{56} = \frac{56c}{56}$$

d. Simplify.
$$\frac{56c}{56} = \frac{\cancel{56}c}{\cancel{56}} = c = \frac{\overset{5}{\cancel{280}}}{\underset{1}{\cancel{56}}} = 5$$
or $c = 5$.

e. To verify, replace the unknown in the original proportion with the answer, and check that the cross products are equal.
$$\frac{5}{20} = \frac{14}{56} \qquad 20 \cdot 14 = 280$$
$$5 \cdot 56 = 280$$
The cross products are equal.

*Problems 55 through 71 are solved using the same steps outlined above. Compute the cross products from the given proportion. Then form an equation by separating the cross products with an equal sign. Divide both sides by the number on the side with the unknown. This guarantees the unknown will be by itself on one side of the equation and the solution to the proportion problem will be on the other. As a last step, check your answer by writing the proportion with the unknown replaced by the value you found and verify the cross products are equal.*

55. $\dfrac{8}{5} = \dfrac{w}{90}$

$$5 \cdot w = 5w$$
$$8 \cdot 90 = 720$$
$$5w = 720$$
$$\frac{5w}{5} = \frac{720}{5}$$
$$w = 144$$
Verify:
$$\frac{8}{5} = \frac{144}{90} \qquad \begin{array}{l} 5 \cdot 144 = 720 \\ 8 \cdot 90 = 720 \end{array}$$

57. $\dfrac{c}{35} = \dfrac{48}{70}$

$$35 \cdot 48 = 1680$$
$$c \cdot 70 = 70c$$
$$70c = 1680$$
$$\frac{70c}{70} = \frac{1680}{70}$$
$$c = 24$$
Verify:
$$\frac{24}{35} = \frac{48}{70} \qquad \begin{array}{l} 35 \cdot 48 = 1680 \\ 24 \cdot 70 = 1680 \end{array}$$

59. $\dfrac{4.4}{22} = \dfrac{6}{m}$

$$22 \cdot 6 = 132$$
$$4.4 \cdot m = 4.4m$$
$$4.4m = 132$$
$$\frac{4.4m}{4.4} = \frac{132}{4.4}$$
$$m = 30$$
Verify:
$$\frac{4.4}{22} = \frac{6}{30} \qquad \begin{array}{l} 22 \cdot 6 = 132 \\ 4.4 \cdot 30 = 132 \end{array}$$

61. $\dfrac{v}{30} = \dfrac{30}{9}$

$30 \cdot 30 = 900$

$v \cdot 9 = 9v$

$9v = 900$

$\dfrac{9v}{9} = \dfrac{900}{9}$

$v = 100$

Verify:

$\dfrac{100}{30} = \dfrac{30}{9}$    $30 \cdot 30 = 900$

            $100 \cdot 9 = 900$

63. $\dfrac{5}{\frac{1}{4}} = \dfrac{60}{j}$

$\dfrac{1}{4} \cdot 60 = 15$

$5 \cdot j = 5j$

$5j = 15$

$\dfrac{5j}{5} = \dfrac{15}{5}$

$j = 3$

Verify:

$\dfrac{5}{\frac{1}{4}} = \dfrac{60}{3}$    $\dfrac{1}{4} \cdot 60 = 15$

            $5 \cdot 3 = 15$

65. $\dfrac{2.8}{1.2} = \dfrac{8.4}{h}$

$1.2 \cdot 8.4 = 10.08$

$2.8 \cdot h = 2.8h$

$2.8h = 10.08$

$\dfrac{2.8h}{2.8} = \dfrac{10.08}{2.8}$

$h = 3.6$

Verify:

$\dfrac{2.8}{1.2} = \dfrac{8.4}{3.6}$    $1.2 \cdot 8.4 = 10.08$

            $2.8 \cdot 3.6 = 10.08$

67. $\dfrac{16}{b} = \dfrac{64}{120}$

$b \cdot 64 = 64b$

$16 \cdot 120 = 1920$

$64b = 1920$

$\dfrac{64b}{64} = \dfrac{1920}{64}$

$b = 30$

Verify:

$\dfrac{16}{30} = \dfrac{64}{120}$    $30 \cdot 64 = 1920$

            $16 \cdot 120 = 1920$

69. $\dfrac{6}{42} = \dfrac{8}{q}$

$42 \cdot 8 = 336$

$6 \cdot q = 6q$

$6q = 336$

$\dfrac{6q}{6} = \dfrac{336}{6}$

$q = 56$

Verify:

$\dfrac{6}{42} = \dfrac{8}{56}$    $42 \cdot 8 = 336$

            $6 \cdot 56 = 336$

71. $\dfrac{y}{1.5} = \dfrac{20}{6}$

$1.5 \cdot 20 = 30$

$y \cdot 6 = 6y$

$6y = 30$

$\dfrac{6y}{6} = \dfrac{30}{6}$

$y = 5$

Verify:

$\dfrac{5}{1.5} = \dfrac{20}{6}$    $1.5 \cdot 20 = 30$

            $5 \cdot 6 = 30$

73. In the English class, the ratio of students who received a B is $\dfrac{18}{40}$. In the economics class, the ratio is $\dfrac{54}{120}$. The cross products of the two ratios are $40 \cdot 54 = 2160$ and $18 \cdot 120 = 2160$ which are equal. Students earn a grade of B at proportional rates in the two classes.

75. The ratio of John's premium to his coverage is $\dfrac{250}{10,000}$. For Sam it is $\dfrac{400}{18,000}$. To find if these are in proportion, compute the cross products: $10,000 \cdot 400 = 4,000,000$ and $250 \cdot 18,000 = 4,500,000$. The cross products are not equal so the insurance rates are not proportional.

77. Let $p$ be the number of passenger flights. There were 32 cargo flights and we know the ratio of cargo flights to passenger flights is 4 to 3. Then $p$ satisfies the proportion $\dfrac{32}{p} = \dfrac{4}{3}$. Solving this proportion gives $4p = 32 \cdot 3 = 96$ and so $p = \dfrac{96}{4} = 24$. There were 24 passenger flights.

79. Let $b$ be the number of buses needed for 588 students. If 3 buses are required for every 126 students, then $b$ satisfies the proportion $\dfrac{b}{588} = \dfrac{3}{126}$. Cross multiplying gives $126b = 588 \cdot 3 = 1764$. Now divide to find $b = \dfrac{1764}{126} = 14$. A total of 14 buses are required for 588 students.

81. If $c$ is the amount of whipped cream, in cups, needed then $c$ satisfies the proportion $\dfrac{c}{8} = \dfrac{1\frac{3}{4}}{2}$ since the recipe calls for $1\frac{3}{4}$ cups per 2 cups of strawberries and we have 8 cups of strawberries. Cross multiplying gives the equation $2c = 8 \cdot 1\frac{3}{4} = 8 \cdot \frac{7}{4} = 14$. Dividing by 2 gives $c = \dfrac{14}{2} = 7$ as the number of cups of whipped cream.

83. Let $i$ be the amount of interest on \$5400. Since \$34.50 is the interest on \$2000, $i$ must satisfy the proportion $\dfrac{34.50}{2000} = \dfrac{i}{5400}$. Solving the proportion gives $2000i = 34.50 \cdot 5400 = 186,300$ and so $i = \dfrac{186,300}{2000} = 93.15$. The interest on \$5400 would be \$93.15.

85. a. Let $c$ be the number of children who attended. There 1500 adults and we know the ratio of children to adults is 8 to 5. Thus $\dfrac{c}{1500} = \dfrac{8}{5}$. Solving this proportion gives $5c = 1500 \cdot 8 = 12,000$ and so $c = \dfrac{12,000}{5} = 2400$. There were 2400 children at the circus.

    b. Let $p$ be the number of popcorn bags sold. Given that 2400 children attended (from part a.) and the rate of 1.75 bags per child, we have the proportion $\dfrac{p}{2400} = \dfrac{1.75}{1}$. Cross multiplying gives $p = 1.75 \cdot 2400 = 4200$ bags of popcorn sold.

    c. From part b., 4200 bags were sold. At \$4.50 per bag the popcorn revenue was $4200 \cdot \$4.50 = \$18,900$.

87. In 2002, the birth rate was 13.9 per 1000 people. To find the births in a population of 150,000, set up the proportion $\dfrac{b}{150,000} = \dfrac{13.9}{1000}$ where $b$ is the number of births. Solving this proportion gives $1000b = 150,000 \cdot 13.9 = 2,085,000$ and $b = \dfrac{2,085,000}{1000} = 2085$. There were 2085 births.

89. Let $p$ denote the city's population. The birth rate in 1990 was 16.7 births per 1000 people. If the city had 5845 births then the proportion $\dfrac{5845}{p} = \dfrac{16.7}{1000}$ can be used to find $p$. Cross multiplying gives $16.7p = 5845 \cdot 1000 = 5,845,000$ and so $p = \dfrac{5,845,000}{16.7} = 350,000$. The population was 350,000.

91. Let $b$ denote the number of births per 1000 people in 2005. Since there were 3,942,000 births per 298,000,000 people, we have the proportion $\dfrac{b}{1000} = \dfrac{3,942,000}{298,000,000}$. Solving this proportion in the usual

way gives $298,000,000b = 1000 \cdot 3,942,000$ and $b = \dfrac{3,942,000,000}{298,000,000} = 13.228....$ To the nearest tenth, the

birth rate in 2005 was 13.2 births per 1000 people.

93. In the similar triangles, the ratios of a side in the smaller triangle to the corresponding side in the bigger

triangle are in proportion. It follows that $\dfrac{6}{44} = \dfrac{4.5}{b}$. Cross multiplying gives $6b = 44 \cdot 4.5 = 198$ and so

$b = \dfrac{198}{6} = 33$. The length, $b$, is 33 inches.

95. Let $x$ be the height of the sailboat mast in feet. The ratio of the height of the mast to the length of its

shadow is $\dfrac{x}{4}$. The ratio of the tree's height to its shadow length is $\dfrac{28}{8}$. Since the mast is next to the tree,

the ratios are in proportion and $\dfrac{x}{4} = \dfrac{28}{8}$. Cross multiply to get $8x = 4 \cdot 28 = 112$ and so $x = \dfrac{112}{8} = 14$. The

mast is 14 feet tall.

## Cumulative Skills Review

1. Given $2.85\overline{)9.2625}$ first move the decimal point two places to the right in both the divisor and dividend so the divisor is a whole number, $285\overline{)926.25}$. Then use long division to find

$$
\begin{array}{r}
3.25 \\
285\overline{)926.25} \\
-855 \phantom{.25} \\
\hline
712 \phantom{5} \\
-570 \phantom{5} \\
\hline
1425 \\
-1425 \\
\hline
0
\end{array}
$$

3. a. $1\dfrac{1}{2} + \dfrac{2}{3} = \dfrac{3}{2} + \dfrac{2}{3} = \dfrac{9}{6} + \dfrac{4}{6} = \dfrac{13}{6} = 2\dfrac{1}{6}$

   b. $1\dfrac{1}{2} - \dfrac{2}{3} = \dfrac{3}{2} - \dfrac{2}{3} = \dfrac{9}{6} - \dfrac{4}{6} = \dfrac{5}{6}$

   c. $1\dfrac{1}{2} \cdot \dfrac{2}{3} = \dfrac{3}{2} \cdot \dfrac{2}{3} = \dfrac{6}{6} = 1$

   d. $1\dfrac{1}{2} \div \dfrac{2}{3} = \dfrac{3}{2} \div \dfrac{2}{3} = \dfrac{3}{2} \cdot \dfrac{3}{2} = \dfrac{9}{4} = 2\dfrac{1}{4}$

5. There are four decimal places in 62.0399 indicating ten-thousandths. Using this information and reading the decimal point as "and", the word form is "sixty-two and three hundred ninety-nine ten-thousandths"

7. The cost of $3\dfrac{2}{3}$ pounds at \$9.00 per pound is

$$3\dfrac{2}{3} \cdot 9 = \dfrac{11}{3} \cdot \dfrac{9}{1} = \dfrac{99}{3} = 33 \text{ or } \$33.$$

9. Three ways to write the ratio of 18 to 25 are

$$18 \text{ to } 25, \ 18{:}25, \ \dfrac{18}{25}$$

## Chapter 4 Numerical Facts Of Life

1. First set up a proportion with $x$ representing the unknown quantity.

$$\dfrac{1230 \text{ calories per hour}}{170 \text{ pounds}} = \dfrac{x \text{ calories per hour}}{185 \text{ pounds}}$$

Then solve the proportion by setting the cross products equal.

$$170x = 1230 \cdot 185$$

$$170x = 227550$$

$$x = \frac{227550}{170} = 1338.529...$$

Mike will burn roughly 1339 calories per hour when running.

3.   In one hour, Mike will burn 1339 calories running for each 588 calories burned playing tennis.  For each calorie burned playing tennis, the number $z$ of calories burned running can be found by solving the proportion

$$\frac{1339}{588} = \frac{z}{1} \text{ or } z = \frac{1339}{588} = 2.277...$$

To the nearest tenth, Mike burns 2.3 calories running for each calorie burned playing tennis.

## Chapter 4 Review Exercises

1.   The ratio of 3 to 8 may be written

   3 to 8      3 : 8      $\dfrac{3}{8}$

2.   The ratio of 62 to 7 may be written

   62 to 7      62 : 7      $\dfrac{62}{7}$

3.   The ratio of 12 to 5.2 may be written

   12 to 5.2      12 : 5.2      $\dfrac{12}{5.2}$

4.   The ratio of 9 to 14.3 may be written

   9 to 14.3      9 : 14.3      $\dfrac{9}{14.3}$

5.   The ratio of 3 to $\dfrac{5}{9}$ may be written

   $3 \text{ to } \dfrac{5}{9}$      $3 : \dfrac{5}{9}$      $\dfrac{3}{\frac{5}{9}}$

6.   The ratio of $2\dfrac{1}{2}$ to $\dfrac{1}{16}$ may be written

   $2\dfrac{1}{2} \text{ to } \dfrac{1}{16}$      $2\dfrac{1}{2} : \dfrac{1}{16}$      $\dfrac{2\frac{1}{2}}{\frac{1}{16}}$

7.   Simplify 10 to 16.

$$\frac{10}{16} = \frac{\overset{5}{\cancel{10}}}{\underset{8}{\cancel{16}}} = \frac{5}{8}$$

8.   Simplify 58 to 6.

$$\frac{58}{6} = \frac{\overset{29}{\cancel{58}}}{\underset{3}{\cancel{6}}} = \frac{29}{3}$$

9.   Simplify 16 to 52.

$$\frac{16}{52} = \frac{\overset{4}{\cancel{16}}}{\underset{13}{\cancel{52}}} = \frac{4}{13}$$

10.  Simplify 24 to 21.

$$\frac{24}{21} = \frac{\overset{8}{\cancel{24}}}{\underset{7}{\cancel{21}}} = \frac{8}{7}$$

11. Simplify 5 to 12.5.

$$\frac{5}{12.5} = \frac{5}{12.5} \cdot \frac{10}{10} = \frac{50}{125} = \frac{\cancel{50}^{2}}{\cancel{125}_{5}} = \frac{2}{5}$$

12. Simplify 1.1 to 15.

$$\frac{1.1}{15} = \frac{1.1}{15} \cdot \frac{10}{10} = \frac{11}{150}$$

13. Simplify 2 to $\frac{7}{8}$.

$$\frac{2}{\frac{7}{8}} = \frac{2}{1} \cdot \frac{8}{7} = \frac{16}{7}$$

14. Simplify $\frac{9}{17}$ to 7.

$$\frac{\frac{9}{17}}{7} = \frac{9}{17} \cdot \frac{1}{7} = \frac{9}{119}$$

15. 110 feet to 16 yards
Smaller unit: feet
16 yards = $16 \cdot 3$ feet = 48 feet

$$\frac{110 \text{ feet}}{16 \text{ yards}} = \frac{110}{48} = \frac{55}{24}$$

16. 3 pounds to 20 ounces
Smaller unit: ounces
3 pounds = $3 \cdot 16$ ounces
$= 48$ ounces

$$\frac{3 \text{ pounds}}{20 \text{ ounces}} = \frac{48}{20} = \frac{12}{5}$$

17. 12 minutes to 220 seconds
Smaller unit: seconds
12 minutes = $12 \cdot 60$ seconds
$= 720$ seconds

$$\frac{12 \text{ minutes}}{220 \text{ seconds}} = \frac{720}{220} = \frac{36}{11}$$

18. 10,000 feet to 4 miles
Smaller unit: feet
4 miles = $4 \cdot 5280$ feet
$= 21{,}120$ feet

$$\frac{10{,}000 \text{ feet}}{4 \text{ miles}} = \frac{10{,}000}{21{,}120} = \frac{125}{264}$$

19. 8500 pounds to 5 tons
Smaller unit: pounds
5 tons = $5 \cdot 2000$ pounds
$= 10{,}000$ pounds

$$\frac{8500 \text{ pounds}}{5 \text{ tons}} = \frac{8500}{10{,}000} = \frac{17}{20}$$

20. 12 quarts to 50 pints
Smaller unit: pints
12 quarts = $12 \cdot 2$ pints
$= 24$ pints

$$\frac{12 \text{ quarts}}{50 \text{ pints}} = \frac{24}{50} = \frac{12}{25}$$

21. 280 days to 16 weeks
Smaller unit: days
16 weeks = $16 \cdot 7$ days
$= 112$ days

$$\frac{280 \text{ days}}{16 \text{ weeks}} = \frac{280}{112} = \frac{5}{2}$$

22. 5.5 gallons to 12 quarts
Smaller unit: quarts
5.5 gallons = $5.5 \cdot 4$ quarts
$= 22$ quarts

$$\frac{5.5 \text{ gallons}}{12 \text{ quarts}} = \frac{22}{12} = \frac{11}{6}$$

23. 75 sprinklers for 6 acres

$$\frac{75 \text{ sprinklers}}{6 \text{ acres}} = \frac{25 \text{ sprinklers}}{2 \text{ acres}}$$

25 sprinklers for every 2 acres

24. 392 avocados for 40 trees

$$\frac{392 \text{ avocados}}{40 \text{ trees}} = \frac{49 \text{ avocados}}{5 \text{ trees}}$$

49 avocados for every 5 trees

25. 38 kittens for 3 pet stores

$$\frac{38 \text{ kittens}}{3 \text{ pet stores}}$$

38 kittens for every 3 pet stores

26. $98 for 4 tires

$$\frac{\$98}{4 \text{ tires}} = \frac{\$49}{2 \text{ tires}}$$

$49 for every 2 tires

27. 30 ponies for 12 trainers

$$\frac{30 \text{ ponies}}{12 \text{ trainers}} = \frac{5 \text{ ponies}}{2 \text{ trainers}}$$

5 ponies for every 2 trainers

28. 120 cheeseburgers for $200

$$\frac{120 \text{ cheeseburgers}}{\$200} = \frac{3 \text{ cheeseburgers}}{\$5}$$

3 cheeseburgers for $5

29. 60 miles in 5 days

$$\frac{60 \text{ miles}}{5 \text{ days}} = \frac{60 \div 5 \text{ miles}}{5 \div 5 \text{ days}} = \frac{12 \text{ miles}}{1 \text{ day}}$$

12 miles per day

30. 1588 pounds in 2 trucks

$$\frac{1588 \text{ pounds}}{2 \text{ trucks}} = \frac{1588 \div 2 \text{ pounds}}{2 \div 2 \text{ trucks}}$$
$$= \frac{794 \text{ pounds}}{1 \text{ truck}}$$

794 pounds per truck

31. 18 yards in 7 minutes

$$\frac{18 \text{ yards}}{7 \text{ minutes}} = \frac{18 \div 7 \text{ yards}}{7 \div 7 \text{ minutes}} = \frac{2.571... \text{ yards}}{1 \text{ minute}}$$

To the nearest tenth, 2.6 yards per minute

32. 9615 jellybeans in 12 bags

$$\frac{9615 \text{ jellybeans}}{12 \text{ bags}} = \frac{9615 \div 12 \text{ jellybeans}}{12 \div 12 \text{ bags}}$$
$$= \frac{801.25 \text{ jellybeans}}{1 \text{ bag}}$$

To the nearest tenth, 801.3 jellybeans per bag

33. 168 cars in 6 lanes

$$\frac{168 \text{ cars}}{6 \text{ lanes}} = \frac{168 \div 6 \text{ cars}}{6 \div 6 \text{ lanes}} = \frac{28 \text{ cars}}{1 \text{ lane}}$$

28 cars per lane

34. 13,005 bees in 9 beehives

$$\frac{13,005 \text{ bees}}{9 \text{ beehives}} = \frac{13,005 \div 9 \text{ bees}}{9 \div 9 \text{ beehives}} = \frac{14458 \text{ bees}}{1 \text{ beehive}}$$

1445 bees per beehive

35. 47 tons of fuels in 3 cruises

$$\frac{47 \text{ tons}}{3 \text{ cruises}} = \frac{47 \div 3 \text{ tons}}{3 \div 3 \text{ cruises}} = \frac{15.666... \text{ tons}}{1 \text{ cruise}}$$

To the nearest tenth, 15.7 tons of fuel per cruise

36. 25 pounds in 8 weeks

$$\frac{25 \text{ pounds}}{8 \text{ weeks}} = \frac{25 \div 8 \text{ pounds}}{8 \div 8 \text{ weeks}} = \frac{3.125 \text{ pounds}}{1 \text{ week}}$$

To the nearest tenth, 3.1 pounds per week

37. 5 tickets for $90

$$\frac{\$90}{5 \text{ tickets}} = \frac{\$90 \div 5}{5 \div 5 \text{ tickets}} = \frac{\$18}{1 \text{ ticket}}$$

The unit price is $18 per ticket.

38. 15 T-shirts for $187.50

$$\frac{\$187.50}{15 \text{ T-shirts}} = \frac{\$187.50 \div 15}{15 \div 15 \text{ T-shirts}} = \frac{\$12.50}{1 \text{ T-shirt}}$$

The unit price is $12.50 per T-shirt.

39. $14 for 2 car washes

$$\frac{\$14}{2 \text{ car washes}} = \frac{\$14 \div 2}{2 \div 2 \text{ car washes}} = \frac{\$7}{1 \text{ car wash}}$$

The unit price is $7 per car wash.

40. $695 for 4 days

$$\frac{\$695}{4 \text{ days}} = \frac{\$695 \div 4}{4 \div 4 \text{ days}} = \frac{\$173.75}{1 \text{ day}}$$

The unit price is $173.75 per day.

41. 6 flight lessons for $510

$$\frac{\$510}{6 \text{ lessons}} = \frac{\$510 \div 6}{6 \div 6 \text{ lessons}} = \frac{\$85.00}{1 \text{ lesson}}$$

The unit price is $85 per flight lesson.

42. 125 sugar cookies for $81.25

$$\frac{\$81.25}{125 \text{ cookies}} = \frac{\$81.25 \div 125}{125 \div 125 \text{ cookies}} = \frac{\$0.65}{1 \text{ cookie}}$$

The unit price is $0.65 per sugar cookie.

43. $4.75 for 3 tennis balls

$$\frac{\$4.75}{3 \text{ balls}} = \frac{\$4.75 \div 3}{3 \div 3 \text{ balls}} = \frac{\$1.5833...}{1 \text{ ball}}$$

The unit price, to the nearest cent, is $1.58 per tennis ball.

44. $12,900 for 3 sales events

$$\frac{\$12,900}{3 \text{ events}} = \frac{\$12,900 \div 3}{3 \div 3 \text{ events}} = \frac{\$4,300}{1 \text{ event}}$$

The unit price is $4,300 per sales event.

45. 9 is to 11 as 36 is to 44.
written as a proportion is
$$\frac{9}{11} = \frac{36}{44}$$

46. 124 graduates is to 3 schools as 248 graduates is to 6 schools.
written as a proportion is
$$\frac{124 \text{ graduates}}{3 \text{ schools}} = \frac{248 \text{ graduates}}{6 \text{ schools}}$$

47. 3 is to 5 as 300 is to 500.
written as a proportion is
$$\frac{3}{5} = \frac{300}{500}$$

48. 2 days is to 95 mail orders as 6 days is to 285 mail orders.
written as a proportion is
$$\frac{2 \text{ days}}{95 \text{ mail orders}} = \frac{6 \text{ days}}{285 \text{ mail orders}}$$

49. 2.1 is to 6.5 as 16.8 is to 52. written as a proportion is
$$\frac{2.1}{6.5} = \frac{16.8}{52}$$

50. 5 concerts is to 7 days as 15 concerts is to 21 days. written as a proportion is
$$\frac{5 \text{ concerts}}{7 \text{ days}} = \frac{15 \text{ concerts}}{21 \text{ days}}$$

51. $\dfrac{30 \text{ violins}}{5 \text{ orchestras}} = \dfrac{90 \text{ violins}}{15 \text{ orchestras}}$
left hand side: 30 violins to 5 orchestras
right hand side: 90 violins to 15 orchestras
Sentence form:
30 violins is to 5 orchestras as 90 violins is to 15 orchestras.

52. $\dfrac{9}{13} = \dfrac{81}{117}$
left hand side: 9 to 13
right hand side: 81 to 117
Sentence form:
9 is to 13 as 81 is to 117.

53. $\dfrac{3 \text{ tours}}{450 \text{ bicycles}} = \dfrac{6 \text{ tours}}{900 \text{ bicycles}}$
left hand side: 3 tours to 450 bicycles
right hand side: 6 tours to 900 bicycles
Sentence form:
3 tours is to 450 bicycles as 6 tours is to 900 bicycles.

54. $\dfrac{12}{33} = \dfrac{36}{99}$
left hand side: 12 to 33
right hand side: 36 to 99
Sentence form:
12 is to 33 as 36 is to 99.

55. $\dfrac{8 \text{ swings}}{3 \text{ playgrounds}} = \dfrac{16 \text{ swings}}{6 \text{ playgrounds}}$
left hand side: 8 swings to 3 playgrounds
right hand side: 16 swings to 6 playgrounds
Sentence form:
8 swings is to 3 playgrounds as 16 swings is to 6 playgrounds.

56. $\dfrac{3.7}{1.2} = \dfrac{37}{12}$
left hand side: 3.7 to 1.2
right hand side: 37 to 12
Sentence form:
3.7 is to 1.2 as 37 is to 12.

57. $\dfrac{18}{17} \overset{?}{=} \dfrac{54}{51}$
$17 \cdot 54 = 918, \ 18 \cdot 51 = 918$
The cross products are equal.
The ratios are proportional.
$$\frac{18}{17} = \frac{54}{51}$$

58. $\dfrac{6}{1.9} \overset{?}{=} \dfrac{18}{5.4}$
$1.9 \cdot 18 = 34.2, \ 6 \cdot 5.4 = 32.4$
The cross products are not equal. The ratios are not proportional.

59. $\dfrac{39}{28} \overset{?}{=} \dfrac{13}{9}$
$28 \cdot 13 = 364, \ 39 \cdot 9 = 351$
The cross products are not equal. The ratios are not proportional.

60. $\dfrac{35}{21} \overset{?}{=} \dfrac{70}{42}$
$21 \cdot 70 = 1470, \ 35 \cdot 42 = 1470$
The cross products are equal.
The ratios are proportional.
$$\frac{35}{21} = \frac{70}{42}$$

61. $\dfrac{7.5}{11} \overset{?}{=} \dfrac{60}{88}$
$11 \cdot 60 = 660, \ 7.5 \cdot 88 = 660$
The cross products are equal.
The ratios are proportional.
$$\frac{7.5}{11} = \frac{60}{88}$$

62. $\dfrac{2.3}{5.5} \overset{?}{=} \dfrac{9.2}{22}$
$5.5 \cdot 9.2 = 50.6, \ 2.3 \cdot 22 = 50.6$
The cross products are equal.
The ratios are proportional.
$$\frac{2.3}{5.5} = \frac{9.2}{22}$$

63. $\dfrac{2}{5} = \dfrac{14}{g}$

$\quad\quad 5 \cdot 14 = 70$

$\quad\quad 2 \cdot g = 2g$

$2g = 70$

$\dfrac{2g}{2} = \dfrac{70}{2}$

$g = 35$

Verify:

$\dfrac{2}{5} = \dfrac{14}{35} \quad\quad 5 \cdot 14 = 70$

$\quad\quad\quad\quad\quad\quad 2 \cdot 35 = 70$

64. $\dfrac{28}{100} = \dfrac{y}{25}$

$\quad 100 \cdot y = 100y$

$\quad 28 \cdot 25 = 700$

$100y = 700$

$\dfrac{100y}{100} = \dfrac{700}{100}$

$y = 7$

Verify:

$\dfrac{28}{100} = \dfrac{7}{25} \quad 100 \cdot 7 = 700$

$\quad\quad\quad\quad\quad 28 \cdot 25 = 700$

65. $\dfrac{m}{3} = \dfrac{46}{2}$

$\quad\quad 3 \cdot 46 = 138$

$\quad\quad m \cdot 2 = 2m$

$2m = 138$

$\dfrac{2m}{2} = \dfrac{138}{2}$

$m = 69$

Verify:

$\dfrac{69}{3} = \dfrac{46}{2} \quad 3 \cdot 46 = 138$

$\quad\quad\quad\quad\quad 69 \cdot 2 = 138$

66. $\dfrac{t}{3} = \dfrac{44}{6}$

$\quad\quad 3 \cdot 44 = 132$

$\quad\quad t \cdot 6 = 6t$

$6t = 132$

$\dfrac{6t}{6} = \dfrac{132}{6}$

$t = 22$

Verify:

$\dfrac{22}{3} = \dfrac{44}{6} \quad 3 \cdot 44 = 132$

$\quad\quad\quad\quad\quad 22 \cdot 6 = 132$

67. $\dfrac{18}{0.5} = \dfrac{a}{3}$

$\quad\quad 0.5 \cdot a = 0.5a$

$\quad\quad 18 \cdot 3 = 54$

$0.5a = 54$

$\dfrac{0.5a}{0.5} = \dfrac{54}{0.5}$

$a = 108$

Verify:

$\dfrac{18}{0.5} = \dfrac{108}{3} \quad 0.5 \cdot 108 = 54$

$\quad\quad\quad\quad\quad\quad 18 \cdot 3 = 54$

68. $\dfrac{1.6}{f} = \dfrac{4}{40}$

$\quad\quad f \cdot 4 = 4f$

$\quad\quad 1.6 \cdot 40 = 64$

$4f = 64$

$\dfrac{4f}{4} = \dfrac{64}{4}$

$f = 16$

Verify:

$\dfrac{1.6}{16} = \dfrac{4}{40} \quad 16 \cdot 4 = 64$

$\quad\quad\quad\quad\quad 1.6 \cdot 40 = 64$

69. $\dfrac{4}{u} = \dfrac{24}{18}$

$\quad\quad u \cdot 24 = 24u$

$\quad\quad 4 \cdot 18 = 72$

$24u = 72$

$\dfrac{24v}{24} = \dfrac{72}{24}$

$u = 3$

Verify:

$\dfrac{4}{3} = \dfrac{24}{18} \quad 3 \cdot 24 = 72$

$\quad\quad\quad\quad\quad 4 \cdot 18 = 72$

70. $\dfrac{b}{2} = \dfrac{30}{12}$

$\quad\quad 2 \cdot 30 = 60$

$\quad\quad b \cdot 12 = 12b$

$12b = 60$

$\dfrac{12b}{12} = \dfrac{60}{12}$

$b = 5$

Verify:

$\dfrac{5}{2} = \dfrac{30}{12} \quad 2 \cdot 30 = 60$

$\quad\quad\quad\quad\quad 5 \cdot 12 = 60$

71. $\dfrac{q}{7} = \dfrac{20}{3.5}$

$\quad\quad 7 \cdot 20 = 140$

$\quad\quad q \cdot 3.5 = 3.5q$

$3.5q = 140$

$\dfrac{3.5q}{3.5} = \dfrac{140}{3.5}$

$q = 40$

Verify:

$\dfrac{40}{7} = \dfrac{20}{3.5} \quad 7 \cdot 20 = 140$

$\quad\quad\quad\quad\quad\quad 40 \cdot 3.5 = 140$

72. $\dfrac{2.5}{10} = \dfrac{r}{12}$

$\quad\quad 10 \cdot r = 10r$

$\quad\quad 2.5 \cdot 12 = 30$

$10r = 30$

$\dfrac{10r}{10} = \dfrac{30}{10}$

$r = 3$

Verify:

$\dfrac{2.5}{10} = \dfrac{3}{12} \quad 10 \cdot 3 = 30$

$\quad\quad\quad\quad\quad\quad 2.5 \cdot 12 = 30$

73. 
$$\frac{2\frac{1}{4}}{h} = \frac{9}{12}$$
$$h \cdot 9 = 9h$$
$$2\frac{1}{4} \cdot 12 = \frac{9}{4} \cdot 12 = 27$$
$$9h = 27$$
$$\frac{9h}{9} = \frac{27}{9}$$
$$h = 3$$
Verify:
$$\frac{2\frac{1}{4}}{3} = \frac{9}{12} \qquad 3 \cdot 9 = 27$$
$$\qquad\qquad\qquad 2\frac{1}{4} \cdot 12 = 27$$

74. 
$$\frac{3\frac{1}{4}}{6\frac{1}{2}} = \frac{4}{x}$$
$$6\frac{1}{2} \cdot 4 = \frac{13}{2} \cdot 4 == 26$$
$$3\frac{1}{4} \cdot x = \frac{13}{4}x$$
$$\frac{13}{4}x = 26$$
$$\frac{\frac{13}{4}x}{\frac{13}{4}} = \frac{26}{\frac{13}{4}}$$
$$x = 26 \cdot \frac{4}{13} = 8$$
Verify:
$$\frac{3\frac{1}{4}}{6\frac{1}{2}} = \frac{4}{8} \qquad \begin{array}{l} 6\frac{1}{2} \cdot 4 = 26 \\ 3\frac{1}{4} \cdot 8 = 26 \end{array}$$

75. a. Given 14 girls and 27 boys, the ratio of girls to boys can be written in the following ways:

$$14 \text{ to } 27,\ 14{:}27,\ \frac{14}{27}.$$

 b. The total number of students is $14 + 27 = 41$. There are 27 boys. The ratio of boys to the total may be written as: $27 \text{ to } 41,\ 27{:}41,\ \dfrac{27}{41}.$

76. a. There are 125 condominium units and 33 townhouses. The ratio of condominium units to townhouses may be written as: $125 \text{ to } 33,\ 125{:}33,\ \dfrac{125}{33}.$

 b. The total number of homes is $125 + 33 = 158$. There are 125 condominiums. The ratio of condominiums to the total may be written as: $158 \text{ to } 125,\ 158{:}125,\ \dfrac{158}{125}.$

77. a. The ratio of Randy's Monday mileage to his Wednesday mileage is $\dfrac{65}{25} = \dfrac{5 \cdot 13}{5 \cdot 5} = \dfrac{13}{5}.$

 b. The ratio of Randy's Tuesday mileage to his Thursday mileage is $\dfrac{40}{50} = \dfrac{4 \cdot 10}{5 \cdot 10} = \dfrac{4}{5}.$

 c. Randy traveled a total of $65 + 40 + 25 + 50 = 180$ miles with 65 of those miles on Monday. The simplified ratio of Monday mileage to the total is $\dfrac{65}{180} = \dfrac{5 \cdot 13}{5 \cdot 36} = \dfrac{13}{36}.$

78. a. The number of associate in science degrees is 225 and the number of bachelor degrees awarded was 60. The simplified ratio of associate in science degrees to bachelor degrees is $\dfrac{225}{60} = \dfrac{15 \cdot 15}{15 \cdot 4} = \dfrac{15}{4}$

b. The number of associates in art degrees and science degrees were 150 and 225 respectively. The ratio of associate arts degrees to science degrees is $\dfrac{150}{225} = \dfrac{75 \cdot 2}{75 \cdot 3} = \dfrac{2}{3}$.

c. The total number of degrees awarded was $150 + 225 + 60 = 435$ of which 60 were bachelor degrees. The simplified ratio of bachelor degrees to total degrees is $\dfrac{60}{435} = \dfrac{15 \cdot 4}{15 \cdot 29} = \dfrac{4}{29}$.

79. a. The coffeemaker can brew 4 cups of coffee in 3 minutes. As a rate, this may be written as $\dfrac{4 \text{ cups of coffee}}{3 \text{ minutes}}$.

b. The unit rate for part a. is $\dfrac{4 \text{ cups of coffee}}{3 \text{ minutes}} = \dfrac{4 \div 3 \text{ cups of coffee}}{3 \div 3 \text{ minute}} = \dfrac{1\frac{1}{3} \text{ cups of coffee}}{1 \text{ minute}}$ or $1\frac{1}{3}$ cups of coffee per minute.

c. The unit price is $\dfrac{\$1.80}{4 \text{ cups}} = \dfrac{\$1.80 \div 4}{4 \div 4 \text{ cups}} = \dfrac{\$0.45}{1 \text{ cup}}$ or $0.45 per cup.

80. a. The machine requires 27 oranges for 5 pints of orange juice. As a rate, this is written as $\dfrac{27 \text{ oranges}}{5 \text{ pints}}$.

b. The unit rate for part a. is $\dfrac{27 \text{ oranges}}{5 \text{ pints}} = \dfrac{27 \div 5 \text{ oranges}}{5 \div 5 \text{ pints}} = \dfrac{5\frac{2}{5} \text{ oranges}}{1 \text{ pint}}$ or $5\frac{2}{5}$ oranges per pint.

c. We are given the cost of 27 oranges is $11.50. Note, we are looking for the unit price per pint not per orange. From part a., 27 oranges produce 5 pints of juice. The unit price is $\dfrac{\$11.50}{5 \text{ pints}} = \dfrac{\$11.50 \div 5}{5 \div 5 \text{ pints}} = \dfrac{\$2.30}{1 \text{ pint}}$ or $2.30 per pint.

81. The 10-ounce bag has a unit price of $\dfrac{\$1.20}{10 \text{ ounces}} = \dfrac{\$0.12}{1 \text{ ounce}}$ or $0.12 per ounce. The 13-ounce bag has a unit price of $\dfrac{\$1.69}{13 \text{ ounces}} = \dfrac{\$0.13}{1 \text{ ounce}}$. The best buy (lowest unit price) is 10-ounces at $0.12 per ounce.

82. The table below shows the unit prices for each number of rides.

| No. of rides | Price | Unit price |
|---|---|---|
| 3 | $9.75 | $9.75 \div 3 = $3.25 per ride |
| 6 | $18.00 | $18.00 \div 6 = $3.00 per ride |
| 12 | $34.80 | $34.80 \div 12 = $2.90 per ride |

The best buy is 12 rides at $2.90 per ride.

83. Two dozen bagels have a unit price of $5.95 \div 2 = $2.975$ per dozen while three dozen bagels have a unit price of $8.75 \div 3 = $2.9166...$ per dozen. The latter unit price is the smaller of the two so the best buy, rounding to the nearest cent, is three dozen bagels at $2.92 per dozen.

84. The table below shows the unit prices for each number of classes.

| No. of classes | Price | Unit price |
|---|---|---|
| 9 | $50.00 | $50.00 ÷ 9 = $5.555... per class |
| 12 | $64.20 | $64.20 ÷ 12 = $5.35 per class |
| 15 | $84.75 | $84.75 ÷ 15 = $5.65 per class |

The best buy is 12 classes at $5.35 per class.

85. We must determine the validity of the proportion $\frac{152{,}000}{1600} \stackrel{?}{=} \frac{185{,}250}{1950}$. Compute the cross products: $1600 \cdot 185{,}250 = 296{,}400{,}000$ and $152{,}000 \cdot 1950 = 296{,}400{,}000$. Since the cross products are equal, the ratios are in proportion and we may write the cost per square feet proportion as $\frac{\$152{,}000}{1600 \text{ sq. ft.}} = \frac{\$185{,}250}{1950 \text{ sq. ft.}}$.

86. The prints-to-time ratio for the first printer is $\frac{5}{7}$ while this ratio for the second printer is $\frac{7}{9}$. To determine if these rates are proportional, compute the cross products: $7 \cdot 7 = 49$ and $5 \cdot 9 = 45$. The cross products are not equal so the rates are not proportional.

87. The question is whether the rates 15 acres for every 3 hours and 19 acres for every 4 hours are proportional rates, or equivalently is $\frac{15}{3} \stackrel{?}{=} \frac{19}{4}$ a valid proportion. Compute the cross products: $3 \cdot 19 = 57$ and $15 \cdot 4 = 60$. The cross products are not equal so the rates are not proportional.

88. Harold's fuel consumption rate, in dollars per days, is $\frac{52}{8}$ while Jenna's is $\frac{32.50}{5}$. The cross products for these two rates are $8 \cdot 32.50 = 260$ and $52 \cdot 5 = 260$. The cross products are equal so the rates are proportional which can be written $\frac{\$52.00}{8 \text{ days}} = \frac{\$32.50}{5 \text{ days}}$.

89. The ratio of puppies to instructors is 5 to 3. If $p$ is the number of puppies that can attend a class with 9 instructors then $\frac{p}{9} = \frac{5}{3}$. Cross multiply to find $3p = 9 \cdot 5 = 45$ and then divide to find $p = \frac{45}{3} = 15$. In a class with 9 instructors, 15 puppies can attend.

90. Let $x$ denote the total square feet in the new corral which will hold 12 horses. If the corrals are proportional then the ratios of each corral's area to the number of horses will be proportional. Thus, $\frac{x}{12} = \frac{1500}{5}$ and so $5x = 12 \cdot 1500 = 18{,}000$ and $x = \frac{18{,}000}{5} = 3600$. The new corral should contain 3600 square feet.

91. The ratio of notebooks to students is $\frac{45}{20}$. Let $n$ be the number of notebooks needed for 60 students, then $\frac{n}{60} = \frac{45}{20}$. Solving this proportion gives $20n = 60 \cdot 45 = 2700$ and so $n = \frac{2700}{20} = 135$. A class of 60 students needs 135 notebooks.

92. Let $p$ the price of the 26-ounce soda. For the prices to be proportional, $p$ must satisfy $\dfrac{p}{26} = \dfrac{2.50}{20}$. Now, cross multiply to find the equation $20p = 26 \cdot 2.50 = 65$. To find $p$, divide by 20 to get $p = \dfrac{65}{20} = 3.25$. The 26-ounce size should sell for \$3.25.

93. Shaun's work minutes-to-break minutes ratio is $\dfrac{50}{5}$. Let $w$ be the amount of time Shaun needs to work to earn 30 minutes break time. Then $w$ satisfies the proportion $\dfrac{50}{5} = \dfrac{w}{30}$. Solving this proportion gives $5w = 50 \cdot 30 = 1500$ and $w = \dfrac{1500}{5} = 300$ minutes. Shaun must work 300 minutes or 5 hours.

94. Let $c$ be the number of classes in 12 semesters. Alicia's class-to-semester ratio is $\dfrac{10}{3}$, so $c$ satisfies the proportion $\dfrac{c}{12} = \dfrac{10}{3}$. Cross multiplying gives $3c = 12 \cdot 10 = 120$ and so $c = \dfrac{120}{3} = 40$. Alicia will take 40 classes in 12 semesters.

95. Let $n$ be the number of nails used in 20 minutes. If the nail gun uses nails at a rate of 65 nails for every $1\frac{1}{2}$ minutes, then $n$ satisfies the proportion $\dfrac{n}{20} = \dfrac{65}{1\frac{1}{2}}$. Solve this proportion in the usual way:

$1\frac{1}{2} \cdot n = 20 \cdot 65 = 1300$ and $n = \dfrac{1300}{1\frac{1}{2}} = \dfrac{1300}{\frac{3}{2}} = 1300 \cdot \dfrac{2}{3} = \dfrac{2600}{3} = 866.666\ldots$. To the nearest whole nail, 867 nails will be used in 20 minutes.

96. Let $p$ be the amount of peaches, in pounds, needed for a recipe having 2 cups of sugar. Since 3 pounds of peaches correspond to $\dfrac{3}{4}$ cup of sugar, $p$ satisfies the proportion $\dfrac{p}{2} = \dfrac{3}{\frac{3}{4}}$. By cross multiplying,

$\dfrac{3}{4}p = 2 \cdot 3 = 6$. Dividing by $\dfrac{3}{4}$, or multiplying by $\dfrac{4}{3}$, gives the value $p = \dfrac{4}{3} \cdot 6 = \dfrac{24}{3} = 8$. Eight pounds of peaches are needed.

97. Let $b$ be the number of pairs of work boots Brad will buy in 2 years. Brad buys 3 pairs every 6 months. When setting up a proportion, the units must agree, so first express 2 years as 24 months and then write the proportion $\dfrac{b}{24} = \dfrac{3}{6}$. Solving this proportion gives $6b = 24 \cdot 3 = 72$ and $b = \dfrac{72}{6} = 12$. Brad will buy 12 pairs of boots in 2 years.

98. Claudia's car gets 28 miles for every 1 gallon. Let $g$ be the gallons of gasoline needed for a 1498 mile trip. Then $g$ satisfies the proportion $\dfrac{1498}{g} = \dfrac{28}{1}$. Cross multiplying yields the equation $28g = 1498$ and dividing by 28 gives $g = \dfrac{1498}{28} = 53.5$. Claudia needs 53.5 gallons of gasoline for the trip.

99. Let $v$ be the number of viewers expected over 7 nights where $v$ is in millions. Since 3 million viewers are expected over 4 nights, $v$ satisfies the proportion $\dfrac{v}{7} = \dfrac{3}{4}$. Solve this proportion in the usual way: $4v = 7 \cdot 3 = 21$ and $v = \dfrac{21}{4} = 5.25$. Thus, 5.25 million or 5,250,000 million viewers are expected.

100. Suppose $w$ is the amount of water, in gallons, used in 1 hour by the sprinkler. The sprinkler uses 20 gallons of water in 5 minutes so x satisfies the proportion $\dfrac{x}{60} = \dfrac{20}{5}$. Note the rate on the right hand side is expressed as gallons over minutes so the denominator on the left must also be in minutes so 1 hour was converted to its equivalent of 60 minutes. Now solve by cross multiplying: $5x = 60 \cdot 20 = 1200$ and $x = \dfrac{1200}{5} = 240$. The sprinkler will use 240 gallons in one hour.

101. Since the pots are similar geometric figures, the ratio of distances across the top is proportional to the ratio of heights of the smaller to larger pot. Let $x$ be the height of the smaller pot, then $\dfrac{6}{9} = \dfrac{x}{12}$. Cross multiplying yields the equation $9x = 6 \cdot 12 = 72$. Dividing by 9 gives $x = \dfrac{72}{9} = 8$. The smaller pot is 8 inches high.

102. Since the new pool is built proportional to the old pool, the ratio of a side in the smaller pool to the corresponding side in the larger pool is in proportion to any other such ratio. Let $z$ be the unknown side of the larger pool. The side of length 12 feet in the smaller pool corresponds to the side of length 30 feet in the new pool. Similarly, the side of length 20 feet in the smaller pool corresponds to the side of length $z$ in the new pool. Thus, $\dfrac{12}{30} = \dfrac{20}{z}$ and, by comparing cross products, $12z = 30 \cdot 20 = 600$. Lastly, divide by 12 to find $z = \dfrac{600}{12} = 50$. The side of the new pool is 50 feet long.

103. Let $x$ be the height of the mast in feet. A nearby piling has a height-to-shadow length ratio of $\dfrac{10}{14}$. Since the mast casts a shadow of length 56 feet, $x$ must satisfy the proportion $\dfrac{x}{56} = \dfrac{10}{14}$. By comparing cross products, $14x = 56 \cdot 10 = 560$ and so $x = \dfrac{560}{14} = 40$. The mast is 40 feet tall.

104. Let $x$ represent the height of the building in feet. We know the building casts a shadow 60 feet long. The ratio of the crane's height to shadow length is 80 to 25. Since the two are next to each other their height-to-shadow ration are in proportion and so $\dfrac{x}{60} = \dfrac{80}{25}$. Solving in the usual manner gives $25x = 60 \cdot 80 = 4800$ and $x = \dfrac{4800}{25} = 192$. The building is 192 feet tall.

105. The wingspan of the Boeing 787 is given as 186 feet. The wingspan of the Airbus A330 is shown as 198 feet. Simplifying the ratio of the Boeing's wingspan to that of the Airbus gives $\dfrac{186}{198} = \dfrac{6 \cdot 31}{6 \cdot 33} = \dfrac{31}{33}$. The three ways to write this simplified ratio are 31 to 33, 31:33, and $\dfrac{31}{33}$.

106. Using the measurements in the diagrams, the ratio of the 787's length to that of the 767 is
$\dfrac{182}{180} = \dfrac{2 \cdot 91}{2 \cdot 90} = \dfrac{91}{90}$. This ratio can be written as 91 to 90, 91:90, and $\dfrac{91}{90}$.

107. The cruising speed of the Airbus A330 is given as 635 mph while that for the Boeing 767 is 530 mph.
First, write their ratio as a fraction and simplify $\dfrac{635}{530} = \dfrac{5 \cdot 127}{5 \cdot 106} = \dfrac{127}{106}$. Next write the simplified ratio in
word form: 127 miles per hour of the Airbus A330 for every 106 miles per hour of the Boeing 767.

108. The passenger capacity of the Boeing 787 is specified as 200 people while that for the model 767 is 218
people. First write the 787-to-767 passenger ratio as a fraction and simplify: $\dfrac{200}{218} = \dfrac{2 \cdot 100}{2 \cdot 109} = \dfrac{100}{109}$. The
word form for the simplified ratio is: 100 passengers on the boeing 787 for every 109 passengers on the
Boeing 767.

109. a. From the drawing, the wingspan of the base model Boeing 787 is 186 feet and the length is 182 feet. The
wingspan-to-length ratio is $\dfrac{186}{182} = \dfrac{2 \cdot 93}{2 \cdot 91} = \dfrac{93}{91}$.

   b. The length of the new "stretch" model will be 198 feet. Let $w$ denote the wingspan of the short model in
feet. Since the wingspan-to-length ratio will be the same as determined in part a., we have the proportion
$\dfrac{w}{198} = \dfrac{93}{91}$. Cross multiply to find $91w = 198 \cdot 93 = 18{,}414$. Then $w = \dfrac{18{,}414}{91} = 202.35\dots$. To the nearest
foot, the wingspan of the new model should be 202 feet.

110. a. From the diagram, the wingspan of the base model A330 is 198 feet and the length is 194 feet. The
wingspan-to-length ratio is $\dfrac{198}{194} = \dfrac{2 \cdot 99}{2 \cdot 97} = \dfrac{99}{97}$.

   b. The wingspan of the new "short" model will be 180 feet. Let $x$ denote the length of the short model in
feet. Since the wingspan-to-length ratio will be the same as determined in part a., we have the proportion
$\dfrac{180}{x} = \dfrac{99}{97}$. Cross multiply to find $99x = 180 \cdot 97 = 17{,}460$. Then $x = \dfrac{17{,}460}{99} = 176.36\dots$. To the nearest
foot, the length of the new model should be 176 feet.

## Chapter 4 Assessment Test

1.  The ratio of 28 to 65 may be written

    28 to 65     28 : 65     $\dfrac{28}{65}$

2.  The ratio of 5.8 to 2.1 may be written

    5.8 to 2.1     5.8 : 2.1     $\dfrac{5.8}{2.1}$

3.  Simplify 16 to 24.

    $\dfrac{16}{24} = \dfrac{\overset{2}{\cancel{16}}}{\underset{3}{\cancel{24}}} = \dfrac{2}{3}$

4.  Simplify 15 to 6.

    $\dfrac{15}{6} = \dfrac{\overset{5}{\cancel{15}}}{\underset{2}{\cancel{6}}} = \dfrac{5}{2}$

5. Simplify 68 to 36.

$$\frac{68}{36} = \frac{\overset{17}{\cancel{68}}}{\underset{9}{\cancel{36}}} = \frac{17}{9}$$

6. Simplify 2.5 to 75.

$$\frac{2.5}{75} = \frac{25}{750} = \frac{\overset{1}{\cancel{25}}}{\underset{30}{\cancel{750}}} = \frac{1}{30}$$

7. 2 days to 15 hours
   Smaller unit: hours
   2 days = 2 · 24 hours = 48 hours

   $$\frac{2 \text{ days}}{15 \text{ hours}} = \frac{48}{15} = \frac{16}{5}$$

8. 5 quarts to 3 pints
   Smaller unit: pints
   5 quarts = 5 · 2 pints = 10 pints

   $$\frac{5 \text{ quarts}}{3 \text{ pints}} = \frac{10}{3}$$

9. 56 apples for 10 baskets
   in fraction form is

   $$\frac{56 \text{ apples}}{10 \text{ baskets}} = \frac{28 \text{ apples}}{5 \text{ baskets}}.$$

   In sentence form, the simplified
   fraction form is "28 apples for
   every 5 baskets."

10. 12 cabinets for 88 files
    in fraction form is

    $$\frac{12 \text{ cabinets}}{88 \text{ files}} = \frac{3 \text{ cabinets}}{22 \text{ files}}.$$

    In sentence form, the simplified
    fraction form is "3 cabinets for
    every 22 files."

11. 385 miles for 12 gallons in fraction

    is $\dfrac{385 \text{ gallons}}{12 \text{ miles}}$. Divide to get a

    denominator of 1,

    $$\frac{385 \text{ gallons}}{12 \text{ miles}} = \frac{32.0833... \text{ gallons}}{1 \text{ mile}}$$

    To the nearest tenth, the unit rate is
    32.1 gallons per mile.

12. 12 birds in 4 cages in fraction form

    is $\dfrac{12 \text{ birds}}{4 \text{ cages}}$. Divide to get a

    denominator of 1,

    $$\frac{12 \text{ birds}}{4 \text{ cages}} = \frac{3 \text{ birds}}{1 \text{ cage}}$$

    The unit rate is 3 birds per cage.

13. $675 for 4 dining room chairs
    Write in fraction form, then divide
    by the denominator:

    $$\frac{\$675}{4 \text{ chairs}} = \frac{\$675 \div 4}{4 \div 4 \text{ chair}} = \frac{\$168.75}{1 \text{ chair}}$$

    The unit price is $168.75 per chair.

14. $5.76 for 12 tropical fish
    Write in fraction form, then divide
    by the denominator:

    $$\frac{\$5.76}{12 \text{ fish}} = \frac{\$5.76 \div 12}{12 \div 12 \text{ fish}} = \frac{\$0.48}{1 \text{ fish}}$$

    The unit price is $0.48 per fish.

15. 3 is to 45 as 18 is to 270
    written as a proportion is

    $$\frac{3}{45} = \frac{18}{270}$$

16. 9 labels is to 4 folders as 45 labels is to 20 folders
    written as a proportion is

    $$\frac{9 \text{ labels}}{4 \text{ folders}} = \frac{45 \text{ labels}}{20 \text{ folders}}$$

17. Given the proportion $\dfrac{2}{17} = \dfrac{6}{51}$

    left hand side: 2 to 17
    right hand side: 6 to 51
    Sentence form:
    2 is to 17 as 6 is to 51.

18. Given the proportion $\dfrac{12 \text{ photos}}{5 \text{ hours}} = \dfrac{24 \text{ photos}}{10 \text{ hours}}$

    left hand side: 12 photos to 5 hours
    right hand side: 24 photos to 10 hours
    Sentence form:
    12 photos is to 5 hours as 24 photos is to 10 hours.

19. $\dfrac{7}{16} \overset{?}{=} \dfrac{35}{80}$

Compute cross products:
$16 \cdot 35 = 560, \quad 7 \cdot 80 = 560$

The cross products are equal so the ratios are proportional.

20. $\dfrac{22}{18} \overset{?}{=} \dfrac{14}{12}$

Compute cross products:
$18 \cdot 14 = 252, \quad 22 \cdot 12 = 264$

The cross products are not equal so the ratios are not proportional.

21. $\dfrac{5}{p} = \dfrac{125}{150}$

Cross multiply: $125p = 5 \cdot 150 = 750$

Solve: $p = \dfrac{125p}{125} = \dfrac{750}{125} = 6$

Verify:
$\dfrac{5}{6} = \dfrac{125}{150} \qquad \begin{array}{l} 6 \cdot 125 = 750 \\ 5 \cdot 150 = 750 \end{array}$

22. $\dfrac{8}{13} = \dfrac{c}{52}$

Cross multiply: $13c = 8 \cdot 52 = 416$

Solve: $c = \dfrac{13c}{13} = \dfrac{416}{13} = 32$

Verify:
$\dfrac{8}{13} = \dfrac{32}{52} \qquad \begin{array}{l} 13 \cdot 32 = 416 \\ 8 \cdot 52 = 416 \end{array}$

23. $\dfrac{22}{19} = \dfrac{55}{m}$

Cross multiply: $22m = 19 \cdot 55 = 1045$

Solve: $m = \dfrac{22m}{22} = \dfrac{1045}{22} = 47.5$

Verify:
$\dfrac{22}{19} = \dfrac{55}{47.5} \qquad \begin{array}{l} 19 \cdot 55 = 1045 \\ 22 \cdot 47.5 = 1045 \end{array}$

24. $\dfrac{t}{3.5} = \dfrac{27}{10.5}$

Cross multiply: $10.5t = 3.5 \cdot 27 = 94.5$

Solve: $t = \dfrac{10.5c}{10.5} = \dfrac{94.5}{10.5} = 9$

Verify:
$\dfrac{9}{3.5} = \dfrac{27}{10.5} \qquad \begin{array}{l} 3.5 \cdot 27 = 94.5 \\ 9 \cdot 10.5 = 94.5 \end{array}$

25. a. The player ran 160 yards in 12 carries. As a fraction, his performance can be expressed as the ratio
$\dfrac{160}{12} = \dfrac{40}{3}$. Two other ways to express this ratio are $40 : 3$ and $40$ to $3$.

b. From part a., the fraction representing his performance is $\dfrac{40}{3} = 13.333.....$ . This fraction to the nearest tenth is 13.3. and represents the ratio of yards to carry. The unit rate is then 13.3 yards per carry.

26. Five pounds of bananas for \$7.25 is a unit rate of $\dfrac{7.25}{5} = 1.45$ dollars per pound. Three pounds of bananas for \$4.20 is a unit rate of $\dfrac{4.20}{3} = 1.40$ dollars per pound which is the smaller unit rate. Three pounds are the better buy.

27. The unit rates are

18 ounces of sugar for \$3.60, the unit rate is $\dfrac{3.60}{18} = 0.20$ dollars per ounce,

24 ounces of sugar for \$4.08, the unit rate is $\dfrac{4.08}{24} = 0.17$ dollars per ounce,

32 ounces of sugar for \$5.76, the unit rate is $\dfrac{5.76}{32} = 0.18$ dollars per ounce.

The smallest unit rate is \$0.17 per ounce so the 24 ounce deal is the best.

28. The ratio representing the first assembly lines production is $\dfrac{45}{4}$, while for the second line it is $\dfrac{78}{7}$. The question is whether $\dfrac{45}{4} \overset{?}{=} \dfrac{78}{8}$ is a valid proportion. Compute cross products to find $4 \cdot 78 = 312$ and $45 \cdot 8 = 360$. Since 312 is not equal to 360, the cross products are not equal and the rates of operation are not proportional.

29. Let $x$ denote the actual distance between the cities. Then solve the proportion $\dfrac{4}{35} = \dfrac{2.6}{x}$. First, cross multiply to find $4x = 2.6 \cdot 35 = 91$. Then divide to find $x$, $x = \dfrac{4x}{4} = \dfrac{91}{4} = 22.75$. The distance is 22.75 miles.

30. Let $x$ denote the height of the building in feet. Then solve the proportion $\dfrac{x}{130} = \dfrac{12}{20}$. First, cross multiply to find $20x = 130 \cdot 12 = 1560$. Then divide to find $x$, $x = \dfrac{20x}{20} = \dfrac{1560}{20} = 78$. The building is 78 feet high.

# Chapter 5    Percents

## Section 5.1    Introduction to Percents

### Concept Check

1.  A __percent__ is a part per 100.

3.  To convert a percent to a fraction, write the number preceding the percent sign over __100__. Simplify, if possible.

5.  To convert a decimal to a percent, multiply the decimal by __100%__. Alternatively, move the decimal point two places to the __right__ and append a percent sign.

7.  a. What percent is represented by the shaded area?  __21%__
    Note the figure is divided into $10 \times 10 = 100$ squares, 21 of which are shaded, hence 21% is shaded.

    b. What percent is represented by the unshaded area?  __79%__
    Note the figure is divided into $10 \times 10 = 100$ squares, 79 of which are unshaded, hence 79% is unshaded.

### Guide Problems

9.  Consider 32%

    a. Convert 32% to a fraction.

    $$32\% = \frac{\underline{32}}{100}$$

    To convert a percent to a fraction, write the value before the percent sign over a denominator of 100.

    b.  Simplify the fraction in part a.

    $$\frac{32}{100} = \frac{\underline{8}}{\underline{25}}$$

    Divide out a factor of 4.

11. Convert 29% to a decimal.

    To convert a percent to a decimal, drop the percent sign and multiply by 0.01.

    $$29\% = 29 \cdot 0.01 = \underline{0.29}$$

*In Problems 13 through 39, percents are converted to fractions as follows. The value to the left of the percent sign becomes the numerator of a fraction with 100 as the denominator. If decimal values are involved, the fraction is multiplied by 1 in the form n/n where n is a power of 10 chosen to clear all decimal points. If a mixed numbers is involved, the mixed number is rewritten as a fraction . Lastly, the fraction is simplified.*

13. $60\% = \dfrac{60}{100} = \dfrac{3}{5}$
   
15. $10\% = \dfrac{10}{100} = \dfrac{1}{10}$

17. $25\% = \dfrac{25}{100} = \dfrac{1}{4}$

19. $42\% = \dfrac{42}{100} = \dfrac{21}{50}$

21. $14\dfrac{3}{4}\% = \dfrac{14\dfrac{3}{4}}{100} = \dfrac{\dfrac{59}{4}}{100}$

    $= \dfrac{59}{4} \cdot \dfrac{1}{100} = \dfrac{59}{400}$

23. $13\dfrac{1}{2}\% = \dfrac{13\dfrac{1}{2}}{100} = \dfrac{\dfrac{27}{2}}{100}$

    $= \dfrac{27}{2} \cdot \dfrac{1}{100} = \dfrac{27}{200}$

25. $4\dfrac{1}{4}\% = \dfrac{4\dfrac{1}{4}}{100} = \dfrac{\dfrac{17}{4}}{100}$

    $= \dfrac{17}{400}$

27. $14.5\% = \dfrac{14.5}{100} = \dfrac{14.5}{100} \cdot \dfrac{10}{10}$

    $= \dfrac{145}{1000} = \dfrac{29}{200}$

29. $10.5\% = \dfrac{10.5}{100} = \dfrac{10.5}{100} \cdot \dfrac{10}{10}$

    $= \dfrac{105}{1000} = \dfrac{21}{200}$

31. $18.5\% = \dfrac{18.5}{100} = \dfrac{18.5}{100} \cdot \dfrac{10}{10}$

    $= \dfrac{185}{1000} = \dfrac{37}{200}$

33. $3.75\% = \dfrac{3.75}{100} = \dfrac{3.75}{100} \cdot \dfrac{100}{100}$

    $= \dfrac{375}{10,000} = \dfrac{3}{80}$

35. $110\% = \dfrac{110}{100} = \dfrac{11}{10} = 1\dfrac{1}{10}$

37. $165\% = \dfrac{165}{100} = \dfrac{33}{20} = 1\dfrac{13}{20}$

39. $260\% = \dfrac{260}{100} = \dfrac{13}{5} = 2\dfrac{3}{5}$

*In Exercises 41 through 67 a percent is converted to a decimal as follows. If the numerical value to the right of the percent sign contains a fraction, the value is first written in decimal form. The percent is converted to a decimal by dropping the percent sign and multiplying by 0.01. Alternatively, drop the percent sign and move the decimal point two places to left.*

41. $45\% = 45 \cdot 0.01 = 0.45$

43. $98\% = 98 \cdot 0.01 = 0.98$

45. $150\% = 150 \cdot 0.01 = 1.5$

47. $115\% = 115 \cdot 0.01 = 1.15$

49. $98\dfrac{1}{5}\% = 98\dfrac{2}{10}\% = 98.2\%$

    $= 98.2 \cdot 0.01 = 0.982$

51. $76\dfrac{1}{5}\% = 76\dfrac{2}{10}\% = 76.2\%$

    $= 76.2 \cdot 0.01 = 0.762$

53. $57\dfrac{2}{5}\% = 57\dfrac{4}{10}\% = 57.4\%$

    $= 57.4 \cdot 0.01 = 0.574$

55. $64\dfrac{4}{5}\% = 64\dfrac{8}{10}\% = 64.8\%$

    $= 64.8 \cdot 0.01 = 0.648$

57. $87.8\% = 87.8 \cdot 0.01 = 0.878$

59. $35.74\% = 35.74 \cdot 0.01 = 0.3574$

61. $4.2\% = 4.2 \cdot 0.01 = 0.042$

63. $0.37\% = 0.37 \cdot 0.01 = 0.0037$

65. $0.72\% = 0.72 \cdot 0.01 = 0.0072$

67. $0.883\% = 0.883 \cdot 0.01 = 0.00883$

## Guide Problems

69. Convert 0.81 to a percent.

    Multiply by 100%.

    $0.81 \cdot 100\% = 81\%$

Alternatively, move the decimal
two places to the right and append
a percent sign.

71. Consider $\dfrac{7}{16}$.

 a. Convert the fraction to a decimal.

   $\dfrac{7}{16} = \underline{\ 0.4375\ }$

 b. Convert the decimal in part a to a percent.
    Multiply by 100%.

   $\underline{\ 0.4375\ } \cdot 100\% = 43.75\%$

*In each of Exercises 73 through 99 a decimal value is converted to a percent. In all cases, the conversion is done by multiplying by 100%. Alternatively, the conversion can be done by moving the decimal point in the given value two places to the right and adding % symbol.*

73. $0.45 = 0.45 \cdot 100\% = 45\%$     75. $0.3 = 0.3 \cdot 100\% = 30\%$     77. $0.01 = 0.01 \cdot 100\% = 1\%$

79. $0.769 = 0.769 \cdot 100\% = 76.9\%$     81. $0.6675 = 0.6675 \cdot 100\% = 66.75\%$

83. $0.072 = 0.072 \cdot 100\% = 7.2\%$     85. $0.00048. = 0.00048 \cdot 100\% = 0.048\%$

87. $10 = 10 \cdot 100\% = 1000\%$     89. $16 = 16 \cdot 100\% = 1600\%$

91. $3.76 = 3.76 \cdot 100\% = 376\%$     93. $2.268 = 2.268 \cdot 100\% = 226.8\%$

95. $6.92 = 6.92 \cdot 100\% = 692\%$     97. $1.99 = 1.99 \cdot 100\% = 199\%$     99. $2.35 = 2.35 \cdot 100\% = 235\%$

*In each of Exercises 101 through 127, a fraction or mixed fraction is converted to a percent. In each case, the fraction is first converted to an equivalent decimal value using the methods of chapter 3. The resulting decimal is then converted to a percent using the techniques of Exercises 73 through 99 above.*

101. $\dfrac{1}{10} = 0.01 = 0.01 \cdot 100\% = 10\%$     103. $\dfrac{1}{5} = 0.2 = 0.2 \cdot 100\% = 20\%$

105. $\dfrac{23}{100} = 0.23 = 0.23 \cdot 100\% = 23\%$     107. $\dfrac{21}{100} = 0.21 = 0.21 \cdot 100\% = 21\%$

109. $\dfrac{3}{25} = 0.12 = 0.12 \cdot 100\% = 12\%$     111. $\dfrac{13}{20} = 0.65 = 0.65 \cdot 100\% = 65\%$

113. $\dfrac{9}{8} = 1.125 = 1.125 \cdot 100\% = 112.5\%$     115. $\dfrac{19}{8} = 2.375 = 2.375 \cdot 100\% = 237.5\%$

117. $\dfrac{33}{16} = 2.0625 = 2.0625 \cdot 100\% = 206.25\%$     119. $\dfrac{45}{16} = 2.8125 = 2.8125 \cdot 100\% = 281.25\%$

121. $1\frac{4}{5} = 1.8 = 1.8 \cdot 100\% = 180\%$                    123. $1\frac{22}{25} = 1.88 = 1.88 \cdot 100\% = 188\%$

125. $2\frac{3}{5} = 2.6 = 2.6 \cdot 100\% = 260\%$                    127. $2\frac{3}{4} = 2.75 = 2.75 \cdot 100\% = 275\%$

129. The percent of campers from out of state is $0.46 = 0.46 \cdot 100\% = 46\%$.

131. 89% as a decimal is 0.89 (drop the % symbol and move the decimal two places to the left).

133. The percent 38.4% is represented by the decimal $38.4\% = 38.4 \cdot 0.01 = 0.384$.

135. The decimal form of 37.5% is $37.5\% = 37.5 \cdot 0.01 = 0.375$.

137. The decimal 1.22 is equivalent to the percent $1.22 = 1.22 \cdot 100\% = 122\%$.

139. The percent 231.4% in decimal form is $231.4\% = 231.4 \cdot 0.01 = 2.314$.

## Cumulative Skills Review

1. In the product $9 \cdot 9 \cdot 9 \cdot 9$, the number 9 is being repeatedly multiplied so 9 is the base in exponential notation. There are four factors of 9 so 4 is the exponent. Thus, in exponential notation $9 \cdot 9 \cdot 9 \cdot 9$ is $9^4$ which is read "nine to the fourth power."

3. Given the ratios $\frac{12}{39}$ and $\frac{4}{13}$ compute the cross products: $12 \cdot 13 = 156$ and $4 \cdot 39 = 156$. Since the cross products are equal, the ratios are proportional and the corresponding proportion may be written $\frac{12}{39} = \frac{4}{13}$.

5. $0.386 = \frac{386}{1000} = \frac{193 \cdot 2}{500 \cdot 2} = \frac{193}{500}$          7. $\frac{8}{6} - \frac{1}{3} = \frac{8}{6} - \frac{1}{3} \cdot \frac{2}{2} = \frac{8}{6} - \frac{2}{6} = \frac{6}{6} = 1$

9. The three different ways to write a ratio of two values are as a fraction with the first value as numerator and the second value as denominator, with a colon separating the values, and with the word *to* separating the values. The ratio of 22 to 17 may be written as

$$\frac{22}{17} \qquad\qquad 22:17 \qquad\qquad 22 \text{ to } 17$$

## Section 5.2    Solve Percent Problems Using Equations

## Concept Check

1. Percent problems can be solved using either __equations__ or __proportions__.

3. When writing a percent problem as an equation, the word *of* indicates __multiplication__ and the word *is* indicates __equality__.

5. In a percent problem, the __amount__ is the number that represents part of a whole.

7. In a percent problem, the __base__ is the number that represents the whole. In a percent problem, it is preceded by the word *of*.

9.  When solving for the base, we can use the formula $\text{Base} = \dfrac{\text{Amount}}{\text{Percent}}$ as a variation of the percent equation.

11. When solving the percent equation for the amount or the base, the percent is converted to a  decimal .

## Guide Problems

13. Fill in the following to complete the percent equation.

    75    is    what percent    of    80?

    ↓    ↓         ↓            ↓    ↓
    75   =        p        ·    80

    Recall that in a percent problem, the word *is* indicates an equal sign and the word *of* indicates multiplication. Phrases such as *what*, *what number*, and *what percent* indicate an unknown quantity represented by a variable. Here we use *p* since the unknown quantity is a percent.

*In Problems 15 through 25 a percent problem is written as an equation. The word " is" typically translates into an equal sign while the word "of" becomes multiplication. The phrases what, what number, and what percent indicate an unknown to be represented by a variable. If the unknown quantity is a percent, the variable p is used, otherwise a and b are used. Note how the resulting equation compares to the percent equation,* Amount = Percent · Base .

15. 195.3   is   30%   of   what number?

    ↓      ↓    ↓    ↓        ↓
    195.3  =   30%   ·        b

17. 10   is   3%   of   what number?

    ↓    ↓   ↓    ↓        ↓
    10   =   3%   ·        b

19. 736   is   86%   of   what number?

    ↓     ↓    ↓    ↓        ↓
    736   =   86%   ·        b

21. 38%   of   what number   is   406?

    ↓     ↓        ↓          ↓    ↓
    38%   ·        b          =    406

23. Find   25%   of   500?

    ↓      ↓     ↓    ↓
    a =    25%   ·    500

25. 87%   of   4350   is   what number?

    ↓     ↓    ↓      ↓        ↓
    87%   ·    4350   =        a

Note, in this case the word "Find" corresponded to a variable followed by an equal sign since you are asked to find an unknown value.

## Guide Problems

27. What number is 20% of 70?

    a. Identify the parts of the percent problem.

    amount: _unknown, a_
    rate: _20%_
    base: _70_

    The number 70 is what you are taking a percent of, so it is the base and the amount is unknown.

    b. Write the problem as a percent equation.

    Use Amount = Percent · Base .
    $$a = 20\% \cdot 70$$

    c. Convert the percent to a decimal.

    $$a = 0.2 \cdot 70$$

    d. Solve the equation.

    Multiply: $a = 0.2 \cdot 70 = 14$

29. What percent of 500 is 25?

    a. Identify the parts of the percent problem.

    amount: _25_
    rate: _unknown, p_
    base: _500_

    The number 500 is what you are taking a percent of, so it is the base and the percent is unknown.

    b. Write the problem as a percent equation.

    Use Amount = Percent · Base .
    $$500 \cdot p = 25$$

    c. Solve the equation.
    Divide to isolate the variable:

    $$\frac{500 \cdot p}{500} = \frac{25}{500}$$
    $$p = \frac{25}{500} = 0.05$$

    d. Convert the answer to a percent.

    $$p = 0.05 = 0.05 \cdot 100\% = 5\%$$

31. 1050 is 10% of what number?

    amount: 1050
    percent: 10%
    base: unknown, b

    Percent · Base = Amount
    $$10\% \cdot b = 1050$$
    $$0.01b = 1050$$
    $$\frac{0.01b}{0.01} = \frac{1050}{0.01}$$
    $$b = 105,000$$

33. 2350 is 25% of what number?

    amount: 2350
    percent: 25%
    base: unknown, b

    Percent · Base = Amount
    $$25\% \cdot b = 2350$$
    $$0.25b = 2350$$
    $$\frac{0.25b}{0.25} = \frac{2350}{0.25}$$
    $$b = 9400$$

35. What is 2% of 500?

   amount: unknown, $a$
   percent: 2%
   base: 500

   $$\text{Amount} = \text{Percent} \cdot \text{Base}$$
   $$a = 2\% \cdot 1500$$
   $$a = 0.02 \cdot 1500$$
   $$a = 30$$

37. 2088 is 36% of what number?

   amount: 2088
   percent: 36%
   base: unknown, $b$

   $$\text{Percent} \cdot \text{Base} = \text{Amount}$$
   $$36\% \cdot b = 2088$$
   $$0.36b = 2088$$
   $$\frac{0.36b}{0.36} = \frac{2088}{0.36}$$
   $$b = 5800$$

39. What number is 32% of 4900?

   amount: unknown, $a$
   percent: 32%
   base: 4900

   $$\text{Amount} = \text{Percent} \cdot \text{Base}$$
   $$a = 32\% \cdot 4900$$
   $$a = 0.32 \cdot 4900$$
   $$a = 1568$$

41. 6120 is what percent of 12,000?

   amount: 6120
   percent: unknown, $p$
   base: 12,000

   $$\text{Percent} \cdot \text{Base} = \text{Amount}$$
   $$p \cdot 12,000 = 6120$$
   $$\frac{p \cdot 12,000}{12,000} = \frac{6120}{12,000}$$
   $$p = 0.51$$
   $$p = 0.51 \cdot 100\%$$
   $$p = 51\%$$

43. 2700 is 30% of what number?

   $$30\% \cdot b = 2700$$
   $$0.3b = 2700$$
   $$\frac{0.3b}{0.3} = \frac{2700}{0.3}$$
   $$b = 9000$$

45. What is 3% of 3900?

   $$a = 3\% \cdot 3900$$
   $$a = 0.03 \cdot 3900$$
   $$a = 117$$

47. 13,248 is 69% of what number?

   $$69\% \cdot b = 13,248$$
   $$0.69b = 13,248$$
   $$\frac{0.69b}{0.69} = \frac{13,248}{0.69}$$
   $$b = 19200$$

49. What percent of 13,000 is 2210?

   $$p \cdot 13,000 = 2210$$
   $$\frac{p \cdot 13,000}{13,000} = \frac{2210}{13,000}$$
   $$p = 0.17$$
   $$p = 17\%$$

51. What number is 82% of 122?

   $$a = 82\% \cdot 122$$
   $$a = 0.82 \cdot 122$$
   $$a = 100.04$$

53. What is 12% of 365?

   $$a = 12\% \cdot 365$$
   $$a = 0.12 \cdot 365$$
   $$a = 43.8$$

55. What number is 44% of 699?

$$a = 44\% \cdot 699$$
$$a = 0.44 \cdot 699$$
$$a = 307.56$$

57. What percent of 751 is 518.19?

$$p \cdot 751 = 518.19$$
$$\frac{p \cdot 518.19}{751} = \frac{518.19}{751}$$
$$p = 0.69$$
$$p = 69\%$$

59. 326.65 is 47% of what number?

$$47\% \cdot b = 326.65$$
$$0.47b = 326.65$$
$$\frac{0.47b}{0.47} = \frac{326.65}{0.47}$$
$$b = 695$$

61. The problem translates to : what percent of 500 is 400? Solving the corresponding percent equation gives

$$p \cdot 500 = 400 \text{ or } p = \frac{400}{500} = 0.8 = 80\% \, .$$

63. The problem translates to the question: 20% of what number is $55. Translating to an equation gives $20\% \cdot b = 55$ and solving this equation tells us $0.2b = 55$ and so $b = \frac{55}{0.2} = 275$. Nancy's goal is $275 in contributions.

65. The favored candidate has won 60% of 25 states or $60\% \cdot 25 = 0.6 \cdot 25 = 15$ states.

67. a. Selma has loaded 30%of 500 songs or $30\% \cdot 500 = 0.3 \cdot 500 = 150$ songs.

   b. Selma has room for $500 - 150 = 350$ songs.

   c. Selma has added 47 songs to the original 150 songs for a total of $150 + 47 = 197$ songs out of a possible 500. The percent 197 out of 500 represents can be found by solving $p \cdot 500 = 197$ which gives

   $$p = \frac{197}{500} = 0.394 = 39.4\% \, . \text{ Rounding to the nearest whole percent yields 39\%.}$$

69. There are 46 defective air conditioners and this represents 4% of the total. The percent problem here can be stated simply as "4% of what number is 46?" Using the percent equation, we have

   $$4\% \cdot b = 46 \text{ so } 0.04b = 46 \text{ and } b = \frac{46}{0.04} = 1150 \, . \text{ A total of 1150 air conditioners were made.}$$

71. John spends 5% of his monthly income, given as $3250, on entertainment. The amount he spends on entertainment is then the answer to the percent problem "what amount is 5% of $3250?" By the percent equation $a = 5\% \cdot 3250 = 0.05 \cdot 3250 = 162.50$. John Spends $162.50 a month on entertainment.

73. The factory has produced 90 lamp shades out of a total of 600 needed for an order. The percent of the order that is complete is $p = \frac{90}{600} = 0.15 = 0.15 \cdot 100\% = 15\% \, .$

75. a. The problem is the same as the percent problem: what percent is 67 of 280? The percent equation is

$p \cdot 280 = 67$ and the solution is $p = \dfrac{67}{280} = 0.2393 = 23.93\%$ or 24% to the nearest whole percent.

b. 280 cars are washed each week and 15% of these get a wax job. The number of cars, on average, that get a wax job each week is $a = 15\% \cdot 280 = 0.15 \cdot 280 = 42$.

77. 60% of the customers get both manicures and haircuts. Since 33 customers got both manicures and haircuts on Wednesday, 33 represents the amount and we seek the base, the total number of customers.
Solving the percent equation $60\% \cdot b = 0.6b = 33$ gives $b = \dfrac{33}{0.6} = 55$ customers.

79. The 146 papers Frank has delivered represent 40% of his route. The total size of his route is unknown and this number represents the base in a percent problem. The corresponding percent equation is $40\% \cdot b = 146$ or $0.4 \cdot b = 146$. Solving gives $b = \dfrac{146}{0.4} = 365$ so Frank delivers 365 papers total.

81. Miguel's mutual fund is worth $5600. His profit was 12.5% of that value or $12.5\% \cdot 5600 = 0.125 \cdot 5600 = 700$. His profit was $700.

83. The 65.1 million people who play basketball represent 23.25% of the population so the total population is found by solving the equation $23.25\% \cdot b = 65.1$ for $b$ where $b$ is in millions. Solving $b = \dfrac{65.1}{0.2325} = 280$. At the time of the analysis the population was 280 million people.

85. Using the total population of 280 million from exercise 83 and the fact that 26.4 million people play tennis, the percent that play tennis $p = \dfrac{26.4}{280} = .09429 = 9.429\%$ or 9.43% to the nearest hundredth.

87. From the graph, 69.1 million people bowl. If $\dfrac{1}{2}\%$ of these average over 200 points per game, then
$\dfrac{1}{2}\% \cdot 69.1 = 0.5\% \cdot 69.1 = 0.005 \cdot 69.1 = 0.3455$ million average over 200 points. Equivalently $0.3455$ million $= 0.3455 \cdot 1,000,000 = 345,500$ people average over 200 points.

## Cumulative Skills Review

1. $72 \div 8 = \dfrac{72}{8} = \dfrac{8 \cdot 9}{8} = \dfrac{\cancel{8} \cdot 9}{\cancel{8}} = 9$

3. The figure is divided into five equal parts, three of which are shaded. The fraction shaded is therefore $\dfrac{3}{5}$.

5. Add vertically:
$$\begin{array}{r} \overset{1\ 1\ 1}{5\,3\,5\,4} \\ +\ \ 7\,7\,7 \\ \hline 6\,1\,3\,1 \end{array}$$

7. $\dfrac{6}{15} + \dfrac{7}{15} = \dfrac{6+7}{15} = \dfrac{13}{15}$

9. Given $\dfrac{3}{9} = \dfrac{y}{36}$, then by cross multiplication, $9 \cdot y = 3 \cdot 36$ or $9 \cdot y = 108$ so $y = \dfrac{108}{9} = 12$.

## Section 5.3    Solve Percent Problems Using Proportions

### Concept Check

1.  Label the parts of a percent proportion $\dfrac{\quad}{\underline{\quad}} = \dfrac{\quad}{100}$.    $\dfrac{\text{Amount}}{\text{Base}} = \dfrac{\text{Part}}{100}$

## Guide Problems

3.  What number is 85% of 358?

    a. Identify the parts of the percent problem.

    amount:  _unknown, a_
    part: _85_
    base: _358_

    Recall the base typically follows *of* in the problem statement. The word *what* indicates the unknown and the part precedes the % symbol.

    b. Write the percent proportion.

    $$\frac{\text{Amount}}{\text{Base}} = \frac{\text{Part}}{100}$$

    c. Substitute the values of proportion, base, and part into the proportion.

    $$\frac{a}{358} = \frac{85}{100}$$

5.  28% of what number is 16?

    a. Identify the parts of the percent problem.

    amount:  _16_
    part: _28_
    base:  _unknown, b_

    Recall the base typically follows *of* in the problem statement. The word *what* indicates the unknown and the part precedes the % symbol.

    b. Write the percent proportion.

    $$\frac{\text{Amount}}{\text{Base}} = \frac{\text{Part}}{100}$$

    c. Substitute the values of proportion, base, and part into the proportion.

    $$\frac{16}{b} = \frac{28}{100}$$

7.  What is 12% of 279?

    amount:  _unknown, a_
    part: _12_
    base: _279_

    $$\frac{\text{Amount}}{\text{Base}} = \frac{\text{Part}}{100}$$

    $$\frac{a}{279} = \frac{12}{100}$$

9.  What number is 88.5% of 190?

    amount:  _unknown, a_
    part: _88.5_
    base: _190_

    $$\frac{\text{Amount}}{\text{Base}} = \frac{\text{Part}}{100}$$

    $$\frac{a}{190} = \frac{88.5}{100}$$

11. 200 is what percent of 1000?

    amount:  _200_
    part: _unknown, p_
    base: _1000_

    $$\frac{\text{Amount}}{\text{Base}} = \frac{\text{Part}}{100}$$

    $$\frac{200}{1000} = \frac{p}{100}$$

## Guide Problems

13. What percent of 128 is 96?

a. Identify the parts of the percent problem.

amount: __96__
part: __unknown, p__
base: __128__

Recall the base follows *of*.

b. Substitute the values of amount, base, and part into the proportion.

$$\frac{\text{Amount}}{\text{Base}} = \frac{\text{Part}}{100}$$

$$\frac{96}{128} = \frac{p}{100}$$

c. Simplify the fractions in the percent proportion, if possible.

$$\frac{3}{4} = \frac{p}{100}$$

d. Cross-multiply and set the cross-products equal.

$$4 \cdot p = 3 \cdot 100$$
$$4p = 300$$

e. Divide both sides of the equation by the number on the side with the unknown.

$$\frac{4p}{4} = \frac{300}{4}$$
$$p = 75$$

__75__% of 128 is 96.

*In Exercises 15 through 41, the same steps are followed to solve the percent problem as a proportion. First, the parts of the percent problem are identified with one part an unknown. The parts are substituted into the percent proportion* $\dfrac{\text{Amount}}{\text{Base}} = \dfrac{\text{Part}}{100}$. *Fractions are simplified, if possible, then the cross-products are computed and set equal to each other. Lastly, each side is divided by the number on the side with the unknown.*

15. What number is 25% of 700?

amount: __unknown, a__
part: __25__
base: __700__

$$\frac{\text{Amount}}{\text{Base}} = \frac{\text{Part}}{100}$$

$$\frac{a}{700} = \frac{25}{100}$$
$$\frac{a}{700} = \frac{1}{4}$$
$$700 \cdot 1 = a \cdot 4$$
$$4a = 700$$
$$\frac{4a}{4} = \frac{700}{4}$$
$$a = 175$$

175 is 25% of 700

17. 330 is what percent of 1100?

amount: __330__
part: __unknown, p__
base: __1100__

$$\frac{\text{Amount}}{\text{Base}} = \frac{\text{Part}}{100}$$

$$\frac{330}{1100} = \frac{p}{100}$$
$$\frac{3}{10} = \frac{p}{100}$$
$$10 \cdot p = 3 \cdot 100$$
$$10p = 300$$
$$\frac{10p}{10} = \frac{300}{10}$$
$$p = 30$$

330 is 30% of 1100

19. 1440 is what percent of 900?

amount: __1440__
part: __unknown, p__
base: __900__

$$\frac{\text{Amount}}{\text{Base}} = \frac{\text{Part}}{100}$$

$$\frac{1440}{900} = \frac{p}{100}$$
$$\frac{8}{5} = \frac{p}{100}$$
$$5 \cdot p = 8 \cdot 100$$
$$5p = 800$$
$$\frac{5p}{5} = \frac{800}{5}$$
$$p = 160$$

1440 is 160% of 900

21.  150% of 622 is what number?

$$\frac{a}{622} = \frac{150}{100}$$

$$\frac{a}{622} = \frac{3}{2}$$

$$622 \cdot 3 = a \cdot 2$$

$$2a = 1866$$

$$a = \frac{1866}{2} = 933$$

150% of 622 is 933.

23.  6975 is 31% of what number?

$$\frac{6975}{b} = \frac{31}{100}$$

$$\frac{6975}{b} = \frac{31}{100}$$

$$b \cdot 31 = 697,500$$

$$31b = 697,500$$

$$b = \frac{697,500}{31} = 22,500$$

6975 is 31% of 22,500

25.  270 is 60% of what number?

$$\frac{270}{b} = \frac{60}{100}$$

$$\frac{270}{b} = \frac{3}{5}$$

$$b \cdot 3 = 270 \cdot 5$$

$$3b = 1350$$

$$b = \frac{1350}{3} = 450$$

270 is 60% of 450.

27.  What is 2.5% of 2000?

$$\frac{a}{2000} = \frac{2.5}{100}$$

$$\frac{a}{2000} = \frac{25}{1000} = \frac{1}{40}$$

$$2000 \cdot 1 = a \cdot 40$$

$$40a = 2000$$

$$a = \frac{2000}{40} = 50$$

50 is 2.5% of 2000.

29.  288 is what percent of 1200?

$$\frac{288}{1200} = \frac{p}{100}$$

$$\frac{6}{25} = \frac{p}{100}$$

$$25 \cdot p = 6 \cdot 100$$

$$25p = 600$$

$$p = \frac{600}{25} = 24$$

288 is 24% of 1200.

31.  300 is 16% of what number?

$$\frac{300}{b} = \frac{16}{100}$$

$$\frac{300}{b} = \frac{4}{25}$$

$$b \cdot 4 = 300 \cdot 25$$

$$4b = 7500$$

$$b = \frac{7500}{4} = 1875$$

300 is 16% of 1875.

33.  25% of 528 is what number?

$$\frac{a}{528} = \frac{25}{100}$$

$$\frac{a}{528} = \frac{1}{4}$$

$$528 \cdot 1 = a \cdot 4$$

$$4a = 528$$

$$a = \frac{528}{4} = 132$$

25% of 528 is 132.

35.  390 is what percent of 300?

$$\frac{390}{300} = \frac{p}{100}$$

$$\frac{13}{10} = \frac{p}{100}$$

$$10 \cdot p = 13 \cdot 100$$

$$10p = 1300$$

$$p = \frac{1300}{10} = 130$$

390 is 130% of 300.

37.  What number is 10% of 3525?

$$\frac{a}{3525} = \frac{10}{100}$$

$$\frac{a}{3525} = \frac{1}{10}$$

$$3525 \cdot 1 = a \cdot 10$$

$$10a = 3525$$

$$a = \frac{3525}{10} = 352.5$$

352.5 is 10% of 3525.

39.  1950 is what percent of 6000?

$$\frac{1950}{6000} = \frac{p}{100}$$

$$\frac{13}{40} = \frac{p}{100}$$

$$40 \cdot p = 13 \cdot 100$$

$$40p = 1300$$

$$p = \frac{1300}{40} = 32.5$$

1950 is 32.5% of 6000.

41.  What is 2% of 263?

$$\frac{a}{263} = \frac{2}{100}$$

$$\frac{a}{263} = \frac{1}{50}$$

$$263 \cdot 1 = a \cdot 50$$

$$50a = 263$$

$$a = \frac{263}{50} = 5.26$$

5.26 is 2% of 263.

43. The total number of students is the unknown base, $b$, of the problem. The amount 1350 and 20% says the part is 20 of 100. Substituting into the percent proportion gives $\dfrac{1350}{b} = \dfrac{20}{100}$. Simplify the fractions and cross-multiply to get $\dfrac{1350}{b} = \dfrac{20}{100} = \dfrac{1}{5}$ and $b = 5 \cdot 1350 = 6750$ students.

45. a. The total number of grams recommended per day is the unknown base $b$. We are told 7 grams is 14% of the total so the part is 14 and the amount is 7. The percent proportion is $\dfrac{7}{b} = \dfrac{14}{100} = \dfrac{7}{50}$. Solving the proportion gives $7b = 7 \cdot 50 = 350$ and $b = \dfrac{350}{7} = 50$. The recommended daily total is 50 grams.

    b. As in part a., the total is $b$, the amount 2 and the part, 8. Thus $\dfrac{2}{b} = \dfrac{8}{100} = \dfrac{2}{25}$ and $2b = 2 \cdot 25 = 50$ and $b$ is 25 grams.

47. The problem asks for the value of your 40% share so the amount, $a$, is unknown. The part is 40 and the base is \$58,000. The percent proportion is $\dfrac{a}{58,000} = \dfrac{40}{100} = \dfrac{2}{5}$. Cross-multiplying gives $5a = 2 \cdot 58,000 = 116,000$ and so $a = \dfrac{116,000}{5} = 23,200$. Your share is \$23,200.

49. The problem is asking what percent of \$15,500 is \$1450? Thus, the base is \$15,500, the amount is \$1450, and the part, $p$, is unknown. The percent proportion is $\dfrac{1450}{15,500} = \dfrac{p}{100}$ or, after reducing, $\dfrac{29}{310} = \dfrac{p}{100}$. Solving the proportion gives $310p = 2900$ and $p = \dfrac{2900}{310} = 9.355$. The down payment is 9.4% to the nearest tenth of a percent.

51. Since 67 babies represents 2% of the deliveries, the total number of deliveries is the unknown base, $b$. The percent proportion is $\dfrac{67}{b} = \dfrac{2}{100} = \dfrac{1}{50}$. Cross-multiplying gives the total as $b = 67 \cdot 50 = 3350$ deliveries.

53. The question is what percent of 12 is 10? The part, p, is unknown and the corresponding percent proportion is $\dfrac{10}{12} = \dfrac{p}{100}$. The proportion is solved as follows: $\dfrac{10}{12} = \dfrac{5}{6} = \dfrac{p}{100}$ and so $6p = 500$ and $p = \dfrac{500}{6} = 83.333$. The percent of the coffee pot, to the nearest whole percent, is 83%.

55. a. The percent proportion is $\dfrac{7000}{40,000} = \dfrac{p}{100}$ or $\dfrac{7}{40} = \dfrac{p}{100}$. Solving the proportion gives $40p = 7 \cdot 100 = 700$ and so $p = \dfrac{700}{40} = 17.5$. The pool was filled to 17.5% of capacity.

    b. From part a., 17.5% was filled so $100\% - 17.5\% = 82.5\%$ remained to be filled.

57. The base here is the mother's weight of 880 pounds and the amount is the baby's weight of 54 pounds. The part, $p$, is unknown. The percent proportion is $\dfrac{54}{880} = \dfrac{p}{100}$. Reducing and solving gives $\dfrac{27}{440} = \dfrac{p}{100}$,

$440p = 2700$ and $p = \dfrac{2700}{440} = 6.136$. To the nearest tenth of a percent, the baby's weight is 6.1% of its mother's weight.

59. If five out of six packages arrive on time, then one out of six does not. The percent that does not is found by solving the percent proportion $\dfrac{1}{6} = \dfrac{p}{100}$. Cross-multiplication gives $6p = 100$ so $p = \dfrac{100}{6} = 16.666$. To the nearest tenth of a percent, 16.7% of packages are late.

61. The question is "what is 5% of 280?" So the base is 280, the part is 5 and the amount $a$ is unknown. The percent proportion is $\dfrac{a}{280} = \dfrac{5}{100} = \dfrac{1}{20}$. Cross-multiplying gives $20a = 280$ and $a = \dfrac{280}{20} = 14$. Matt would have 14 vacation days after working 280 days.

*In Problems 63-65, the problem is rephrased into one the standard forms, the parts of the percent problem are identified and substituted into the percent proportion which is then solved.*

63. What percent is 117.5 million of 166.5 million? Base: 166.5, amount: 117.5, part: $p$.

$\dfrac{117.5}{166.5} = \dfrac{p}{100}$, $\quad 166.5p = 100 \cdot 117.5 = 11{,}750$, $\quad p = \dfrac{11{,}750}{166.5} = 70.571$

Elvis Presley's album sales are about 71% of that of the Beatles.

65. What number is 8.5% of 105 million? Base: 105, amount: $a$, part: 8.5.

$\dfrac{a}{105} = \dfrac{8.5}{100} = \dfrac{85}{1000} = \dfrac{17}{200}$, $\quad 200a = 105 \cdot 17 = 1785$, $\quad a = \dfrac{1785}{200} = 8.925$

Internet sales account for about 9 million albums of Garth Brook's total sales.

## Cumulative Skills Review

1. $85\% = \dfrac{85}{100} = \dfrac{17 \cdot 5}{20 \cdot 5} = \dfrac{17}{20}$; $\;85\% = 85 \cdot 0.01 = 0.85$

3. The digits of 51.2523 agree with those of 51.252478 through the thousandths place. In the ten-thosuandths place 4 is greater than 3 so $51.2523 < 51.252478$.

5. $22.5\% = 22.5 \cdot 0.01 = 0.225$

7. Given the ratios $\dfrac{3}{13}$ and $\dfrac{21}{45}$ compute the cross products $3 \cdot 45 = 135$ and $13 \cdot 21 = 273$. Since the cross products are not equal, the ratios are not proportional.

9. Since the recipe calls for one cup of tomato sauce and Brian is using 4 cups of sauce, the recipe is being quadrupled and so four times the amount of mushrooms are needed or $4 \cdot \dfrac{1}{4} = \dfrac{4}{1} \cdot \dfrac{1}{4} = \dfrac{4}{4} = 1$ cup.

## Section 5.4    Solve Percent Application Problems

## Concept Check

1.   A common application of percents is expressing how much a particular quantity has   changed  .

3.   When numbers go down, the percent change is referred to as a percent   decrease  .

5.    Sales    tax   is a state tax based is the retail price or rental cost of certain items.

7.   Write the tip equation. Tip = Tip rate · Bill amount

9.   Write the commission equation.    Commission = Commission rate · Sales amount

## Guide Problems

11.  If a number changes from 100 to 123, what is the percent change?

a. What is the change amount?

The new amount is greater than the original amount, so the amount has *increased* and the change amount is found by subtracting the original amount from the new amount.

$$123 - 100 = 23$$

b. Set up the percent change formula.

$$\text{Percent change} = \frac{\text{Change amount}}{\text{Original amount}}$$
$$= \frac{23}{100}$$

c. Solve and state the percent change as an increase or decrease.

$$\frac{23}{100} = 0.23 = 23\%$$

The percent change is a 23% increase.

*In completing the table for exercises 13 – 21, the following strategy is used. To determine the Amount Of Change, compute* New Amount – Original Amount *, if the new amount is greater than the original amount. If the new amount is less than the original amount, compute* Original Amount – New Amount *. The Percent*

*Change is found by computing* $\dfrac{\text{Change amount}}{\text{Original amount}}$ *and writing the result as a percent. Lastly, the percent change*

*is an increase if the new amount is greater than the original amount and a decrease if the new amount is less than the original amount. Percents are rounded to the nearest whole percent.*

|      | Original Amount | New Amount | Amount Of Change | Percent Change |
|------|-----------------|------------|------------------|----------------|
| 13.  | 50              | 65         | New – Orig. = 65 – 50 = 15 | $\frac{15}{50} = 0.30 = 30\%$  increase |

| 15. | 18 | 22 | New − Orig. = 22 − 18 = 4 | $\dfrac{4}{18} = 0.22 = 22\%$ increase |
|---|---|---|---|---|
| 17. | 1000 | 1260 | New − Orig. = 1260 − 1000 = 260 | $\dfrac{260}{1000} = 0.26 = 26\%$ increase |
| 19. | $150 | $70 | Orig. − New = $150 − $70 = $80 | $\dfrac{80}{150} = 0.53 = 53\%$ decrease |
| 21. | 68 | 12 | Orig. − New = 68 − 12 = 56 | $\dfrac{56}{68} = 0.82 = 82\%$ decrease |

## Guide Problems

23. A toaster sells for $55. If the sales tax rate is 5%, determine the sales tax and the total purchase price.

   a. Write the sales tax formula.
      Sales tax = Sales tax rate · Item cost

   b. Substitute the values in the formula.

      $t = 5\% \cdot \$55$

   c. Calculate the sales tax.

      $t = 5\% \cdot \$55 = 0.05 \cdot \$55$
      $\phantom{t} = \$2.75$

   d. Determine the total purchase price.

      Item Cost + Sales Tax =
          Total purchase price
      $55.00 + $2.75 = $57.75

25. Bernie earned a 20% commission on $600 in magazine sales. How much commission will he receive?

   a. Write the commission formula.
      Commission =
          Commission rate · Sales amount

   b. Substitute the values in the formula.

      $c = 20\% \cdot \$600$

   c. Calculate the commission.

      $c = 20\% \cdot \$600 = 0.2 \cdot \$600 = \$120$

   Bernie earned a commission of $120.

27. Item cost: $65.40   Sales tax rate: 6%
    Sales tax = Sales tax rate · Item cost
       $= 6\% \cdot \$65.40 = 0.06 \cdot \$65.40$
       $= \$3.924 \text{ or } \$3.92$

29. a. Item cost: $327.19   Sales tax rate: 4.3%
       Sales tax = Sales tax rate · Item cost
          $= 4.3\% \cdot \$327.19 = 0.043 \cdot \$327.19$
          $= \$14.06917 \text{ or } \$14.07$

    b. Total purchase price
       = Item Cost + Sales Tax
       = $327.19 + $14.07 = $341.26

31. Bill amount: $13.00   Tip rate: 15%
    Tip = Tip rate · Bill amount

    $\quad = 15\% \cdot \$13.00 = 0.15 \cdot \$13.00$

    $\quad = \$1.95$

    $\quad\quad$ Total = Bill amount + Tip

    $\quad\quad\quad = \$13.00 + \$1.95 = \$14.95$

33. The bill amount is $120 while the total is
    $150.  The tip is then $\$150 - \$120 = \$30$.

    Let $r$ represent the tip rate. Then from the
    tip equation
    $\quad\quad$ Tip = Tip rate · Bill amount

    $\quad \$30 = r \cdot \$120$

    $\quad \$120r = \$30$

    Solving for $r$ give a tip rate of

    $$r = \frac{30}{120} = \frac{1}{4} = 0.25 = 25\% \ .$$

35. Commission rate: 2.5%
    Total sales: $500,000
    Commission

    $\quad$ = Commission rate · Sales amount

    $\quad = 2.5\% \cdot \$500,000 = 0.025 \cdot \$500,000$

    $\quad = \$12,500$

    Ramsey earns a commission of $12,500.

37. Commission: $139.50
    Total sales: $900
    Let $r$ denote the unknown commission rate.
    Then from the commission equation
    $\quad\quad \$139.50 = r \cdot \$900$
    Solving for $r$ gives a commission rate of
    $$r = \frac{139.50}{900} = 0.155 = 15.5\%$$

39. a.  Discount = Discount rate · Original cost

    $\quad\quad = 30\% \cdot \$90 = 0.3 \cdot \$90 = \$27$

    b.  Sale price = Original price − Discount

    $\quad\quad = \$90 - \$27 = \$63$

    The sale price of the racket is $63.

41. The original cost is $350 and the discount
    is $44. Let $r$ denote the unknown discount
    rate. From the discount equation,

    $\quad$ Discount = Discount rate · Original cost

    $\quad\quad \$44 = r \cdot \$350$

    Solve for $r$,  $r = \dfrac{44}{350} = 0.126 = 12.6\%$ .

    The stereo was discounted 12.6%.

43. The jacket has an original price of $500 and a sale price of $300.  The discount is then
    $\$500 - \$300 = \$200$ . If $r$ is the unknown discount rate, the discount equation gives

    $\quad$ Discount = Discount rate · Original cost

    $\quad\quad \$200 = r \cdot \$500$

    Solving for r,  $r = \dfrac{200}{500} = \dfrac{2}{5} = 0.4 = 40\%$ .

    The summer discount rate is 40%.

## Guide Problems

45. A refrigerator weighed 400 pounds empty. After putting in food. the weight increased by 20%. How much
    does the refrigerator weigh after the increase?

    a. Because the weight increased, add the percent change to 100%.

    $\quad\quad \underline{\ \ 20\%\ \ } + 100\% = \underline{\ \ 120\%\ \ }$

b. Substitute the original weight and the percent from part a. into the percent equation. Recall the percent equation

$\quad$ Amount = Percent · Base

Here the Amount is an unknown, $a$, while the percent is 120% and the base is 400 pounds.

$$a = 120\% \cdot 400 \text{ pounds}$$

c. Determine the new weight.

$$a = 120\% \cdot 400 \text{ pounds}$$
$$= 1.20 \cdot 400 \text{ pounds}$$
$$= 480 \text{ pounds}$$

*In exercises 47-59, the table is completed as follows. In each exercise, the percent equation is solved to find either the unknown original amount or the unknown new amount. The original amount corresponds to the base, b, in the percent equation, while the new amount is the amount, a. The percent is* $100\% + $ Percent Change *if an increase is indicated or* $100\% - $ Percent Change *if a decrease is indicated.*

| | Original Amount | Percent Change | Increase or Decrease | New Amount |
|---|---|---|---|---|
| 47. | $290 | 57% | increase | Amount = Percent · Base<br>$a = (100\% + 57\%) \cdot \$290 = 157\% \cdot \$290$<br>$a = 1.57 \cdot \$290 = \$455.30$<br>or $455 rounded |
| 49. | Amount = Percent · Base<br>$4000 = (100\% - 35\%) \cdot b$<br>$4000 = 65\% \cdot b = 0.65b$<br>$b = \dfrac{4000}{0.65} = 6153.846$<br>or 6154 rounded | 35% | decrease | 4000 |
| 51. | $13,882 | 15% | decrease | $a = (100\% - 15\%) \cdot \$13,882$<br>$a = 85\% \cdot \$13,882$<br>$a = 0.85 \cdot \$13,882 = \$11,799.70$<br>or $11,800 rounded |
| 53. | $20,000 | 19% | increase | $a = (100\% + 19\%) \cdot \$20,000$<br>$a = 119\% \cdot \$20,000$<br>$a = 1.19 \cdot \$20,000 = \$23,800$ |
| 55. | 90 | 9% | decrease | $a = (100\% - 9\%) \cdot 90 = 91\% \cdot 90$<br>$a = 0.91 \cdot 90 = 81.9$<br>or 82 rounded. |
| 57. | $100 | 4% | decrease | $a = (100\% - 4\%) \cdot \$100 = 96\% \cdot \$100$<br>$a = 0.96 \cdot \$100 = \$96$ |
| 59. | 16 | 6% | decrease | $a = (100\% - 6\%) \cdot 16 = 94\% \cdot 16$<br>$a = 0.94 \cdot 16 = 15.04$<br>or 15 rounded. |

61. The value of $43,500 is the new price of the motor home and we seek the original or base price, $b$. Further, a decrease of 20% means the new price is $100\% - 20\% = 80\%$ of the original price. Solving $\$43,500 = 80\% \cdot b$

gives $b = \dfrac{\$43,500}{0.8} = \$54,375$.

63. a. The change amount in customers is $10,000 - 8500 = 1500$. The percent change is given by

$\text{Percent change} = \dfrac{\text{Change amount}}{\text{Original amount}} = \dfrac{1500}{10,000} = 0.15 = 15\%$. There was a 15% decrease in customers.

   b. The original amount is the current months total which is 8500 customers. The percent of this month expected next month is $100\% + 10\% = 110\%$ or $a = 110\% \cdot 8500 = 1.1 \cdot 8500 = 9350$ customers.

65. The number of June weddings planned will be $100\% + 40\% = 140\%$ will be $140\% \cdot 5 = 1.4 \cdot 5 = 7$.

67. The problem asks what percent of 1850 is 500. From the percent equation $500 = p \cdot 1850$ and so

$p = \dfrac{500}{1850} = 0.27$ or 27%.

69. A decrease of 19.1% means the sales in 2004 were $100\% - 19.1\% = 80.9\%$ of sales in 2000. From the percent equation, the sales in 2004 were $80.9\% \cdot 948 = 0.809 \cdot 948 = 766.932$ million or about 767 million.

71. a. The increase in length is $18 - 6 = 12$ inches which is a percent change of $\dfrac{12}{6} = 2 = 200\%$.

   b. A change 300% means a total length of $100\% + 300\% = 400\%$ of the original length of 6 inches or $400\% \cdot 6 = 4 \cdot 6 = 24$ inches.

73. The expected increase in the GDP is $\$12 - \$9.4 = \$2.6$ trillion dollars. This represents a percent increase

from 2002 of $\dfrac{2.6}{9.4} = 0.277 = 27.7\%$.

75. The predicted increase in NASA spending is $\$18 - \$14.3 = \$3.7$ billion dollars from 1993 to 2009. This

represents a percent increase from 1993 of $\dfrac{3.7}{14.3} = 0.2587 = 25.87\%$.

77. The percent change was a 17% increase, so the new pressure is $100\% + 17\% = 117\%$ of the old pressure of 27 psi. The new pressure is then $117\% \cdot 27 = 31.59$ or about 32 psi.

79. The change in the amount of water is $2.4 - 1 = 1.4$ gallons. This is a percent change of $\dfrac{1.4}{2.4} = 0.583$ or 58%

to the nearest whole percent.

81. The change in the number of cats showed an increase of $90.5 - 59.1 = 31.4$ million. This represents a

percent increase of $\dfrac{31.4}{59.1} = 0.531 = 53.1\%$ from 1996 to 2006.

83. The change in mobile users from 2003 to 2004 was $154.2 - 147.6 = 6.6$ million users which represents a

percent increase of $\dfrac{6.6}{147.6} = 0.04472 = 4.472\%$ or 4.5% to the nearest tenth of a percent.

85. An 8% increase means the 2003 amount of 147.6 users is $100\% + 8\% = 108\%$ of the usage, $b$, in 2002.

From the percent equation, $147.6 = 108\% \cdot b$ so $b = \dfrac{147.6}{1.08} = 136.7$ million users in 2002.

## Cumulative Skills Review

1.  The percent proportion is $\dfrac{\text{Amount}}{\text{Base}} = \dfrac{\text{Part}}{100}$. In the problem "25 is what percent of 500?", the amount is 25, the base is 500 and the percent, p, is unknown. Substituting these values into the percent proportion gives $\dfrac{25}{500} = \dfrac{p}{100}$.

3.  By the order of operations, evaluate the exponent first, followed by the multiplication and lastly the addition. So $22.8 + 3.95 \times 2.1^3 = 22.8 + 3.95 \times 9.261 = 22.8 + 36.58095 = 59.38095$.

5.  Given the ratios $\dfrac{25}{7}$ and $\dfrac{75}{21}$ compute the cross products $7 \cdot 75 = 525$ and $25 \cdot 21 = 525$. Since the cross products are equal, the ratios are proportional.

7.  Multiply vertically. Since the two numbers being multiplied each have three decimal places, the result must have a total of six decimal places.

$$
\begin{array}{r}
0.895 \\
\times\ 0.322 \\
\hline
1790 \\
1790 \\
2685 \\
\hline
0.288190
\end{array}
$$

9.  $92\% = \dfrac{92}{100} = \dfrac{23 \cdot 4}{25 \cdot 4} = \dfrac{23}{25}$; $\quad 92\% = 92 \cdot 0.01 = 0.92$

## Chapter 5 Numerical Facts Of Life

First summarize Toby Kaluzny's income and his current obligations.

Monthly income: $4650.00     Non-housing monthly obligations: $615.00

If Toby qualifies for the mortgage, his monthly housing expense will be $1230.00.

a.  Toby's housing expense ratio is

$$\text{Housing expense ratio} = \frac{\text{Monthly housing expense}}{\text{Monthly gross income}} = \frac{1230.00}{4650.00} = 0.2645 = 26.5\%$$

to a tenth of a percent.

b.  With the mortgage, Toby's total monthly financial obligations will be $615 + \$1230 = \$1845$. Toby's obligations ratio is

$$\text{Total obligations ratio} = \frac{\text{Total monthly financial obligations}}{\text{Monthly gross income}} = \frac{1845.00}{4650.00} = 0.3967 = 39.7\%$$

to a tenth of a percent.

c.  Toby's housing expense ratio of 26.5% is below the 29% maximum required for an FHA mortgage and below the 28% percent maximum for a conventional mortgage.  However, his 39.7% total obligations ratio exceeds the allowable maximum of 36% for a conventional mortgage and so Toby would be denied a conventional mortgage.  Fortunately, 39.7% does not exceed the maximum total obligations ratio of 41% for an FHA mortgage and so Toby qualifies for an FHA mortgage.

## Chapter 5 Review Exercises

1.  $4.5\% = \dfrac{4.5}{100} = \dfrac{4.5}{100} \cdot \dfrac{10}{10}$
    $= \dfrac{45}{1000} = \dfrac{9}{200}$

2.  $284\% = \dfrac{284}{100} = 2\dfrac{84}{100}$
    $= 2\dfrac{21}{25}$

3.  $8\% = \dfrac{8}{100} = \dfrac{2}{25}$

4.  $75\% = \dfrac{75}{100} = \dfrac{3}{4}$

5.  $32.5\% = \dfrac{32.5}{100} = \dfrac{32.5}{100} \cdot \dfrac{10}{10}$
    $= \dfrac{325}{1000} = \dfrac{13}{40}$

6.  $0.25\% = \dfrac{0.25}{100} = \dfrac{0.25}{100} \cdot \dfrac{100}{100}$
    $= \dfrac{25}{10,000} = \dfrac{1}{400}$

7.  $37.5\% = 37.5 \cdot 0.01 = 0.375$

8.  $56\dfrac{4}{5}\% = 56.8\%$
    $= 56.8 \cdot 0.01$
    $= 0.568$

9.  $95\% = 95 \cdot 0.01 = 0.95$

10.  $40.01\% = 40.01 \cdot 0.01$
     $= 0.4001$

11.  $88\% = 88 \cdot 0.01 = 0.88$

12.  $77\dfrac{1}{2}\% = 77.5\% = 77.5 \cdot 0.01$
     $= 0.775$

13.  $1.65 = 1.65 \cdot 100\% = 165\%$

14.  $0.2 = 0.2 \cdot 100\% = 20\%$

15.  $9.0 = 9.0 \cdot 100\% = 900\%$

16.  $0.45 = 0.45 \cdot 100\% = 45\%$

17.  $0.0028 = 0.0028 \cdot 100\%$
     $= 0.28\%$

18.  $0.31 = 0.31 \cdot 100\% = 31\%$

19.  $2\dfrac{1}{5} = 2.2 = 2.2 \cdot 100\%$
     $= 220\%$

20.  $1\dfrac{1}{2} = 1.5 = 1.5 \cdot 100\%$
     $= 150\%$

21.  $\dfrac{21}{25} = 0.84 = 0.84 \cdot 100\%$
     $= 84\%$

22.  $\dfrac{17}{50} = 0.34 = 0.34 \cdot 100\%$
     $= 34\%$

23.  $\dfrac{7}{8} = 0.875 = 0.875 \cdot 100\%$
     $= 87.5\%$

24.  $3\dfrac{2}{5} = 3.4 = 3.4 \cdot 100\%$
     $= 340\%$

| | FRACTION | DECIMAL | PERCENT |
|---|---|---|---|
| 25. | $60\% = \dfrac{60}{100} = \dfrac{3}{5}$ | $60\% = 60 \cdot 0.01 = 0.60$ | $60\%$ (given) |
| 26. | $1\dfrac{5}{8}$ (given) | $1\dfrac{5}{8} = 1.625$ | $1\dfrac{5}{8} = 1.625 = 1.625 \cdot 100\% = 162.5\%$ |
| 27. | $0.81 = \dfrac{81}{100}$ | $0.81$ (given) | $0.81 = 0.81 \cdot 100\% = 81\%$ |
| 28. | $\dfrac{2}{5}$ (given) | $\dfrac{2}{5} = 0.4$ | $\dfrac{2}{5} = 0.4 = 0.4 \cdot 100\% = 40\%$ |

| | FRACTION | DECIMAL | PERCENT |
|---|---|---|---|
| 29. | $68\% = \dfrac{68}{100} = \dfrac{17}{25}$ | $68\% = 68 \cdot 0.01 = 0.68$ | $68\%$ (given) |
| 30. | $0.14 = \dfrac{14}{100} = \dfrac{7}{50}$ | $0.14$ (given) | $0.14 = 0.14 \cdot 100\% = 14\%$ |
| 31. | $\dfrac{17}{400}$ (given) | $\dfrac{17}{400} = 0.0425$ | $\dfrac{17}{400} = 0.0425 = 0.0425 \cdot 100\% = 4.25\%$ |
| 32. | $79\% = \dfrac{79}{100}$ | $79\% = \dfrac{79}{100} = 0.79$ | $79\%$ (given) |

33. What number is 22% of 1980?

$$\text{Amount} = \text{Percent} \cdot \text{Base}$$
$$a = 0.22 \cdot 1980$$
$$a = 435.6$$

34. 120 is 80% of what number?

$$\text{Base} = \frac{\text{Amount}}{\text{Percent}}$$
$$b = \frac{120}{0.8}$$
$$b = 150$$

35. What percent of 50 is 7.5?

$$\text{Percent} = \frac{\text{Amount}}{\text{Base}}$$
$$p = \frac{7.5}{50}$$
$$= \frac{15}{100} = 15\%$$

36. 245 is 49% of what number?

$$\text{Base} = \frac{\text{Amount}}{\text{Percent}}$$
$$b = \frac{245}{0.49}$$
$$b = 500$$

37. 2200 is what percent of 4400?

$$\text{Percent} = \frac{\text{Amount}}{\text{Base}}$$
$$p = \frac{2200}{4400}$$
$$= 0.5 = 50\%$$

38. 70% of 690 is what number?

$$\text{Amount} = \text{Percent} \cdot \text{Base}$$
$$a = 0.7 \cdot 690$$
$$a = 483$$

39.  392 is 28% of what number?

$$\text{Base} = \frac{\text{Amount}}{\text{Percent}}$$
$$b = \frac{392}{0.28}$$
$$b = 1400$$

40.  What is 60% of 300?

$$\text{Amount} = \text{Percent} \cdot \text{Base}$$
$$a = 0.60 \cdot 300$$
$$a = 180$$

41.  What percent is 64 of 400?

$$\text{Percent} = \frac{\text{Amount}}{\text{Base}}$$
$$p = \frac{64}{400}$$
$$= \frac{16}{100} = 16\%$$

42.  What is 30% of 802?

$$\text{Amount} = \text{Percent} \cdot \text{Base}$$
$$a = 0.3 \cdot 802$$
$$a = 240.6$$

43.  13 is what percent of 65?

$$\text{Percent} = \frac{\text{Amount}}{\text{Base}}$$
$$p = \frac{13}{65}$$
$$= \frac{1}{5} = 20\%$$

44.  64.08 is 72% of what number?

$$\text{Base} = \frac{\text{Amount}}{\text{Percent}}$$
$$b = \frac{64.08}{0.72}$$
$$b = 89$$

45.  What number is 7% of 2000?

$$\frac{\text{Amount}}{\text{Base}} = \frac{\text{Part}}{100}$$
$$\frac{a}{2000} = \frac{7}{100}$$
$$100a = 14,000$$
$$a = 140$$

46.  261 is what percent of 450?

$$\frac{\text{Amount}}{\text{Base}} = \frac{\text{Part}}{100}$$
$$\frac{261}{450} = \frac{p}{100}$$
$$\frac{29}{50} = \frac{p}{100}$$
$$50p = 2900$$
$$p = 58$$
261 is 58% of 450.

47.  30 is 20% of what number?

$$\frac{\text{Amount}}{\text{Base}} = \frac{\text{Part}}{100}$$
$$\frac{30}{b} = \frac{20}{100}$$
$$\frac{30}{b} = \frac{1}{5}$$
$$b = 150$$

48.  50 is what percent of 50?

$$\frac{\text{Amount}}{\text{Base}} = \frac{\text{Part}}{100}$$
$$\frac{50}{50} = \frac{p}{100}$$
$$\frac{1}{1} = \frac{p}{100}$$
$$p = 100$$
50 is 100% of 50.

49. What is 15% of 8000?

$$\frac{\text{Amount}}{\text{Base}} = \frac{\text{Part}}{100}$$

$$\frac{a}{8000} = \frac{15}{100}$$

$$\frac{a}{8000} = \frac{3}{20}$$

$$20a = 24,000$$

$$a = 1200$$

50. 136 is 34% of what number?

$$\frac{\text{Amount}}{\text{Base}} = \frac{\text{Part}}{100}$$

$$\frac{136}{b} = \frac{34}{100}$$

$$\frac{136}{b} = \frac{17}{50}$$

$$17b = 6800$$

$$b = 400$$

51. What is 9% of 540?

$$\frac{\text{Amount}}{\text{Base}} = \frac{\text{Part}}{100}$$

$$\frac{a}{540} = \frac{9}{100}$$

$$100a = 4860$$

$$a = 48.6$$

52. 136 is 40% of what number?

$$\frac{\text{Amount}}{\text{Base}} = \frac{\text{Part}}{100}$$

$$\frac{136}{b} = \frac{40}{100}$$

$$\frac{136}{b} = \frac{2}{5}$$

$$2b = 680$$

$$b = 340$$

53. 516 is what percent of 645?

$$\frac{\text{Amount}}{\text{Base}} = \frac{\text{Part}}{100}$$

$$\frac{516}{645} = \frac{p}{100}$$

$$\frac{4}{5} = \frac{p}{100}$$

$$5p = 400$$

$$p = 80$$

516 is 80% of 645.

54. 210 is 56% of what number?

$$\frac{\text{Amount}}{\text{Base}} = \frac{\text{Part}}{100}$$

$$\frac{210}{b} = \frac{56}{100}$$

$$\frac{210}{b} = \frac{14}{25}$$

$$14b = 5250$$

$$b = 375$$

55. 245 is what percent of 100?

$$\frac{\text{Amount}}{\text{Base}} = \frac{\text{Part}}{100}$$

$$\frac{245}{100} = \frac{p}{100}$$

$$\frac{49}{20} = \frac{p}{100}$$

$$20p = 4900$$

$$p = 245$$

245 is 245% of 100.

56. What number is 70% of 833?

$$\frac{\text{Amount}}{\text{Base}} = \frac{\text{Part}}{100}$$

$$\frac{a}{833} = \frac{70}{100}$$

$$\frac{a}{833} = \frac{7}{10}$$

$$10a = 5831$$

$$a = 583.1$$

57. The change amount is an increase of $\$3,000,000 - \$2,500,000 = \$500,000$. The percent change is given by

$$\text{Percent change} = \frac{\text{Change amount}}{\text{Original amount}} = \frac{500,000}{2,500,000} = \frac{1}{5} = 20\%$$

so a 20% increase in sales is expected.

58. The change amount is a decrease of $\$11,000 - \$8400 = \$2600$. The percent change is given by

$$\text{Percent change} = \frac{\text{Change amount}}{\text{Original amount}} = \frac{2600}{11,000} = 0.236 = 23.6\%$$

so a 23.6% decrease in monthly long distance charges resulted from changing providers.

59. The amount of change in staffing is a decrease of $600 - 550 = 50$. The percent change is given by

$$\text{Percent change} = \frac{\text{Change amount}}{\text{Original amount}} = \frac{50}{600} = 0.0833... = 8\%$$

so there was a 8% decrease in employees.

60. The change in sales is an increase of $120 - 40 = 80$ surfboards. The percent change is given by

$$\text{Percent change} = \frac{\text{Change amount}}{\text{Original amount}} = \frac{80}{40} = 2 = 200\%$$

so there is a 200% increase in sales from the regular season to the summer months.

61. a. The tip is found from the tip equation, $\text{Tip} = \text{Tip rate} \cdot \text{Bill amount} = 0.20 \cdot \$79.50 = \$15.90$.

    b. The total bill is $\$79.50 + \$15.90 = \$95.40$.

62. a. The sales tax on *one* car is found from the sales tax equation
    $$\text{Sales tax} = \text{Sales tax rate} \cdot \text{Item cost} = 0.058 \cdot \$12,500 = \$725.$$
    The total sales tax for three cars is $3 \cdot \$725 = \$2175$.

    b. The total purchase price is the total item cost, $3 \cdot \$12,500 = \$37,500$, plus the total sales tax or
    $$\$37,500 + \$2175 = \$39,675.$$

63. Clay's commission is given by the commission equation
    $$\text{Commission} = \text{Commission rate} \cdot \text{Sales amount} = 0.075 \cdot \$1,500,000 = \$112,500.$$

64. a. The discount amount is $\text{Discount rate} \cdot \text{Original cost} = 0.10 \cdot \$16 = \$1.60$.

    b. The sale price is $\$16 - \$1.60 = \$14.40$.

65. The number of jobs increased by 15% so the new amount is 115% of last month's figure. Last month's figure forms the base of the percent equation and is found to be

    $$\text{Base} = \frac{\text{Amount}}{\text{Percent}} = \frac{144,900}{1.15} = 126,000 \text{ employees.}$$

66. Since car rental rates have decreased 16%, the current rate is $100\% - 16\% = 84\%$ of the original or base rate. The original rate can be found using the percent equation

$$\text{Base} = \frac{\text{Amount}}{\text{Percent}} = \frac{42}{0.84} = 50 \text{ or } \$50 \text{ per day.}$$

67. The new mortgage payment is 22% less than the rental amount of $900. Equivalently, the mortgage payment is $100\% - 22\% = 78\%$ *of* the rental amount. The mortgage payment is $\$900 \cdot 0.78 = \$702$.

68. Tracy's weekly gasoline cost has decreased by $36 - 26 = 10$ dollars. The percent change is $\frac{10}{36} = 0.28 = 28\%$. Her weekly cost has decreased 28%.

69. From 2001 to 2002, snack food sales grew by $\$22.5 - \$21.8 = \$0.7$ billion. This represents a percent increase of
$$\frac{0.7}{21.8} = 0.032 = 3.2\%.$$

70. If sales in 2007 represent a 10% increase from 2002, then 2007 sales are 110% of 2002 sales or $1.1 \cdot 22.5 = 24.75$ billion. To the nearest tenth of a billion, 2007 sales were $24.8 billion.

71. Nut sales make up 8.4% of the snack food market so the question is what number is 8.4% of $20.7 billion, the total revenue in 2000. Since $0.084 \cdot 20.7 = 1.7388$ billion, the revenue from nut sales is $1,738,800,000.

72. Tortilla chip sales make up 19.9% of the snack food market so the question is what number is 19.9% of $21.8 billion, the total revenue in 2001. Since $0.199 \cdot 21.8 = 4.3382$ billion, the revenue from tortilla chips in 2001 was $4,338,200,000.

## Chapter 5 Assessment Test

1. $76\% = \frac{76}{100} = \frac{19}{25}$

2. $3\% = \frac{3}{100}$

3. $13.5\% = 13.5 \cdot 0.01 = 0.135$

4. $68.8\% = 68.8 \cdot 0.01 = 0.688$

5. $0.57 = 0.57 \cdot 100\% = 57\%$

6. $6.45 = 6.45 \cdot 100\% = 645\%$

7. $\frac{11}{8} = 1\frac{3}{8} = 1.375 = 1.375 \cdot 100\% = 137.5\%$

8. $10\frac{1}{2} = 10.5 = 10.5 \cdot 100\% = 1050\%$

9. What percent of 610 is 106.75?
$$\text{Percent} = \frac{\text{Amount}}{\text{Base}}$$
$$p = \frac{106.75}{610} = \frac{10,675}{61,000}$$
$$= \frac{7}{40} = 0.175 = 17.5\%$$

10. 186 is 62% of what number?
$$\text{Base} = \frac{\text{Amount}}{\text{Percent}}$$
$$b = \frac{186}{0.62}$$
$$b = 300$$

11. What is 47% of 450?

$$\text{Amount} = \text{Percent} \cdot \text{Base}$$
$$a = 0.47 \cdot 450$$
$$a = 211.5$$

12. 367 is 20% of what number?

$$\text{Base} = \frac{\text{Amount}}{\text{Percent}}$$
$$b = \frac{367}{0.2}$$
$$b = 1835$$

13. What number is 80% of 4560?

$$\text{Amount} = \text{Percent} \cdot \text{Base}$$
$$a = 0.80 \cdot 4560$$
$$a = 3648$$

14. 77 is what percent of 280?

$$\text{Percent} = \frac{\text{Amount}}{\text{Base}}$$
$$p = \frac{77}{280} = \frac{11}{40}$$
$$= 0.275 = 27.5\%$$

15. 560 is 14% of what number?

$$\frac{\text{Amount}}{\text{Base}} = \frac{\text{Part}}{100}$$
$$\frac{560}{b} = \frac{14}{100}$$
$$\frac{560}{b} = \frac{7}{50}$$
$$7b = 28,000$$
$$b = 4000$$

16. 95 is what percent of 380?

$$\frac{\text{Amount}}{\text{Base}} = \frac{\text{Part}}{100}$$
$$\frac{95}{380} = \frac{p}{100}$$
$$\frac{1}{4} = \frac{p}{100}$$
$$4p = 100$$
$$p = 25$$

95 is 25% of 380.

17. What is 83% of 180?

$$\frac{\text{Amount}}{\text{Base}} = \frac{\text{Part}}{100}$$
$$\frac{a}{180} = \frac{83}{100}$$
$$100a = 14,940$$
$$a = 149.4$$

18. What percent of 490 is 196?

$$\frac{\text{Amount}}{\text{Base}} = \frac{\text{Part}}{100}$$
$$\frac{196}{490} = \frac{p}{100}$$
$$\frac{2}{5} = \frac{p}{100}$$
$$5p = 200$$
$$p = 40$$

196 is 40% of 490.

19. What number is 29% of 158?                      20. 245 is 35% of what number?

$$\frac{\text{Amount}}{\text{Base}} = \frac{\text{Part}}{100}$$                      $$\frac{\text{Amount}}{\text{Base}} = \frac{\text{Part}}{100}$$

$$\frac{a}{158} = \frac{29}{100}$$                      $$\frac{245}{b} = \frac{35}{100}$$

$$100a = 4582$$                      $$\frac{245}{b} = \frac{7}{20}$$

$$a = 45.82$$                      $$7b = 4900$$

$$b = 700$$

21. The change in the number of small pizzas is $5600 - 3500 = 2100$ and represents a decrease. The percent change is $\frac{2100}{5600} = \frac{3}{8} = 0.375 = 37.5\%$. Sales of small pizzas have decreased $37.5\%$.

22. Gilbert has increased his rates by $\$575 - \$500 = \$75$. The percent change is $\frac{75}{500} = \frac{3}{20} = 0.15 = 15\%$. Gilbert's rate has increased by $15\%$.

23. The tax payment has decreased by $\$16,800 - \$13,600 = \$3200$. The percent change is $\frac{3200}{16,800} = \frac{4}{21} = 0.19 = 19\%$. Their taxes decreased by $19\%$.

24. The change in rent is $\$1500 - \$800 = \$700$. The percent change is $\frac{700}{800} = \frac{7}{8} = 0.875$. Andrew's rent has increased approximately $88\%$.

25. From the sales tax formula, the tax on $20 is $\text{Sales tax rate} \cdot \text{Item cost} = 0.07 \cdot \$20 = \$1.40$. Adding the tax to the cost of the item gives a total purchase price of $\$20 + \$1.40 = \$21.40$.

26. Since the lunch cost $52.50 and the sales tax is $3.10, the sales tax equation, with $p$ representing the unknown tax rate, gives $52.50p = 3.10$. Thus $p = \frac{3.10}{52.50} = 0.059 = 5.9\%$. The sales tax rate in that area is $5.9\%$.

27. Since Sylvan lost 8% of his original weight, his new weight represents 92% of his original weight of 190 pounds. His new weight is given by the formula $\text{Amount} = \text{Percent} \cdot \text{Base} = 0.92 \cdot 190 = 175$ pounds.

28. A 30% increase means the number of packages today is 130% of the number yesterday. The number of packages delivered today is $54 \cdot 130\% = 54 \cdot 1.3 = 70.2$ or 70 packages.

29. If the pump lost 40% of its pressure, the new pressure, 168 pounds per square inch, is 60% of the original pressure. The original pressure is given by the formula $\text{Base} = \frac{\text{Amount}}{\text{Percent}} = \frac{168}{0.6} = 280$ pounds per square inch.

30. A 22% increase means the current temperature, 90°F, is 122% of what it was. The original temperature is given by the formula $\text{Base} = \frac{\text{Amount}}{\text{Percent}} = \frac{90}{1.22} = 73.77$ degrees or, to the nearest degree, 74°F.

# Chapter 6   Measurement

## Section 6.1    The U.S. Customary System

### Concept Check

1.  A number together with a unit assigned to something to represent its size or magnitude is known as a __measure__ .

3.  Name the basic measurement units for length for the U.S. Customary System. __inch, foot, yard, mile__

5.  The __numerator__ of a unit ratio will contain the new units and the __denominator__ will contain the original units.

7.  __Weight__ is a measure of an object's heaviness.

### Guide Problems

9.  Convert 2640 yards to miles.

    a. Write an appropriate unit ratio.
       We know $1\,\text{mi} = 1760\,\text{yd}$ . The
       unit ratio must have the new unit
       (miles) in the numerator and the
       original unit (yards) in the
       denominator. Thus

       $$\text{Unit ratio} = \frac{\text{New units}}{\text{Original units}}$$

       $$= \frac{1\,\text{mi}}{1760\,\text{yd}}$$

    b. Multiply the original measure by
       the unit fraction.

       $$2640\,\text{yd} \cdot \frac{1\,\text{mi}}{1760\,\text{yd}}$$

       $$= 2640\ \cancel{\text{yd}} \cdot \frac{1\,\text{mi}}{1760\ \cancel{\text{yd}}}$$

       $$= \frac{2640\,\text{mi}}{1760} = \underline{1.5\,\text{mi}}$$

*In Problems 11 through 21, the same procedure is followed to perform the indicated conversion. First find the appropriate unit ratio. The unit ratio will have the desired new unit in the numerator and the original or given unit in the denominator. Convert the given measure by multiplying it by the unit ratio. The units in the denominator will always cancel with those in the original measure.*

11. 25 yards to feet

    Unit ratio: $\dfrac{\text{New unit}}{\text{Old unit}} = \dfrac{3\,\text{feet}}{1\,\text{yard}}$

    Convert:

    $25\,\text{yards} \cdot \dfrac{3\,\text{feet}}{1\,\text{yard}}$

    $= 25\ \cancel{\text{yards}} \cdot \dfrac{3\,\text{feet}}{1\ \cancel{\text{yard}}}$

    $= 25 \cdot 3\,\text{feet} = 75\,\text{feet}$

13. 6000 feet to yards

    Unit ratio: $\dfrac{\text{New unit}}{\text{Old unit}} = \dfrac{1\,\text{yard}}{3\,\text{feet}}$

    Convert:

    $6000\,\text{feet} \cdot \dfrac{1\,\text{yard}}{3\,\text{feet}}$

    $= 6000\ \cancel{\text{feet}} \cdot \dfrac{1\,\text{yard}}{3\ \cancel{\text{feet}}}$

    $= \dfrac{6000}{3}\,\text{yards}$

    $= 2000\,\text{yards}$

15. 5 yards to inches

    Unit ratio: $\dfrac{36\,\text{inches}}{1\,\text{yard}}$

    Convert:

    $5\,\text{yards} \cdot \dfrac{36\,\text{inches}}{1\,\text{yard}}$

    $= 5 \cdot 36\,\text{inches}$

    $= 180\,\text{inches}$

17. 13,200 feet to miles

Unit ratio: $\dfrac{1 \text{ mile}}{5280 \text{ feet}}$

Convert:

$13,200 \text{ feet} \cdot \dfrac{1 \text{ mile}}{5280 \text{ feet}}$

$= \dfrac{13,200}{5280} \text{ miles} = 2.5 \text{ miles}$

19. 360 inches to yards

Unit ratio: $\dfrac{1 \text{ yard}}{36 \text{ inches}}$

Convert:

$360 \text{ inches} \cdot \dfrac{1 \text{ yard}}{36 \text{ inches}}$

$= \dfrac{360}{36} \text{ yards}$

$= 10 \text{ yards}$

21. $\dfrac{2}{3}$ yards to inches

Unit ratio: $\dfrac{36 \text{ inches}}{1 \text{ yard}}$

Convert:

$\dfrac{2}{3} \text{ yards} \cdot \dfrac{36 \text{ inches}}{1 \text{ yard}}$

$= \dfrac{2 \cdot 36}{3} \text{ inches}$

$= 24 \text{ inches}$

## Guide Problems

23. Convert 152 pounds to ounces.

a. Write an appropriate unit ratio. We know $1 \text{ lb} = 16 \text{ oz}$. The unit ratio must have the new unit (oz) in the numerator and the original unit (lb) in the denominator. Thus

Unit ratio $= \dfrac{\text{New units}}{\text{Original units}}$

$= \dfrac{16 \text{ oz}}{1 \text{ lb}}$

b. Multiply the original measure by the unit fraction.

$152 \text{ lb} \cdot \dfrac{16 \text{ oz}}{1 \text{ lb}}$

$= 152 \text{ lb} \cdot \dfrac{16 \text{ oz}}{1 \text{ lb}}$

$= 152 \cdot 16 \text{ oz} = \underline{2432 \text{ oz}}$

25. 12 pounds to ounces

Unit ratio: $\dfrac{\text{New unit}}{\text{Old unit}} = \dfrac{16 \text{ oz}}{1 \text{ lb}}$

Convert:

$12 \text{ lb} \cdot \dfrac{16 \text{ oz}}{1 \text{ lb}}$

$= 12 \text{ lb} \cdot \dfrac{16 \text{ oz}}{1 \text{ lb}}$

$= 12 \cdot 16 \text{ oz} = 192 \text{ ounces}$

27. 20 pounds to ounces

Unit ratio: $\dfrac{\text{New unit}}{\text{Old unit}} = \dfrac{16 \text{ oz}}{1 \text{ lb}}$

Convert:

$20 \text{ lb} \cdot \dfrac{16 \text{ oz}}{1 \text{ lb}}$

$= 20 \text{ lb} \cdot \dfrac{16 \text{ oz}}{1 \text{ lb}}$

$= 20 \cdot 16 \text{ oz} = 320 \text{ ounces}$

29. 48 pounds to ounces

Unit ratio: $\dfrac{16 \text{ oz}}{1 \text{ lb}}$

Convert:

$48 \text{ lb} \cdot \dfrac{16 \text{ oz}}{1 \text{ lb}}$

$= 48 \cdot 16 \text{ oz}$

$= 768 \text{ ounces}$

31. 500 ounces to pounds

Unit ratio: $\dfrac{1 \text{ lb}}{16 \text{ oz}}$

Convert:

$500 \text{ oz} \cdot \dfrac{1 \text{ lb}}{16 \text{ oz}}$

$= \dfrac{500}{16} \text{ lb} = 31.25 \text{ pounds}$

33. 55 tons to pounds

Unit ratio: $\dfrac{2000 \text{ lb}}{1 \text{ t}}$

Convert:

$55 \text{ t} \cdot \dfrac{2000 \text{ lb}}{1 \text{ t}}$

$= 55 \cdot 2000 \text{ lb}$

$= 110,000 \text{ pounds}$

35. 1000 pounds to tons

Unit ratio: $\dfrac{1 \text{ t}}{2000 \text{ lb}}$

Convert:

$1000 \text{ lb} \cdot \dfrac{1 \text{ t}}{2000 \text{ lb}}$

$= \dfrac{1000}{2000} \text{ t} = 0.5 \text{ tons}$

## Guide Problems

37. Convert 12 cups to fluid ounces.

   a. Write an appropriate unit ratio. We know $1\,c = 8\,fl\,oz$. The unit ratio must have the new unit (fl oz) in the numerator and the original unit (c) in the denominator. Thus

   $$\text{Unit ratio} = \frac{\text{New units}}{\text{Original units}}$$

   $$= \frac{8\,fl\,oz}{1\,c}$$

   b. Multiply the original measure by the unit fraction.

   $$12\,c \cdot \frac{8\,fl\,oz}{1\,c}$$

   $$= 12\,\cancel{c} \cdot \frac{8\,fl\,oz}{1\,\cancel{c}}$$

   $$= 12 \cdot 8\,fl\,oz = \underline{96\,fl\,oz}$$

---

39. 2 gallons to quarts

   Unit ratio: $\dfrac{\text{New unit}}{\text{Old unit}} = \dfrac{4\,qt}{1\,gal}$

   Convert:

   $$2\,gal \cdot \frac{4\,qt}{1\,gal}$$

   $$= 2\,\cancel{gal} \cdot \frac{4\,qt}{1\,\cancel{gal}}$$

   $$= 2 \cdot 4\,qt = 8\,\text{quarts}$$

41. $3\frac{1}{2}$ quarts to pints

   Unit ratio: $\dfrac{\text{New unit}}{\text{Old unit}} = \dfrac{2\,pt}{1\,qt}$

   Convert:

   $$3\frac{1}{2}\,qt \cdot \frac{2\,pt}{1\,qt}$$

   $$= 3\frac{1}{2}\,\cancel{qt} \cdot \frac{2\,pt}{1\,\cancel{qt}}$$

   $$= \frac{7}{2} \cdot 2\,pt$$

   $$= 7\,\text{pints}$$

43. $\dfrac{1}{4}$ cup to pints

   Unit ratio: $\dfrac{1\,pt}{2\,c}$

   Convert:

   $$\frac{1}{4}\,c \cdot \frac{1\,pt}{2\,c}$$

   $$= \frac{1}{4} \cdot \frac{1}{2}\,pt$$

   $$= \frac{1}{8}\,\text{pint}$$

---

45. 42 cups to fluid ounces

   Unit ratio: $\dfrac{8\,fl\,oz}{1\,c}$

   Convert:

   $$42\,c \cdot \frac{8\,fl\,oz}{1\,c}$$

   $$= 42 \cdot 8\,fl\,oz$$

   $$= 336\,\text{fluid ounces}$$

47. 2 cups to fluid ounces

   Unit ratio: $\dfrac{8\,fl\,oz}{1\,c}$

   Convert:

   $$2\,c \cdot \frac{8\,fl\,oz}{1\,c}$$

   $$= 2 \cdot 8\,fl\,oz$$

   $$= 16\,\text{fluid ounces}$$

49. 10 quarts to fluid ounces
   We need to multiply by several unit fractions to convert.

   $$10\,qt \cdot \frac{8\,fl\,oz}{1\,c} \cdot \frac{2\,c}{1\,pt} \cdot \frac{2\,pt}{1\,qt}$$

   $$= 10\,qt \cdot \frac{8\,fl\,oz}{1\,\cancel{c}} \cdot \frac{2\,\cancel{c}}{1\,\cancel{pt}} \cdot \frac{2\,\cancel{pt}}{1\,qt}$$

   $$= 10\,qt \cdot \frac{32\,fl\,oz}{1\,qt}$$

   $$= 10 \cdot 32\,fl\,oz$$

   $$= 320\,\text{fluid ounces}$$

---

51. To convert 3 miles to feet, multiply by the unit ratio $\dfrac{5280\,ft}{1\,mi}$ to find

   $$3\,mi \cdot \frac{5280\,ft}{1\,mi} = 3 \cdot 5280\,ft = 15{,}840\,ft.$$ The fence is 15,840 feet long.

53. At a speed of 55 miles per hour, Shari and Chris will travel 55 miles in one hour. Since one mile is equivalent to 1760 yards, multiply by the appropriate unit ratio to find they will travel

$$55 \text{ mi} \cdot \frac{1760 \text{ yd}}{1 \text{ mi}} = 55 \cdot 1760 \text{ yd} = 96,800 \text{ yards each hour.}$$

55. a. Convert 2.5 pounds to ounces using the unit ratio $\frac{16 \text{ oz}}{1 \text{ lb}}$. The dog gained

$$2.5 \text{ lb} \cdot \frac{16 \text{ oz}}{1 \text{ lb}} = 16 \cdot 2.5 \text{ oz} = 40 \text{ ounces.}$$

  b. The dog weighed 48 ounces a week ago. From part a. the dog has gained 40 ounces for a total weight of $48 + 40 = 88$ ounces. Converting from ounces to pounds, the dog's current weight is

$$88 \text{ oz} \cdot \frac{1 \text{ lb}}{16 \text{ oz}} = \frac{88}{16} \text{ lb} = 5.5 \text{ pounds.}$$

57. A 15-pound dumbbell weighs $15 \text{ lb} \cdot \frac{16 \text{ oz}}{1 \text{ lb}} = 15 \cdot 16 \text{ oz} = 240 \text{ ounces.}$

59. In ounces, 1.5 pounds is equivalent to $1.5 \text{ lb} \cdot \frac{16 \text{ oz}}{1 \text{ lb}} = 1.5 \cdot 16 \text{ oz} = 24 \text{ ounces.}$

61. Using the appropriate unit ratio, 2 tablespoons is equivalent to $2 \text{ tbs} \cdot \frac{3 \text{ tsp}}{1 \text{ tbs}} = 2 \cdot 3 \text{ tsp} = 6 \text{ tsp or 6 teaspoons.}$

63. Multiply by a series of unit ratios to convert 15 gallons to pints:

$$15 \text{ gal} \cdot \frac{4 \text{ qt}}{1 \text{ gal}} \cdot \frac{2 \text{ pt}}{1 \text{ qt}} = 15 \text{ gal} \cdot \frac{4 \text{ qt}}{1 \text{ gal}} \cdot \frac{2 \text{ pt}}{1 \text{ qt}} = 15 \cdot 4 \cdot 2 \text{ gal} = 120 \text{ pt. So, 15 gallons is the same as 120 pints.}$$

## Cumulative Skills Review

1. To multiply by 1000, move the decimal point three places to the right.
$$3.264 \cdot 1000 = 3264$$

3. In fraction form, the ratio is
$$\frac{6 \text{ trainers}}{15 \text{ dogs}} = \frac{2 \cdot 3 \text{ trainers}}{5 \cdot 3 \text{ dogs}} = \frac{2 \text{ trainers}}{5 \text{ dogs}}$$
so the simplified ratio is
2 trainers for every 5 dogs.

5.
$$\begin{array}{r} 2\,5\,2\,.4\,8 \\ -\ \ 9\,9\,.9\,0 \\ \hline 1\,5\,2\,.5\,8 \end{array}$$

7. The word "of" implies multiplication so 45% of 3000 is
$$45\% \cdot 3000 = 0.45 \cdot 3000 = 1350.$$

9. 44 to 11 in fraction form is $\frac{44}{11}$.

Simplifying gives
$$\frac{44}{11} = \frac{\overset{4}{\cancel{44}}}{\underset{1}{\cancel{11}}} = \frac{4}{1}.$$

## Section 6.2    Denominate Numbers

## Concept Check

1.  A number together with a unit of measure is called a  <u>denominate</u>  number.

3.  Two or more denominate numbers that are combined are called  <u>compound</u>  denominate numbers.

## Guide Problems

5.  Express 26 feet in terms of yards and feet.

    a. How many feet in a yard?  3
        There are three feet in a yard.

    b. Divide 26 by the answer to part a.

$$\begin{array}{r} 8 \ R \ 2 \\ 3\overline{)26} \\ \underline{24} \\ 2 \end{array}$$

    c. 26 feet = <u>8</u> yards <u>2</u> feet

        The quotient, 8, is the number of yards.
        The remainder 2 is the number of feet
        left over.

7.  There are 3 feet in a yard. Since 16 divided by 3 is 5 R 1, 16 feet = 5 yards 1 foot.

9.  There are 4 quarts in a gallon. Since 13 divided by 4 is 3 R 1, 13 quarts = 3 gallons 1 quart.

11. There are 2000 pounds in a ton. Since
$5800 \div 2000 = 2 \ R \ 1800$,
5800 pounds
      = 2 tons 1800 pounds.

## Guide Problems

13. Simplify 3 pounds 20 ounces.

    a. How many ounces are in a pound?  16

    b. Since 20 ounces is greater than 16 ounces, express
        20 ounces as pounds and ounces.
        20 ounces = 16 ounces + 4 ounces

               = <u>1</u> pounds + <u>4</u> ounces

      Then,
        3 pounds 20 ounces

           = 3 pounds + <u>1</u> pounds + <u>4</u> ounces

           = <u>4</u> pounds + <u>4</u> ounces

15. 3 ft = 1 yd

    12 yards 4 feet
      = 12 yards + 4 feet
      = 12 yards + 3 feet + 1 foot
      = 12 yards + 1 yard + 1 foot
      = 13 yards + 1 foot
      = 13 yards 1 foot

17. 1 lb = 16 oz

    8 pounds 43 ounces
      = 8 pounds + 43 ounces
      = 8 pounds + 32 ounces + 11 ounces
      = 8 pounds + 2 pounds + 11 ounces
      = 10 pounds + 11 ounces
      = 10 pounds 11 ounces

19. 2 pt = 1 qt

    5 quarts 5 pints
      = 5 quarts + 5 pints
      = 5 quarts + 4 pints + 1 pint
      = 5 quarts + 2 quarts + 1 pint
      = 7 quarts + 1 pint
      = 7 quarts 1 pint

## Guide Problems

21. Add.
Add along each column. Then simplify the result if necessary as in exercises 13 through 19.

$$\begin{array}{r} 9 \text{ yd } 2 \text{ ft} \\ + \ 7 \text{ yd } 2 \text{ ft} \\ \hline 16 \text{ yd } 4 \text{ ft} \end{array}$$

$= \underline{16} \text{ yd} + \underline{4} \text{ ft} = \underline{16} \text{ yd} + \underline{3} \text{ ft} + \underline{1} \text{ ft}$

$= \underline{16} \text{ yd} + \underline{1} \text{ yd} + \underline{1} \text{ ft} = \underline{17} \text{ yd} + \underline{1} \text{ ft}$

23.
$$\begin{array}{r} 4 \text{ ft } 6 \text{ in.} \\ + \ 3 \text{ ft } 2 \text{ in.} \\ \hline 7 \text{ ft } 8 \text{ in.} \end{array}$$

25.
$$\begin{array}{r} 3 \text{ c } 3 \text{ fl oz} \\ + \ 9 \text{ c } 6 \text{ fl oz} \\ \hline 12 \text{ c } 9 \text{ fl oz} \end{array}$$

$= 12 \text{ c} + 8 \text{ fl oz} + 1 \text{ fl oz}$

$= 12 \text{ c} + 1 \text{ c} + 1 \text{ fl oz}$

$= 13 \text{ c } 1 \text{ fl oz}$

27.
$$\begin{array}{r} 6 \text{ ft } 10 \text{ in.} \\ 2 \text{ ft } 11 \text{ in.} \\ + \ 9 \text{ ft } 7 \text{ in.} \\ \hline 17 \text{ ft } 28 \text{ in.} \end{array}$$

$= 17 \text{ ft} + 24 \text{ in.} + 4 \text{ in.}$

$= 17 \text{ ft} + 2 \text{ ft} + 4 \text{ in.}$

$= 19 \text{ ft } 4 \text{ in.}$

29.
$$\begin{array}{r} 8 \text{ gal } 3 \text{ qt} \\ - \ 3 \text{ gal } 2 \text{ qt} \\ \hline 5 \text{ gal } 1 \text{ qt} \end{array}$$

31.
$$\begin{array}{r} 8 \text{ c} \\ - \ 2 \text{ c } 3 \text{ fl oz} \end{array}$$

$$\begin{array}{r} \overset{7 \text{ c}}{\cancel{8 \text{ c}}} \quad 8 \text{ fl oz} \\ - \quad 2 \text{ c} \quad 3 \text{ fl oz} \\ \hline 5 \text{ c} \quad 5 \text{ fl oz} \end{array}$$

33.
$$\begin{array}{r} 7 \text{ lb } 1 \text{ oz} \\ - \ 2 \text{ lb } 3 \text{ oz} \end{array}$$

$$\begin{array}{r} \overset{6 \text{ lb}}{\cancel{7 \text{ lb}}} \quad \overset{17 \text{ oz}}{\cancel{1 \text{ oz}}} \\ - \quad 2 \text{ lb} \quad 3 \text{ oz} \\ \hline 4 \text{ lb } 14 \text{ oz} \end{array}$$

## Guide Problems

35. Multiply 8 ft 4 in. by 5.
Multiply each part by 5. Then simplify the result if necessary as in exercises 13 through 19.

$$\begin{array}{r} 8 \text{ ft } 4 \text{ in.} \\ \times \quad\quad 5 \\ \hline 40 \text{ ft } 20 \text{ in.} \end{array}$$

$= \underline{40} \text{ ft} + \underline{20} \text{ in.} = \underline{40} \text{ ft} + \underline{12} \text{ in.} + \underline{8} \text{ in.}$

$= \underline{40} \text{ ft} + \underline{1} \text{ ft} + \underline{8} \text{ in.} = \underline{41} \text{ ft } \underline{8} \text{ in.}$

37.
$$\begin{array}{r} 2 \text{ c } 7 \text{ fl oz} \\ \times \quad\quad 4 \\ \hline 8 \text{ c } 28 \text{ fl oz} \end{array}$$

$= 8 \text{ c} + 24 \text{ fl oz} + 4 \text{ fl oz}$

$= 8 \text{ c} + 3 \text{ c} + 4 \text{ fl oz}$

$= 11 \text{ c } 4 \text{ fl oz}$

39.
$$\begin{array}{r} 9 \text{ lb } 4 \text{ oz} \\ \times \quad\quad 5 \\ \hline 45 \text{ lb } 20 \text{ oz} \end{array}$$

$= 45 \text{ lb} + 16 \text{ oz} + 4 \text{ oz}$

$= 45 \text{ lb} + 1 \text{ lb} + 4 \text{ oz}$

$= 46 \text{ lb } 4 \text{ oz}$

41.
$$\begin{array}{r} 4 \text{ yd } 2 \text{ ft} \\ \times \quad\quad 7 \\ \hline 28 \text{ yd } 14 \text{ ft} \end{array}$$

$= 28 \text{ yd} + 12 \text{ ft} + 2 \text{ ft}$

$= 28 \text{ yd} + 4 \text{ yd} + 2 \text{ ft}$

$= 32 \text{ yd } 2 \text{ ft}$

43.  $\dfrac{27\ \text{ft}\ \ 3\ \text{in.}}{3)81\ \text{ft}\ 9\ \text{in.}}$

$\phantom{3)}\underline{81\ \text{ft}}$

$\phantom{3)81\ \text{ft}}9\ \text{in.}$

45.  $\dfrac{13\ \text{ft}\ \ 4\ \text{in.}}{2)26\ \text{ft}\ 8\ \text{in.}}$

$\phantom{2)}\underline{26\ \text{ft}}$

$\phantom{2)26\ \text{ft}}8\ \text{in.}$

47.  $\dfrac{2\ \text{lb}\ \ 5\ \text{oz}}{4)9\ \text{lb}\ 4\ \text{oz}}$

$\phantom{4)}\underline{8\ \text{lb}}$

$\phantom{4)}1\ \text{lb}\ \ 4\ \text{oz}\ \ \ (= 20\ \text{oz})$

49. Since there are 3 feet in one yard, divide 80 by 3.

$\phantom{3)}26\ \ R\ \ 2$

$3)\overline{80}$

$\phantom{3)}\underline{78}$

$\phantom{3)}2$

so Jasmine will need 80 feet = 26 yards 2 feet of edging.

51. There are 16 ounces in one pound, so divide Michael's weight in ounces, 2300, by 16.

$\phantom{16)2}143\ \ R\ \ 12$

$16)\overline{2300}$

$\phantom{16)}\underline{2288}$

$\phantom{16)2}12$

so Michael weighs 143 pounds 12 ounces.

53. a. Aaron administers a total number of $38\ \text{tsp} + 24\ \text{tsp} = 62$ per day.

 b. There are 3 teaspoons in a tablespoon. Since $62 \div 3 = 20\ R\ 2$, 62 teaspoons is equivalent to 20 tablespoons 2 teaspoons.

55. a. $\phantom{+\ }6\ \text{yd}\ \ 5\ \text{in.}$

$\underline{+\ \ 3\ \text{yd}\ \ 4\ \text{in.}}$

$\phantom{+\ }9\ \text{yd}\ \ 9\ \text{in.}$

The skirt and sleeve require 9 yards 9 inches.

 b. $\overset{14\ \text{yd}}{\cancel{15\ \text{yd}}}\ \overset{39\ \text{in.}}{\cancel{3\ \text{in.}}}$

$\underline{-\ \ 9\ \text{yd}\ \ \ 9\ \text{in.}}$

$\phantom{-\ \ }5\ \text{yd}\ \ 30\ \text{in.}$

There will be 5 yards 30 inches of fabric remaining.

57. To make 12 servings, the recipe must be doubled:

$\phantom{\times\ }2\ \text{c}\ \ \ 6\ \text{fl oz}$

$\underline{\times\ \ \ \ \ \ \ \ \ \ \ 2}$

$4\ \text{c}\ \ 12\ \text{fl oz}$

$= 4\ \text{c} + 8\ \text{fl oz} + 4\ \text{fl oz}$

$= 4\ \text{c} + 1\ \text{c} + 4\ \text{fl oz}$

$= 5\ \text{c}\ \ 4\ \text{fl oz}$

The sauce will require 5 cups 4 fluid ounces of heavy cream.

59. a. The total weight of the cookies is $28 \cdot 1.5\ \text{oz} = 42\ \text{oz}$. Since $42 \div 16 = 2\ R\ 10$, the cookies weigh 2 pounds 10 ounces.

 b. Adding in the weight of the tin gives $42\ \text{oz} + 7\ \text{oz} = 49\ \text{oz}$. Since $49 \div 16 = 3\ R\ 1$, the tin with cookies weighs 3 pounds 1 ounce.

61. The amount of chocolates per box is computed as

$\phantom{20)5}2\ \text{lb}\ \ \ 15\ \text{oz}$

$20)\overline{58\ \text{lb}\ \ \ 12\ \text{oz}}$

$\phantom{20)}\underline{40\ \text{lb}}$

$\phantom{20)}18\ \text{lb}\ \ \ 12\ \text{oz}\ \ \ (= 300\ \text{oz})$

or 2 pounds 15 ounces.

## Cumulative Skills Review

1.  $60\ \text{in.} \cdot \dfrac{1\ \text{ft}}{12\ \text{in.}} = \dfrac{60}{12}\ \text{ft} = 5\ \text{ft}$

$60\ \text{inches} = 5\ \text{feet}$

3. In 5495.89478 the digit in the ten thousandths place is 7. The digit to the right is 8 which is greater than 5 so round the 7 up to get 5495.8948.

5.  $42\% = \dfrac{42}{100} = \dfrac{21}{50}$

7.  Write the fractions over a common denominator

$$\frac{8}{13} = \frac{8}{13} \cdot \frac{31}{31} = \frac{248}{403}$$

$$\frac{9}{31} = \frac{9}{31} \cdot \frac{13}{13} = \frac{117}{403}$$

and compare the numerators. The first numerator is larger so

$$\frac{8}{13} > \frac{9}{31}.$$

9.  The total of Paula's checks is
$\$45.50 + \$22.35 + \$16.75 = \$84.60$
so the balance remaining in the account is
$\$3200.67 - \$84.60 = \$3116.07.$

## Section 6.3    The Metric System

## Concept Check

1.  The basic unit of length in the metric system is the  _meter_ .

3.  The basic unit of capacity in the metric system is the  _liter_ .

5.  Deka-, hecto-, and kilo- represent multiples  _larger_  than the basic unit.

7.  A  _gram_  is the amount of mass of water contained in a cube whose sides measure 1 centimeter each.

## Guide Problems

9.  Convert 8.9 meters to kilometers.

a. Write an appropriate unit ratio.
Since there are 1000 meters in a kilometer and kilometers are the new unit, the unit ratio is

$$\text{Unit ratio} = \frac{\text{New units}}{\text{Original units}}$$

$$= \frac{1 \text{ km}}{1000 \text{ m}}$$

b. Multiply the original measure by the unit fraction.

$$8.9 \text{ m} \cdot \frac{1 \text{ km}}{1000 \text{ m}}$$

$$= 8.9 \text{ m} \cdot \frac{1 \text{ km}}{1000 \text{ m}}$$

$$= \frac{8.9 \text{ m}}{1000} = \underline{0.0089} \text{ m}$$

Alternatively, move the decimal point  _3_  places to the  _left_ . Note km is three units to the left of m in the list below.

km  hm  dam  m  dm  cm  mm

Thus 8.9 meters =  _0.0089_  kilometers.

*Each of Problems 11 through 19 can be done in two ways. First by multiplying by the appropriate unit ration or by moving the decimal point the number of places indicated by examining the list of metric units with prefixes*
*kilo-  hector-  deka-  base  deci-  centi-  milli-*
*where base represents the base unit of measurement.*

11. 1000 meters to dekameters

Unit ratio: $\dfrac{1 \text{ dam}}{10 \text{ m}}$

Convert: $1000 \text{ m} \cdot \dfrac{1 \text{ dam}}{10 \text{ m}} = \dfrac{1000}{10} \text{ dam}$

$= 100 \text{ dam}$

1000 meters = 100 dekameters

13. 72,500 millimeters to meters

Unit ratio: $\dfrac{1 \text{ m}}{1000 \text{ mm}}$

Convert: $72,500 \text{ mm} \cdot \dfrac{1 \text{ m}}{1000 \text{ mm}} = \dfrac{72,500}{1000} \text{ m}$

$= 72.5 \text{ m}$

72,500 millimeters = 72.5 meters

15. 633 dekameters to kilometers
In the list
    <u>km</u>  hm  <u>dam</u>  m  dm  cm  mm
km is two places to the *left* of dam so move the
decimal two places to the left.
      633 dekameters = 6.33 kilometers

17. 99 decimeters to centimeters
In the list
    km  hm  dam  m  <u>dm</u>  <u>cm</u>  mm
cm is one place to the *right* of dm so move the
decimal one place to the right.
      99 decimeters = 990 centimeters

19. 938 dekameters to decimeters
In the list
    km  hm  <u>dam</u>  m  <u>dm</u>  cm  mm
dm is two places to the *right* of dam so move the
decimal two place to the right.
      938 dekameters = 93,800 decimeters

## Guide Problems

21. Convert 5000 centigrams to dekagrams.
Move the decimal point <u> 3 </u> places to the <u> left </u>. Note dag is three units to the left of cg
in the list below.
    kg  hg  <u>dag</u>  m  dg  <u>cg</u>  mg
Thus, 5000 centigrams = <u> 5 </u> dekagrams.

23. 0.33 grams to milligrams

Unit ratio: $\dfrac{1000 \text{ mg}}{1 \text{ g}}$

Convert: $0.33 \text{ g} \cdot \dfrac{1000 \text{ mg}}{1 \text{ g}} = 0.33 \cdot 1000 \text{ mg}$

$= 330 \text{ mg}$

0.33 grams = 330 milligrams

25. 6226 grams to kilograms

Unit ratio: $\dfrac{1 \text{ kg}}{1000 \text{ g}}$

Convert: $6226 \text{ g} \cdot \dfrac{1 \text{ kg}}{1000 \text{ g}} = \dfrac{6226}{1000} \text{ kg}$

$= 6.226 \text{ kg}$

6226 grams = 6.226 kilograms

27. 7753 dekagrams to decigrams
In the list
    kg  hg  <u>dag</u>  g  <u>dg</u>  cg  mg
dg is two places to the *right* of dag so move the
decimal two places to the right.
      7753 dekagrams = 775,300 decigrams

29. 4200 hectograms to kilograms
In the list
    <u>kg</u>  <u>hg</u>  dag  g  dg  cg  mg
kg is one place to the *left* of hg so move the
decimal one place to the left.
      4200 hectograms = 420 kilograms

## Guide Problems

31.  Convert 2500 kiloliters to liters.

a. Write an appropriate unit ratio.
Since there are 1000 liters in a kiloliter and liters are the new unit, the unit ratio is

$$\text{Unit ratio} = \frac{\text{New units}}{\text{Original units}}$$

$$= \frac{1000 \text{ L}}{1 \text{ kL}}$$

b. Multiply the original measure by the unit fraction.

$$2500 \text{ kL} \cdot \frac{1000 \text{ L}}{1 \text{ kL}}$$

$$= 2500 \text{ kL} \cdot \frac{1000 \text{ L}}{1 \text{ kL}}$$

$$= 2500 \cdot 1000 \text{ L}$$

$$= 2,500,000 \text{ L}$$

Alternatively, move the decimal point __3__ places to the __right__ . Note L is three units to the right of kL in the list below.

<u>kL</u>  hL  daL  <u>L</u>  dL  cL  mL

Thus 2500 kiloliters = __2,500,000__ liters.

33.  230 deciliters to liters

Unit ratio: $\dfrac{1 \text{ L}}{10 \text{ dL}}$

Convert:  $230 \text{ dL} \cdot \dfrac{1 \text{ L}}{10 \text{ dL}} = \dfrac{230}{10} \text{ L}$

$$= 23 \text{ L}$$

230 deciliters = 23 liters

35.  6.26 deciliters to liters

Unit ratio: $\dfrac{1 \text{ L}}{10 \text{ dL}}$

Convert:  $6.26 \text{ dL} \cdot \dfrac{1 \text{ L}}{10 \text{ dL}} = \dfrac{6.26}{10} \text{ L}$

$$= 0.626 \text{ L}$$

6.26 deciliters = 0.626 liters

37.  1800 kiloliters to dekaliters
In the list

<u>kL</u>  hL  <u>daL</u>  L  dL  cL  mL

daL is two places to the *right* of kL so move the decimal two places to the right.

1800 kiloliters = 180,000 dekaliters

39.  3.9 kiloliters to centiliters
In the list

<u>kL</u>  hL  daL  L  dL  <u>cL</u>  mL

cL is five places to the *right* of kL so move the decimal five places to the right.

3.9 kiloliters = 390,000 centiliters

41.  2 liters to kiloliters
In the list

<u>kL</u>  hL  daL  <u>L</u>  dL  cL  mL

kL is three places to the *left* of L so move the decimal three places to the left.

2 liters = 0.002 kiloliters

43.  In the list: km  hm  dam  m  dm  <u>cm</u>  <u>mm</u>, the unit mm is one place to the right of cm. Thus, by moving the decimal point one place to the right, an incision of length 0.3 centimeters is equivalent to one of 3 millimeters.

45.  Using a unit ratio, we have  $400 \text{ dL} \cdot \dfrac{1 \text{ L}}{10 \text{ dL}} = \dfrac{400}{10} \text{ L} = 40 \text{ L}$.  The take requires 40 liters of water.

47. a. The rate $\dfrac{250 \text{ mg}}{100 \text{ lb}}$ specifies the dosage rate. For a 225 pound person, the dosage is

$$225 \text{ lb} \cdot \dfrac{250 \text{ mg}}{100 \text{ lb}} = \dfrac{225 \cdot 250}{100} \text{ mg} = 562.5 \text{ mg}.$$

b. From part a. the dosage for a 225 pound person is 562.5 mg. This dosage in grams is

$$562.5 \text{ mg} \cdot \dfrac{1 \text{ g}}{1000 \text{ mg}} = \dfrac{562.5}{1000} \text{ g} = 0.5625 \text{ grams}.$$

49. First, convert 150 mg to grams: $150 \text{ mg} \cdot \dfrac{1 \text{ g}}{1000 \text{ mg}} = \dfrac{150}{1000} \text{ g} = 0.15 \text{ grams}$. Since this dosage is given every

12 hours, it will be given twice in a day, so a daily dosage is $2 \cdot 0.15 = 0.3$ grams.

## Cumulative Skills Review

1.  $\dfrac{16.4}{18} = \dfrac{16.4}{18} \cdot \dfrac{10}{10} = \dfrac{164}{180}$

    $= \dfrac{41 \cdot 4}{45 \cdot 4} = \dfrac{41}{45}$

3.  First write
    $$55 \text{ ft } 2 \text{ in.} = 54 \text{ ft} + 1 \text{ ft} + 2 \text{ in.}$$
    $$= 54 \text{ ft} + 12 \text{ in.} + 2 \text{ in.}$$
    $$= 54 \text{ ft } 14 \text{ in.}$$
    Then subtract

    $$
    \begin{array}{r}
    54 \text{ ft } 14 \text{ in.} \\
    - \ 42 \text{ ft } \ \ 5 \text{ in.} \\
    \hline
    12 \text{ ft } \ \ 9 \text{ in.}
    \end{array}
    $$

5.  Tim worked an average of
    $$\dfrac{125 + 215 + 180}{3} = \dfrac{520}{3} = 173.333...$$
    or about 173 hours per month.

7.  $22,000 \text{ lb} \cdot \dfrac{1 \text{ t}}{2000 \text{ lb}} = \dfrac{22,000}{2000} \text{ t} = 11 \text{ t}$
    $22,000 \text{ pounds} = 11 \text{ tons}$

9.  Create a proportion using corresponding sides of the rectangles.
    $$\dfrac{h}{11} = \dfrac{30}{6}$$
    Then equate the cross products and solve
    $$h \cdot 6 = 11 \cdot 30$$
    $$6h = 330$$
    $$h = \dfrac{330}{6} = 55$$

## Section 6.4    Converting between the U.S. System and the Metric System

## Concept Check

1.  In converting from the U.S. Customary System to the metric system, all conversion values are approximations except " inches  to centimeters ."

## Guide Problems

3.  Convert 5 feet to meters.

    a. Write an appropriate unit ratio.
       Since there are roughly 0.31 meters in a foot
       and meters are the new unit, the unit ratio
       is

       $$\text{Unit ratio} = \frac{\text{New units}}{\text{Original units}}$$

       $$\approx \frac{0.31 \text{ m}}{1 \text{ ft}}$$

    b. Multiply the original measure by
       the unit fraction.

       $$5 \text{ ft} \cdot \frac{0.31 \text{ m}}{1 \text{ ft}}$$

       $$= 5 \text{ ft} \cdot \frac{0.31 \text{ m}}{1 \text{ ft}}$$

       $$= \frac{5 \cdot 0.31 \text{ m}}{1} = \underline{1.55} \text{ m}$$

       Thus 5 feet $\approx$ $\underline{1.55}$ meters.

*In Exercises 5 through 21 a conversion is made from a denominate number in the U.S. system to an equivalent
number in the metric system or vice versa. In each case, the given number is multiplied by an appropriate unit
ratio based on the conversion table in the text. Since most U.S.-to-metric conversion factors are
approximations, the symbol $\approx$ is used to indicate when a conversion is approximate.*

5.  600 feet to meters
    Unit ratio:

    $$\frac{\text{New units}}{\text{Original units}} \approx \frac{0.31 \text{ m}}{1 \text{ ft}}$$

    Convert:

    $$600 \text{ ft} \cdot \frac{0.31 \text{ m}}{1 \text{ ft}} = 600 \cdot 0.31 \text{ m}$$

    $$= 186 \text{ m}$$

    600 feet $\approx$ 186 meters

7.  33 inches to centimeters
    Unit ratio:

    $$\frac{\text{New units}}{\text{Original units}} = \frac{2.54 \text{ cm}}{1 \text{ in}}$$

    Convert:

    $$33 \text{ in} \cdot \frac{2.54 \text{ cm}}{1 \text{ in}} = 33 \cdot 2.54 \text{ cm}$$

    $$= 83.82 \text{ cm}$$

    33 inches = 83.82 centimeters

9.  60 kilometers to miles

    $$\text{Unit ratio: } \frac{0.62 \text{ mi}}{1 \text{ km}}$$

    Convert:

    $$60 \text{ km} \cdot \frac{0.62 \text{ mi}}{1 \text{ km}} = 60 \cdot 0.62 \text{ mi}$$

    $$= 37.2 \text{ mi}$$

    60 kilometers $\approx$ 37.2 miles

11. 135 pounds to kilograms

    $$\text{Unit ratio: } \frac{0.45 \text{ kg}}{1 \text{ lb}}$$

    Convert:

    $$135 \text{ lb} \cdot \frac{0.45 \text{ kg}}{1 \text{ lb}} = 135 \cdot 0.45 \text{ kg}$$

    $$= 60.75 \text{ kg}$$

    135 pounds $\approx$ 60.75 kilograms

13. 109 kilograms to pounds

    $$\text{Unit ratio: } \frac{2.2 \text{ lb}}{1 \text{ kg}}$$

    Convert:

    $$109 \text{ kg} \cdot \frac{2.2 \text{ lb}}{1 \text{ kg}} = 109 \cdot 2.2 \text{ lb}$$

    $$= 239.8 \text{ lb}$$

    109 kilograms $\approx$ 239.8 pounds

15. 15 ounces to grams

    $$\text{Unit ratio: } \frac{28.35 \text{ g}}{1 \text{ oz}}$$

    Convert:

    $$15 \text{ oz} \cdot \frac{28.35 \text{ g}}{1 \text{ oz}} = 15 \cdot 28.35 \text{ g}$$

    $$= 425.25 \text{ g}$$

    15 ounces $\approx$ 425.25 grams

17. 58 pints to liters

    $$\text{Unit ratio: } \frac{0.47 \text{ liters}}{1 \text{ pt}}$$

    Convert:

    $$58 \text{ pt} \cdot \frac{0.47 \text{ liter}}{1 \text{ pt}} = 58 \cdot 0.47 \text{ liter}$$

    $$= 27.26 \text{ liter}$$

    58 pints $\approx$ 27.26 liters

19. 45 gallons to liters

    $$\text{Unit ratio: } \frac{3.78 \text{ liters}}{1 \text{ gal}}$$

    Convert:

    $$45 \text{ gal} \cdot \frac{3.78 \text{ liters}}{1 \text{ gal}} = 45 \cdot 3.78 \text{ liters}$$

    $$= 170.1 \text{ liters}$$

    45 gallons $\approx$ 107.1 liters

21. 13 liters to quarts

    $$\text{Unit ratio: } \frac{1.06 \text{ qt}}{1 \text{ liter}}$$

    Convert:

    $$13 \text{ liters} \cdot \frac{1.06 \text{ qt}}{1 \text{ liter}} = 13 \cdot 1.06 \text{ qt}$$

    $$= 13.78 \text{ qt}$$

    13 liters $\approx$ 13.78 quarts

23. Convert 75 miles to an equivalent number of kilometers: $75 \text{ mi} \cdot \dfrac{1.61 \text{ km}}{1 \text{ mi}} = 75 \cdot 1.61 \text{ km} = 120.75 \text{ km}.$ To stay at the speed limit, John and Bonnie should travel at 120.75 kilometers per hour.

25. A 5 kilogram ball weighs approximately $5 \text{ kg} \cdot \dfrac{2.2 \text{ lb}}{1 \text{ kg}} = 5 \cdot 2.2 \text{ lb} = 11 \text{ pounds}.$

27. At birth Teeny weighed roughly $150 \text{ lb} \cdot \dfrac{0.45 \text{ kg}}{1 \text{ lb}} = 150 \cdot 0.45 \text{ kg} = 67.5 \text{ kilograms}.$

29. Convert 10 pounds and 10 ounces separately to kilograms:

$$10 \text{ lb} \approx 10 \text{ lb} \cdot \frac{0.45 \text{ kg}}{1 \text{ lb}} = 10 \cdot 0.45 \text{ kg} = 4.5 \text{ kilograms}$$

$$10 \text{ oz} \approx 10 \text{ oz} \cdot \frac{28.35 \text{ g}}{1 \text{ oz}} \cdot \frac{1 \text{ kg}}{1000 \text{ g}} = \frac{10 \cdot 28.35}{1000} \text{ kg} = 0.2835 \text{ kilograms}$$

so a 10 pound 10 ounce baby weighs about $4.5 + 0.2835 = 4.7835$ or about 5 kilograms. If the dosage is 1 milligram for each kilogram of body weight, the baby should receive 5 mg of the medication.

31. Since 1 gram contains 1000 milligrams, 1 microgram can be converted to grams as follows:

$$1 \text{ microgram} = \frac{1}{1000} \text{ milligram} \cdot \frac{1 \text{ gram}}{1000 \text{ milligram}} = \frac{1}{1,000,000} \text{ grams} = 0.000001 \text{ grams}.$$

33. There are $50 \text{ mg} \cdot \dfrac{1000 \text{ micrograms}}{1 \text{ mg}} = 50 \cdot 1000 \text{ micrograms} = 50,000 \text{ micrograms}$ in one dose. Since two doses are taken in a day, the daily dosage is $2 \cdot 50,000 \text{ micrograms} = 100,000 \text{ micrograms}.$

## Cumulative Skills Review

1.  $14.25 - 2.5 + 3.3^2 \cdot 100$
    $= 14.25 - 2.5 + 10.89 \cdot 100$
    $= 14.25 - 2.5 + 1089$
    $= 11.75 + 1089$
    $= 1100.75$

3.  $45 \text{ miles} = 45 \text{ mi} \cdot \dfrac{5280 \text{ ft}}{1 \text{ mi}}$
    $= 45 \cdot 5280 \text{ ft}$
    $= 237,600 \text{ feet}$

5.  $0.55 + \dfrac{3}{10} = \dfrac{55}{100} + \dfrac{3}{10}$
    $= \dfrac{11}{20} + \dfrac{3}{10}$
    $= \dfrac{11}{20} + \dfrac{6}{20} = \dfrac{17}{20}$

7.
    $\quad\;\; 48 \text{ ft} \quad 2 \text{ in.}$
    $\quad\;\; 12 \text{ ft} \quad 10 \text{ in.}$
    $\underline{+ \; 7 \text{ ft} \quad 6 \text{ in.}}$
    $\quad\;\; 67 \text{ ft} \quad 18 \text{ in.}$

    $67 \text{ ft } 18 \text{ in.} = 67 \text{ ft} + 12 \text{ in.} + 6 \text{ in.}$
    $= 67 \text{ ft} + 1 \text{ ft} + 6 \text{ in.}$
    $= 68 \text{ ft } 6 \text{ in.}$

9.  In fraction form the rate is

    $\dfrac{21 \text{ swings}}{49 \text{ children}}$. Simplifying gives $\dfrac{\overset{3}{\cancel{21}} \text{ swings}}{\underset{7}{\cancel{49}} \text{ children}} = \dfrac{3 \text{ swings}}{7 \text{ children}}$.

## Section 6.5    Time and Temperature

## Concept Check

1.  One minute is equivalent to __60__ seconds.

3.  One day is equivalent to __24__ hours.

5.  __Temperature__ is a measure of the warmth or coldness of an object, substance, or environment..

7.  The metric system measures temperature in degrees __Celsius__ .

## Guide Problems

9.  Convert 12 weeks to days.

a. Write an appropriate unit ratio.
   Since there are 7 days in one week.

$$\text{Unit ratio} = \frac{\text{New units}}{\text{Original units}}$$

$$= \frac{7 \text{ days}}{1 \text{ wk}}$$

b. Multiply the original measure by
   the unit fraction.

$$12 \text{ wk} \cdot \frac{7 \text{ days}}{1 \text{ wk}}$$

$$= 12 \text{ wk} \cdot \frac{7 \text{ days}}{1 \text{ wk}}$$

$$= \frac{12 \cdot 7 \text{ days}}{1} = \underline{84 \text{ days}}$$

Thus 12 weeks ≈ __84__ days.

11. 12 hours to minutes
    Unit ratio:

$$\frac{\text{New units}}{\text{Original units}} = \frac{60 \text{ min}}{1 \text{ hr}}$$

Convert:

$$12 \text{ hr} \cdot \frac{60 \text{ min}}{1 \text{ hr}} = 12 \cdot 60 \text{ min}$$

$$= 720 \text{ min}$$

13. 3 years to days
    Unit ratio:

$$\frac{\text{New units}}{\text{Original units}} = \frac{365 \text{ days}}{1 \text{ yr}}$$

Convert:

$$3 \text{ yr} \cdot \frac{365 \text{ days}}{1 \text{ yr}} = 3 \cdot 365 \text{ days}$$

$$= 1095 \text{ days}$$

15. 32 years to months

$$\text{Unit ratio: } \frac{12 \text{ months}}{1 \text{ yr}}$$

Convert:

$$32 \text{ yr} \cdot \frac{12 \text{ months}}{1 \text{ yr}}$$

$$= 32 \cdot 12 \text{ months}$$

$$= 384 \text{ months}$$

17. 9200 decades to centuries

$$\text{Unit ratio: } \frac{1 \text{ century}}{10 \text{ decades}}$$

Convert:

$$9200 \text{ decades} \cdot \frac{1 \text{ century}}{10 \text{ decades}}$$

$$= \frac{9200}{10} \text{ centuries}$$

$$= 920 \text{ centuries}$$

19. 1 day to seconds
    Multiply by a sequence of unit
    ratios:

$$1 \text{ day} \cdot \frac{24 \text{ hr}}{1 \text{ day}} \cdot \frac{60 \text{ min}}{1 \text{ hr}} \cdot \frac{60 \text{ sec}}{1 \text{ min}}$$

$$= 1 \text{ day} \cdot \frac{24 \text{ hr}}{1 \text{ day}} \cdot \frac{60 \text{ min}}{1 \text{ hr}} \cdot \frac{60 \text{ sec}}{1 \text{ min}}$$

$$= 24 \cdot 60 \cdot 60 \text{ sec}$$

$$= 86,400 \text{ seconds}$$

21. Henry makes $\dfrac{2 \text{ round trip}}{1 \text{ month}} \cdot \dfrac{2 \text{ flights}}{1 \text{ round trip}} \cdot \dfrac{12 \text{ months}}{1 \text{ year}} \cdot 1 \text{ year} = 48$ flights in a year. At 18 hours per flight,

    Henry travels a total of $48 \text{ flights} \cdot \dfrac{18 \text{ hr}}{1 \text{ flight}} = 48 \cdot 18 \text{ hr} = 864 \text{ hr.}$ Lastly, convert 864 hours to days:

    $864 \text{ hr} \cdot \dfrac{1 \text{ day}}{24 \text{ hr}} = \dfrac{864}{24} \text{ days} = 36 \text{ days.}$

23. The battery charges for $3 \text{ hr} \cdot \dfrac{60 \text{ min}}{1 \text{ hr}} \cdot \dfrac{60 \text{ sec}}{1 \text{ min}} = 3 \cdot 60 \cdot 60 \text{ sec} = 10{,}800$ seconds.

## Guide Problems

25. Convert 77 degrees Fahrenheit to degrees Celsius.

    a. What is the formula to convert Fahrenheit to Celsius?

    $$C = \frac{5}{9}(F - 32)$$

    b. Substitute the degrees Fahrenheit in the formula and simplify.

    $$C = \frac{5}{9}(F - 32) = \frac{5}{9}(77 - 32)$$
    $$= \frac{5}{9} \cdot 45 = 25$$

    Thus, $77^\circ \text{F} = \underline{25} \ ^\circ\text{C}.$

27. $85^\circ$ Fahrenheit to Celsius

    $C = \dfrac{5}{9}(F - 32) = \dfrac{5}{9}(85 - 32)$

    $= \dfrac{5}{9} \cdot 53 = 29.444... \approx 29$

    $85^\circ$ Fahrenheit $\approx 29^\circ$ Celsius

29. $125^\circ$ Fahrenheit to Celsius

    $C = \dfrac{5}{9}(F - 32) = \dfrac{5}{9}(125 - 32)$

    $= \dfrac{5}{9} \cdot 93 = 51.666... \approx 52$

    $125^\circ$ Fahrenheit $\approx 52^\circ$ Celsius

31. $90^\circ$ Celsius to Fahrenheit

    $F = \dfrac{9}{5}C + 32 = \dfrac{9}{5} \cdot 90 + 32$

    $= 162 + 32 = 194$

    $90^\circ$ Celsius $\approx 194^\circ$ Fahrenheit

33. $3^\circ$ Celsius to Fahrenheit

    $F = \dfrac{9}{5}C + 32 = \dfrac{9}{5} \cdot 3 + 32$

    $= 5.4 + 32 = 37.4 \approx 37$

    $3^\circ$ Celsius $\approx 37^\circ$ Fahrenheit

35. $75^\circ$ Celsius to Fahrenheit

    $F = \dfrac{9}{5}C + 32 = \dfrac{9}{5} \cdot 75 + 32$

    $= 135 + 32 = 167$

    $75^\circ$ Celsius $\approx 167^\circ$ Fahrenheit

37. $43^\circ$ Celsius to Fahrenheit

    $F = \dfrac{9}{5}C + 32 = \dfrac{9}{5} \cdot 43 + 32$

    $= 77.4 + 32 = 109.4 \approx 109$

    $43^\circ$ Celsius $\approx 109^\circ$ Fahrenheit

39. From the formula, $F = \dfrac{9}{5}C + 32 = \dfrac{9}{5} \cdot 580 + 32 = 1044 + 32 = 1076$, so the temperature of the lava is $1076^\circ$ Fahrenheit.

41. Substituting 1 for $C$ in the formula gives $F = \dfrac{9}{5}C + 32 = \dfrac{9}{5} \cdot 1 + 32 = 1.8 + 32 = 33.8 \approx 34$. The ice cream is at about $34^\circ$ Fahrenheit.

## Cumulative Skills Review

1. Let $p$ be the unknown percent. By the percent proportion
$$\frac{84}{160} = \frac{p}{100}$$
Equating the cross products gives
$$160\,p = 8400$$
and $p = \dfrac{8400}{160} = 52.5$.
84 is 52.5% of 160.

3. In expanded notation 6301 is
$$6000 + 300 + 1$$
or
$$6 \text{ thousands} + 3 \text{ hundreds} + 1 \text{ one}$$

5. In 314.398 the digit in the hundredths place is 9. The digit to the right is 8 which is greater than 5 so round up to 314.40.

7. $\dfrac{2}{5} + 2\dfrac{1}{3} + 4\dfrac{1}{6} = \dfrac{2}{5} + \dfrac{7}{3} + \dfrac{25}{6}$
$$= \frac{12}{30} + \frac{70}{30} + \frac{125}{30}$$
$$= \frac{207}{30} = \frac{69}{10} = 6\frac{9}{10}$$

9. $\dfrac{3}{42} = \dfrac{5}{70}$ as a sentence is
"3 is to 42 as 5 is to 70"

## Chapter 6 Numerical Facts Of Life

a. To convert 0.002 cents to dollars, first write a unit fraction with the new units, dollars, in the numerator, and the original units, cents, in the denominator. Since 1 dollar = 100 cents, the unit fraction is
$$\text{Unit ratio} = \frac{\text{New units}}{\text{Original units}} = \frac{1 \text{ dollar}}{100 \text{ cents}}.$$ Now multiply 0.002 cents by the unit fraction to obtain
$$0.002 \text{ cents} = 0.002 \text{ cents} \cdot \frac{1 \text{ dollar}}{100 \text{ cents}} = \frac{0.002}{100} \text{ dollars} = 0.00002 \text{ dollars}.$$

b. Given usage of 35,893 kilobytes and an actual rate of 0.002 dollars per kilobyte, we have
$$35{,}893 \text{ kilobytes} \cdot \frac{0.002 \text{ dollar}}{1 \text{ kilobyte}} = 35{,}893 \cdot 0.002 \text{ dollars} = 71.786 \text{ dollars}.$$ The customer was charged $71.786 or $71.79.

c. With the incorrectly quoted rate of 0.002 cents per kilobyte, we have
$$35{,}893 \text{ kilobytes} \cdot \frac{0.002 \text{ cents}}{1 \text{ kilobyte}} = 35{,}893 \cdot 0.002 \text{ cents} = 71.786 \text{ cents}.$$ The customer should have been charged 71.786 cents or about 72¢.

## Chapter 6 Review Exercises

1. 36 inches to feet
   Unit ratio:
   $$\frac{\text{New unit}}{\text{Old unit}} = \frac{1 \text{ ft}}{12 \text{ inches}}$$
   Convert:
   $$36 \text{ inches} \cdot \frac{1 \text{ ft}}{12 \text{ inches}}$$
   $$= 36 \; \cancel{\text{inches}} \cdot \frac{1 \text{ ft}}{12 \; \cancel{\text{inches}}}$$
   $$= \frac{36}{12} \text{ ft}$$
   $$= 3 \text{ feet}$$

2. 2 miles to yards
   Unit ratio:
   $$\frac{\text{New unit}}{\text{Old unit}} = \frac{1760 \text{ yards}}{1 \text{ miles}}$$
   Convert:
   $$2 \text{ miles} \cdot \frac{1760 \text{ yards}}{1 \text{ miles}}$$
   $$= 2 \; \cancel{\text{miles}} \cdot \frac{1760 \text{ yards}}{1 \; \cancel{\text{miles}}}$$
   $$= 2 \cdot 1760 \text{ yards}$$
   $$= 3520 \text{ yards}$$

3. 48 feet to yards
   Unit ratio: $\dfrac{1 \text{ yard}}{3 \text{ feet}}$
   Convert:
   $$48 \text{ feet} \cdot \frac{1 \text{ yard}}{3 \text{ feet}}$$
   $$= \frac{48}{3} \text{ yards}$$
   $$= 16 \text{ yards}$$

4. 552 feet to yards
   Unit ratio: $\dfrac{1 \text{ yard}}{3 \text{ feet}}$
   Convert:
   $$552 \text{ feet} \cdot \frac{1 \text{ yard}}{3 \text{ feet}}$$
   $$= \frac{552}{3} \text{ yards}$$
   $$= 184 \text{ yards}$$

5. 16,720 yards to miles
   Unit ratio: $\dfrac{1 \text{ mile}}{1760 \text{ yards}}$
   Convert:
   $$16,720 \text{ miles} \cdot \frac{1 \text{ mile}}{1760 \text{ yards}}$$
   $$= \frac{16,720}{1760} \text{ miles} = 9.5 \text{ miles}$$

6. 5 tons to ounces
   First convert tons to pounds:
   $$5 \text{ tons} \cdot \frac{2000 \text{ lb}}{1 \text{ tons}}$$
   $$= 5 \cdot 2000 \text{ lb} = 10,000 \text{ lb}$$
   Next convert pounds to ounces:
   $$10,000 \text{ lb} \cdot \frac{16 \text{ ounces}}{1 \text{ lb}}$$
   $$= 10,000 \cdot 16 \text{ ounces}$$
   $$= 160,000 \text{ ounces.}$$

7. 3 tons to pounds
   Unit ratio: $\dfrac{2000 \text{ lb}}{1 \text{ ton}}$
   Convert:
   $$3 \text{ ton} \cdot \frac{2000 \text{ lb}}{1 \text{ ton}}$$
   $$= 3 \cdot 2000 \text{ lb}$$
   $$= 6000 \text{ pounds}$$

8. 496 quarts to gallons
   Unit ratio: $\dfrac{1 \text{ gal}}{4 \text{ qt}}$
   Convert:
   $$496 \text{ qt} \cdot \frac{1 \text{ gal}}{4 \text{ qt}}$$
   $$= \frac{496}{4} \text{ gal}$$
   $$= 124 \text{ gallons}$$

9. 42 cups to fluid ounces
   Unit ratio: $\dfrac{8 \text{ fl oz}}{1 \text{ c}}$
   Convert:
   $$42 \text{ c} \cdot \frac{8 \text{ fl oz}}{1 \text{ c}}$$
   $$= 42 \cdot 8 \text{ fl oz}$$
   $$= 336 \text{ fluid ounces}$$

10. 399 teaspoons to tablespoons
    Unit ratio: $\dfrac{1 \text{ tbs}}{3 \text{ tsp}}$
    Convert:
    $$399 \text{ tsp} \cdot \frac{1 \text{ tbs}}{3 \text{ tsp}}$$
    $$= \frac{399}{3} \text{ tbs}$$
    $$= 133 \text{ tablespoons}$$

11. 64 fluid ounces to cups
    Unit ratio: $\dfrac{1 \text{ c}}{8 \text{ fl oz}}$
    Convert:
    $$64 \text{ fl oz} \cdot \frac{1 \text{ c}}{8 \text{ fl oz}}$$
    $$= \frac{64}{8} \text{ c}$$
    $$= 8 \text{ cups}$$

12. 52 pints to quarts
    Unit ratio: $\dfrac{1 \text{ qt}}{2 \text{ pt}}$
    Convert:
    $$52 \text{ pt} \cdot \frac{1 \text{ qt}}{2 \text{ pt}}$$
    $$= \frac{52}{2} \text{ qt}$$
    $$= 26 \text{ quarts}$$

13. 12 in. = 1 ft
    2 ft 43 in.
      = 2 ft + 43 in.
      = 2 ft + 36 in. + 7 in.
      = 2 ft + 3 ft + 7 in.
      = 5 ft + 7 in.
      = 5 ft 7 in.

14. 3 ft = 1 yd
    12 yd 16 ft
      = 12 yd + 16 ft
      = 12 yd + 15 ft + 1 ft
      = 12 yd + 5 yd + 1 ft
      = 17 yd + 1 ft
      = 17 yd 1 ft

15. 16 fl oz = 1 pt
    2 pt 35 fl oz
      = 2 pt + 35 fl oz
      = 2 pt + 32 fl oz + 3 fl oz
      = 2 pt + 2 pt + 3 fl oz
      = 4 pt + 3 fl oz
      = 4 pt 3 fl oz

16. 12 in. = 1 ft
    18 ft 19 in.
      = 18 ft + 19 in.
      = 18 ft + 12 in. + 7 in.
      = 18 ft + 1 ft + 7 in.
      = 19 ft + 7 in.
      = 19 ft 7 in.

17. 4 qt = 1 gal
    4 gal 82 qt
      = 4 gal + 82 qt
      = 4 gal + 80 qt + 2 qt
      = 4 gal + 20 gal + 2 qt
      = 24 gal + 2 qt
      = 24 gal 2 qt

18. 16 oz = 1 lb
    2 lb 20 oz
      = 2 lb + 20 oz
      = 2 lb + 16 oz + 4 oz
      = 2 lb + 1 lb + 4 oz
      = 3 lb + 4 oz
      = 3 lb 4 oz

19.     12 ft  10 in
    +  8 ft   7 in
      20 ft  17 in
        = 20 ft + 12 in. + 5 in.
        = 20 ft + 1 ft + 5 in.
        = 21 ft  5 in.

20.      4 c   7 fl oz
      12 c   6 fl oz
    + 16 c   5 fl oz
      32 c  18 fl oz
        = 32 c + 16 fl oz + 2 fl oz
        = 32 c + 2 c + 2 fl oz
        = 34 c  2 fl oz

21.     18 lb  12 oz
    + 26 lb  15 oz
      44 lb  27 oz
        = 44 lb + 16 oz + 11 oz
        = 44 lb + 1 lb + 11 oz
        = 45 lb  11 oz

22.    3 t  18 lb
   − 1 t   2 lb
     2 t  16 lb

23.    5 ft  3 in.
   − 4 ft  8 in.

        4 ft   15 in.
       ~~5 ft~~  ~~3 in.~~
    − 4 ft   8 in.
              7 in.

24.    16 c  3 fl oz
   − 12 c  4 fl oz

        15 c   11 fl oz
      ~~16 c~~  ~~3 fl oz~~
    − 12 c   4 fl oz
         3 c   7 fl oz

25.    3 yd  2 ft
    ×       3
     9 yd  6 ft
       = 9 yd + 6 ft
       = 9 yd + 2 yd
       = 11 yd

26.    3 yd   2 ft
    ×       8
    24 yd  16 ft
       = 24 yd + 15 ft + 1 ft
       = 24 yd + 5 yd + 1 ft
       = 29 yd 1 ft

27.    3 qt  1 pt
    ×       5
   15 qt  5 pt
       = 15 qt + 4 pt + 1 pt
       = 15 qt + 2 qt + 1 pt
       = 17 qt  1 pt

28.     2 tbs 1 tsp
   2)4 tbs 2 tsp
     4 tbs
           2 tsp

29.     2 ft 11 in.
   3)8 ft  9 in.
     6 ft
     2 ft   9 in.  (= 33 in.)

30.     3 lb 6 oz
   6)20 lb 4 oz
    18 lb
    2 lb  4 oz  (= 36 oz)

31. 65 meters to centimeters

Unit ratio:  $\dfrac{100 \text{ cm}}{1 \text{ m}}$

Convert:  $65 \text{ m} \cdot \dfrac{100 \text{ cm}}{1 \text{ m}} = 65 \cdot 100 \text{ cm}$

$= 6500 \text{ cm}$

65 meters $= 6500$ centimeters

32. 26 centimeters to meters

Unit ratio:  $\dfrac{1 \text{ m}}{100 \text{ cm}}$

Convert:  $26 \text{ cm} \cdot \dfrac{1 \text{ m}}{100 \text{ cm}} = \dfrac{26}{100} \text{ m}$

$= 0.26 \text{ m}$

26 centimeters $= 0.26$ meters

33. 37,498 millimeters to dekameters

In the list

km  hm  <u>dam</u>  m  dm  cm  <u>mm</u>

dam is four places to the *left* of mm so move the decimal four places to the left.

37,498 millimeters $= 3.7498$ dekameters

34. 14,774 hectometers to meters

Unit ratio:  $\dfrac{100 \text{ m}}{1 \text{ hm}}$

Convert:  $14,774 \text{ hm} \cdot \dfrac{100 \text{ m}}{1 \text{ hm}} = 14,774 \cdot 100 \text{ m}$

$= 1,477,400 \text{ m}$

14,774 hectometers $= 1,477,400$ meters

35. 1.87 grams to centigrams

Unit ratio:  $\dfrac{100 \text{ cg}}{1 \text{ g}}$

Convert:  $1.87 \text{ g} \cdot \dfrac{100 \text{ cg}}{1 \text{ g}} = 1.87 \cdot 100 \text{ cg}$

$= 187 \text{ cg}$

1.87 grams $= 187$ centigrams

36. 8575 milligrams to grams

Unit ratio:  $\dfrac{1 \text{ g}}{1000 \text{ mg}}$

Convert:  $8575 \text{ mg} \cdot \dfrac{1 \text{ g}}{1000 \text{ mg}} = \dfrac{8575}{1000} \text{ g}$

$= 8.575 \text{ g}$

8575 milligrams $= 8.575$ grams

37. 199,836 grams to kilograms

Unit ratio:  $\dfrac{1 \text{ kg}}{1000 \text{ g}}$

Convert:  $199,836 \text{ g} \cdot \dfrac{1 \text{ kg}}{1000 \text{ g}} = \dfrac{199,836}{1000} \text{ kg}$

$= 199.836 \text{ kg}$

199,836 grams $= 199.836$ kilograms

38. 55 dekagrams to milligrams

In the list

kg  hg  <u>dag</u>  g  dg  cg  <u>mg</u>

mg is four places to the *right* of dag so move the decimal four places to the right.

55 dekagrams $= 550,000$ milligrams

39. 37,345 milligrams to kilograms

In the list

<u>kg</u>  hg  dag  g  dg  cg  <u>mg</u>

kg is six places to the *left* of mg so move the decimal six places to the left.

37,345 milligrams $= 0.037345$ kilograms

40. 7000 liters to kiloliters

Unit ratio:  $\dfrac{1 \text{ kL}}{1000 \text{ L}}$

Convert:  $7000 \text{ L} \cdot \dfrac{1 \text{ kL}}{1000 \text{ L}} = \dfrac{7000}{1000} \text{ kL}$

$= 7 \text{ kL}$

7000 liters $= 7$ kiloliters

41. 58 liters to deciliters

Unit ratio:  $\dfrac{10 \text{ dL}}{1 \text{ L}}$

Convert:  $58 \text{ L} \cdot \dfrac{10 \text{ dL}}{1 \text{ L}} = 58 \cdot 10 \text{ dL}$

$= 580 \text{ dL}$

58 liters $= 580$ deciliters

42. 128 deciliters to liters

Unit ratio:  $\dfrac{1 \text{ L}}{10 \text{ dL}}$

Convert:  $128 \text{ dL} \cdot \dfrac{1 \text{ L}}{10 \text{ dL}} = \dfrac{128}{10} \text{ L}$

$= 12.8 \text{ L}$

128 deciliters $= 12.8$ liters

43. 88 kilometers to dekameters
   In the list
   $$\underline{km} \quad hm \quad \underline{dam} \quad m \quad dm \quad cm \quad mm$$
   dam is two places to the *right* of km so move the
   decimal two places to the right.
   $$88 \text{ kilometers} = 8800 \text{ dekameters}$$

44. 10 deciliters to milliliters
   In the list
   $$kL \quad hL \quad daL \quad L \quad \underline{dL} \quad cL \quad \underline{mL}$$
   mL is two places to the *right* of dL so move the
   decimal two places to the right.
   $$10 \text{ deciliters} = 1000 \text{ milliliters}$$

45. 4697 dekaliters to hectoliters
   In the list
   $$kL \quad \underline{hL} \quad \underline{daL} \quad L \quad dL \quad cL \quad mL$$
   hL is one place to the *left* of daL so move the decimal one place to the left.
   $$4697 \text{ dekaliters} = 469.7 \text{ hectoliters}$$

46. 4 meters to feet
   Unit ratio: $\dfrac{3.3 \text{ ft}}{1 \text{ m}}$
   Convert:
   $$4 \text{ m} \cdot \frac{3.3 \text{ ft}}{1 \text{ m}} = 4 \cdot 3.3 \text{ ft}$$
   $$= 13.2 \text{ ft}$$

   4 meters ≈ 13.2 feet

47. 12 miles to kilometers
   Unit ratio: $\dfrac{1.61 \text{ km}}{1 \text{ mi}}$
   Convert:
   $$12 \text{ mi} \cdot \frac{1.61 \text{ km}}{1 \text{ mi}} = 12 \cdot 1.61 \text{ km}$$
   $$= 19.32 \text{ km}$$

   12 miles ≈ 19.32 kilometers

48. 14 inches to centimeters
   Unit ratio: $\dfrac{2.54 \text{ cm}}{1 \text{ in}}$
   Convert:
   $$14 \text{ in} \cdot \frac{2.54 \text{ cm}}{1 \text{ in}} = 14 \cdot 2.54 \text{ cm}$$
   $$= 35.56 \text{ cm}$$

   14 inches = 35.56 centimeters

49. 90 yards to meters
   Unit ratio: $\dfrac{0.91 \text{ m}}{1 \text{ yd}}$
   Convert:
   $$90 \text{ yd} \cdot \frac{0.91 \text{ m}}{1 \text{ yd}} = 90 \cdot 0.91 \text{ m}$$
   $$= 81.9 \text{ m}$$

   90 yards ≈ 81.9 meters

50. 60 kilograms to pounds
   Unit ratio: $\dfrac{2.2 \text{ lb}}{1 \text{ kg}}$
   Convert:
   $$60 \text{ kg} \cdot \frac{2.2 \text{ lb}}{1 \text{ kg}} = 60 \cdot 2.2 \text{ lb}$$
   $$= 132 \text{ lb}$$

   60 kilograms ≈ 132 pounds

51. 250 pounds to kilograms
   Unit ratio: $\dfrac{0.45 \text{ kg}}{1 \text{ lb}}$
   Convert:
   $$250 \text{ lb} \cdot \frac{0.45 \text{ kg}}{1 \text{ lb}} = 250 \cdot 0.45 \text{ kg}$$
   $$= 112.5 \text{ kg}$$

   250 pounds ≈ 112.5 kilograms

52. 16 ounces to grams
   Unit ratio: $\dfrac{28.35 \text{ g}}{1 \text{ oz}}$
   Convert:
   $$16 \text{ oz} \cdot \frac{28.35 \text{ g}}{1 \text{ oz}} = 16 \cdot 28.35 \text{ g}$$
   $$= 453.6 \text{ g}$$

   16 ounces ≈ 453.6 grams

53. 500 fluid ounces to liters
   Unit ratio: $\dfrac{0.03 \text{ liters}}{1 \text{ fl oz}}$
   Convert:
   $$500 \text{ fl oz} \cdot \frac{0.03 \text{ liters}}{1 \text{ fl oz}}$$
   $$= 500 \cdot 0.03 \text{ liters}$$
   $$= 15 \text{ liters}$$

   500 fluid ounces ≈ 15 liters

54. 32 quarts to liters
   Unit ratio: $\dfrac{0.95 \text{ liters}}{1 \text{ qt}}$
   Convert:
   $$32 \text{ qt} \cdot \frac{0.95 \text{ liters}}{1 \text{ qt}} = 32 \cdot 0.95 \text{ liter}$$
   $$= 30.4 \text{ liter}$$

   32 quarts ≈ 30.4 liters

55. 105 liters to gallons

Unit ratio: $\dfrac{0.26 \text{ gal}}{1 \text{ liter}}$

Convert:

$105 \text{ liters} \cdot \dfrac{0.26 \text{ gal}}{1 \text{ liter}} = 105 \cdot 0.26 \text{ gal}$

$\phantom{105 \text{ liters} \cdot \dfrac{0.26 \text{ gal}}{1 \text{ liter}}} = 27.3 \text{ gal}$

105 liters ≈ 27.3 gallons

56. 8 liters to pints

Unit ratio: $\dfrac{2.11 \text{ pt}}{1 \text{ liter}}$

Convert:

$8 \text{ liters} \cdot \dfrac{2.11 \text{ pt}}{1 \text{ liter}} = 8 \cdot 2.11 \text{ pt}$

$\phantom{8 \text{ liters} \cdot \dfrac{2.11 \text{ pt}}{1 \text{ liter}}} = 16.88 \text{ pt}$

8 liters ≈ 16.88 pints

57. 18 gallons to liters

Unit ratio: $\dfrac{3.78 \text{ liters}}{1 \text{ gal}}$

Convert:

$18 \text{ gal} \cdot \dfrac{3.78 \text{ liters}}{1 \text{ gal}} = 18 \cdot 3.78 \text{ liters}$

$\phantom{18 \text{ gal} \cdot \dfrac{3.78}{1}} = 68.04 \text{ liters}$

18 gallons ≈ 68.04 liters

58. 12 minutes to seconds

Unit ratio: $\dfrac{60 \text{ sec}}{1 \text{ min}}$

Convert:

$12 \text{ min} \cdot \dfrac{60 \text{ sec}}{1 \text{ min}}$

$= 12 \cdot 60 \text{ sec}$

$= 720 \text{ seconds}$

59. 52 weeks to days

Unit ratio: $\dfrac{7 \text{ days}}{1 \text{ week}}$

Convert:

$52 \text{ weeks} \cdot \dfrac{7 \text{ days}}{1 \text{ week}}$

$= 52 \cdot 7 \text{ days}$

$= 364 \text{ days}$

60. 3 centuries to years

Unit ratio: $\dfrac{100 \text{ yr}}{1 \text{ century}}$

Convert:

$3 \text{ centuries} \cdot \dfrac{100 \text{ yr}}{1 \text{ century}}$

$= 3 \cdot 100 \text{ yr}$

$= 300 \text{ years}$

61. 88° Fahrenheit to Celsius

$C = \dfrac{5}{9}(F - 32) = \dfrac{5}{9}(88 - 32)$

$= \dfrac{5}{9} \cdot 56 = 31.111... \approx 31$

88° F ≈ 31° C

62. 50° Celsius to Fahrenheit

$F = \dfrac{9}{5}C + 32 = \dfrac{9}{5} \cdot 50 + 32$

$= 90 + 32 = 122$

50° C = 122° F

63. 100° Fahrenheit to Celsius

$C = \dfrac{5}{9}(F - 32) = \dfrac{5}{9}(100 - 32)$

$= \dfrac{5}{9} \cdot 68 = 37.777... \approx 38$

100° F ≈ 38° C

64. First, express 3 yards as the equivalent 9 feet. Then subtract 3 feet 8 inches from 9 feet.

```
        8 ft   12 in.
       9 ft
   −   3 ft    8 in.
   ─────────────────
       5 ft    4 in.
```

There is 5 feet 4 inches left of the stock piece.

65. Convert two quarters to months:

$2 \text{ quarters} \cdot \dfrac{3 \text{ months}}{1 \text{ quarter}}$

$= 2 \cdot 3 \text{ months}$

$= 6 \text{ months}$

and so 6 months have passed.

66. a. Convert 6000 feet to meters

$6000 \text{ ft} \cdot \dfrac{0.31 \text{ m}}{1 \text{ ft}} = 6000 \cdot 0.31 \text{ ft}$

$= 1860 \text{ m}$

Anthony runs roughly 1860 meters.

   b. From part a., Anthony will run $4 \cdot 1860 = 7440$ meters per week. Converting to kilometers

$7440 \text{ m} \cdot \dfrac{1 \text{ km}}{1000 \text{ m}} = \dfrac{7440}{1000} \text{ km}$

$= 7.44 \text{ km}$

Anthony will run about 7.44 kilometers in a week.

c. Convert the answer from b. to miles:

$7.44 \text{ km} \cdot \dfrac{0.62 \text{ mi}}{1 \text{ km}} = 7.44 \cdot 0.62 \text{ mi}$

$\approx 4.6 \text{ mi}$

Anthony runs about 4.6 miles.

Note that Anthony runs 6000 feet per day so in 4 days, Anthony runs 24,000 feet. Converting this number to miles gives

$24,000 \text{ ft} \cdot \dfrac{1 \text{ mi}}{5280 \text{ ft}} = \dfrac{24,000}{5280} \text{ mi}$

$\approx 4.5 \text{ mi}$

The answers differ slightly because of the approximate conversion factors used in U.S. metric conversions.

67. Using the Celsius to Fahrenheit formula with $C = 15$, we have
$$F = \frac{9}{5}C + 32 = \frac{9}{5} \cdot 15 + 32$$
$$= 27 + 32 = 59$$
The temperature at Heathrow is $59^\circ$ Fahrenheit.

68. Use unit ratios to convert 3 days per week to a rate involving quarters.
$$\frac{3 \text{ days}}{1 \text{ week}} \cdot \frac{52 \text{ weeks}}{1 \text{ yr}} \cdot \frac{1 \text{ yr}}{4 \text{ quarters}}$$
$$= \frac{3 \text{ days}}{1 \text{ week}} \cdot \frac{\overset{13}{\cancel{52}} \text{ weeks}}{1 \text{ yr}} \cdot \frac{1 \text{ yr}}{\underset{1}{\cancel{4}} \text{ quarters}}$$
$$= \frac{3 \cdot 13 \text{ days}}{\text{quarter}} = \frac{39 \text{ days}}{\text{quarter}}$$
Firefighters get 39 days off each quarter.

69. a. 30 years is equivalent to
$$30 \text{ yr} \cdot \frac{1 \text{ decade}}{10 \text{ yr}} = \frac{30}{10} \text{ decade} = 3 \text{ decades.}$$
The 30-year reunion attendees went to school 3 decades ago.

b. 10 years is equivalent to
$$10 \text{ yr} \cdot \frac{12 \text{ months}}{1 \text{ yr}} = 10 \cdot 12 \text{ months}$$
$$= 120 \text{ months.}$$
The 10-year reunion attendees went to school 120 months ago.

c. Convert 20 years to quarters.
$$20 \text{ yr} \cdot \frac{4 \text{ quarters}}{1 \text{ yr}} = 20 \cdot 4 \text{ quarters}$$
$$= 80 \text{ quarters}$$
The 20-year reunion attendees went to school 80 quarters ago.

70. If one pill contains 500 milligrams, then four pills contain $4 \cdot 500 = 2000$ milligrams. Convert this amount to grams.
$$2000 \text{ mg} \cdot \frac{1 \text{ g}}{1000 \text{ mg}} = \frac{2000}{1000} \text{ g} = 2 \text{ g}$$
Four tablets contain 2 grams of medication.

71. If each tank holds 1000 gallons then three tanks hold 3000 gallons. Convert this amount to liters.
$$3000 \text{ gal} \cdot \frac{3.78 \text{ L}}{1 \text{ gal}} = 3000 \cdot 3.78 \text{ L}$$
$$= 11,340 \text{ L}$$
The tanks can hold roughly 11,340 liters.

72. To convert 100 milligrams to centigrams, note that in the list
kg  hg  dag  g  dg  <u>cg</u>  <u>mg</u>
cg is one place to the left of mg, so move the decimal point one place to the left to find 100 milligrams = 10 centigrams.
Thus tablets contain 20 centigrams.

73. Convert 1.5 tablespoons to teaspoons.
$$1.5 \text{ tbs} \cdot \frac{3 \text{ tsp}}{1 \text{ tbs}} = 1.5 \cdot 3 \text{ tsp} = 4.5 \text{ tsp}$$
The patient takes 4.5 teaspoons of fiber.

74. First multiply the rate, 30 milliliters per hour, by the time, 24 hours.
$$24 \text{ hr} \cdot \frac{30 \text{ mL}}{1 \text{ hr}} = 720 \text{ mL}$$
Next, convert this amount to liters.
$$720 \text{ mL} \cdot \frac{1 \text{ L}}{1000 \text{ mL}} = \frac{720}{1000} \text{ L} = 0.72 \text{ L}$$
The patient receives 0.72 liters in 24 hours.

75. a. Compute $239^\circ$ Fahrenheit to degrees Celsius.
$$C = \frac{5}{9}(F - 32) = \frac{5}{9}(239 - 32) = \frac{5}{9} \cdot 207 = 115$$
The solution boils at $115^\circ$ Celsius.

b. Convert 4.5 minutes to seconds.
$$4.5 \text{ min} \cdot \frac{60 \text{ sec}}{1 \text{ min}} = 4.5 \cdot 60 \text{ sec} = 270 \text{ sec}$$
The solution boils in 270 seconds.

76. The time increased from 3 hours, 19 minutes, 5 seconds in 1970 to 3 hours, 42 minutes, 42 seconds in 1980. Subtracting hours: $3-3=0$, minutes: $42-19=23$, and seconds: $42-5=37$ gives an increase of 23 minutes 37 seconds.

77. The time increased from 4 hours, 12 minutes, 3 seconds in 1990 to 4 hours, 21 minutes, 32 seconds. Subtracting hours: $4-4=0$, minutes: $21-12=9$, and seconds: $32-3=29$ gives an increase of 9 minutes 29 seconds.

78. The time increased from 3 hours, 19 minutes, 5 seconds in 1970 to 4 hours, 21 minutes, 32 seconds. Subtracting hours: $4-3=1$, minutes: $21-19=2$, and seconds: $32-5=27$ gives an increase of 1 hour 2 minutes 27 seconds.

79. The average increase over the thirty year period is found by dividing 30 into the total increase found in exercise 78 which was 1 hour 2 minutes 27 seconds. The division is shown below. Since 30 does not divide 1, convert 1 hour to 60 minutes and write the dividend as 62 minutes 27 seconds.

$$\begin{array}{r} 2\ \text{min}\phantom{aaaaaaa} \\ 30\overline{)62\ \text{min}\quad 27\ \text{sec}} \\ \underline{60\ \text{min}\phantom{aaaaaa}} \\ 2\ \text{min}\quad 27\ \text{sec}\quad (=147\ \text{sec}) \end{array}$$

Dividing 147 seconds by 30 gives 4.9 seconds, so, to the nearest second, the average increase was 2 minutes 5 seconds.

## Chapter 6 Assessment Test

1. 938 yards to feet
Unit ratio:
$$\frac{\text{New unit}}{\text{Old unit}}=\frac{3\ \text{ft}}{1\ \text{yd}}$$
Convert:
$$938\ \text{yd}\cdot\frac{3\ \text{ft}}{1\ \text{yd}}$$
$$=938\ \cancel{\text{yd}}\cdot\frac{3\ \text{ft}}{1\ \cancel{\text{yd}}}$$
$$=938\cdot 3\ \text{ft}$$
$$=2814\ \text{feet}$$

2. 272 quarts to gallons
Unit ratio:
$$\frac{\text{New unit}}{\text{Old unit}}=\frac{1\ \text{gal}}{4\ \text{qt}}$$
Convert:
$$272\ \text{qt}\cdot\frac{1\ \text{gal}}{4\ \text{qt}}$$
$$=272\ \cancel{\text{qt}}\cdot\frac{1\ \text{gal}}{4\ \cancel{\text{qt}}}$$
$$=\frac{272}{4}\ \text{gal}$$
$$=68\ \text{gallons}$$

3. 3 pounds to ounces
Unit ratio:
$$\frac{\text{New unit}}{\text{Old unit}}=\frac{16\ \text{oz}}{1\ \text{lb}}$$
Convert:
$$3\ \text{lb}\cdot\frac{16\ \text{oz}}{1\ \text{lb}}$$
$$=3\ \cancel{\text{lb}}\cdot\frac{16\ \text{oz}}{1\ \cancel{\text{lb}}}$$
$$=3\cdot 16\ \text{oz}$$
$$=48\ \text{ounces}$$

4. 160 fluid ounces to cups
Unit ratio:
$$\frac{\text{New unit}}{\text{Old unit}}=\frac{1\ \text{c}}{8\ \text{fl oz}}$$
Convert:
$$160\ \text{fl oz}\cdot\frac{1\ \text{c}}{8\ \text{fl oz}}$$
$$=160\ \cancel{\text{fl oz}}\cdot\frac{1\ \text{c}}{8\ \cancel{\text{fl oz}}}$$
$$=\frac{160}{8}\ \text{c}$$
$$=20\ \text{cups}$$

5. There are 4 quarts in a gallon so divide 15 by 4.
$$\begin{array}{r} 3\ R\ 3 \\ 4\overline{)15} \\ \underline{12} \\ 3 \end{array}$$
15 qt = 3 gal 3 qt

6. There are 3 feet in a yard so divide 74 by 3.
$$\begin{array}{r} 24\ R\ 2 \\ 3\overline{)74} \\ \underline{6} \\ 14 \\ \underline{12} \\ 2 \end{array}$$
74 ft = 24 yd 2 ft

7. $\quad\quad\quad$ 2 cups = 1 pint

   15 pints 3 cups = 15 pints + 2 cups + 1 cup

   $\quad\quad\quad\quad\quad$ = 15 pints + 1 pint + 1 cup

   $\quad\quad\quad\quad\quad$ = 16 pt 1 c

8. 5280 feet = 1 mile

   3 miles 6700 feet

   $\quad$ = 3 miles + 5280 feet + 1420 feet

   $\quad$ = 3 miles + 1 mile + 1420 feet

   $\quad$ = 4 mi 1420 ft

9. $\quad$ 15 ft  10 in.

   $\quad$ 20 ft $\quad$ 2 in.

   + $\quad$ 1 ft $\quad$ 8 in.

   ―――――――

   $\quad$ 36 ft $\quad$ 20 in.

   $\quad\quad$ = 36 ft + 12 in. + 8 in.

   $\quad\quad$ = 36 ft + 1 ft + 8 in.

   $\quad\quad$ = 37 ft $\quad$ 8 in.

10. $\quad$ 80 t  100 lb

    $-$ $\quad$ 26 t  500 lb

    $\quad\quad\quad\quad\quad$ 79 t $\quad$ 2100 lb

    $\quad\quad\quad\quad$ ~~80 t~~ $\quad$ ~~100 lb~~

    $\quad\quad$ $-$ $\quad$ 26 t $\quad\quad$ 500 lb

    $\quad\quad$ ―――――――――

    $\quad\quad\quad\quad$ 53 t $\quad$ 1600 lb

11. $\quad$ 12 ft $\quad$ 5 in.

    × $\quad\quad\quad\quad$ 4

    ―――――――

    $\quad$ 48 ft $\quad$ 20 in.

    $\quad\quad$ = 48 ft + 12 in. + 8 in.

    $\quad\quad$ = 48 ft + 1 ft + 8 in.

    $\quad\quad$ = 49 ft $\quad$ 8 in.

12. $\quad\quad$ 1 lb  12 oz

    7)‾12 lb  4 oz

    $\quad\quad$ 7 lb

    $\quad$ ―――――――

    $\quad\quad$ 5 lb $\quad$ 4 oz $\quad$ (= 84 oz)

13. 987 centiliters to milliliters

    In the list

    $\quad\quad$ kL $\quad$ hL $\quad$ daL $\quad$ L $\quad$ dL $\quad$ <u>cL</u> $\quad$ <u>mL</u>

    mL is one place to the *right* of cL so move the decimal one place to the right.

    $\quad\quad$ 987 centiliters = 9870 milliliters

14. 84 grams to decigrams

    Unit ratio: $\dfrac{10\text{ dg}}{1\text{ g}}$

    Convert: $84\text{ g}\cdot\dfrac{10\text{ dg}}{1\text{ g}} = 84\cdot 10\text{ dg}$

    $\quad\quad\quad\quad\quad\quad\quad$ = 840 dg

    84 grams = 840 decigrams

15. 32 hectometers to centimeters

    In the list

    $\quad\quad$ km $\quad$ <u>hm</u> $\quad$ dam $\quad$ m $\quad$ dm $\quad$ <u>cm</u> $\quad$ mm

    cm is four places to the *right* of hm so move the decimal four places to the right.

    $\quad\quad$ 32 hectometers = 320,000 centimeters

16. 3155 centigrams to dekagrams

    In the list

    $\quad\quad$ kg $\quad$ hg $\quad$ <u>dag</u> $\quad$ g $\quad$ dg $\quad$ <u>cg</u> $\quad$ mg

    dag is three places to the *left* of cg so move the decimal three places to the left.

    $\quad\quad$ 3155 centigrams = 3.155 dekagrams

17. 64 feet to meters

    Unit ratio: $\dfrac{0.31\text{ m}}{1\text{ ft}}$

    Convert:

    $64\text{ ft}\cdot\dfrac{0.31\text{ m}}{1\text{ ft}} = 64\cdot 0.31\text{ m}$

    $\quad\quad\quad\quad\quad\quad$ = 19.84 m

    64 feet ≈ 19.84 meters

18. 3 kilometers to miles

    Unit ratio: $\dfrac{0.62\text{ mi}}{1\text{ km}}$

    Convert:

    $3\text{ km}\cdot\dfrac{0.62\text{ mi}}{1\text{ km}} = 3\cdot 0.62\text{ mi}$

    $\quad\quad\quad\quad\quad\quad$ = 1.86 mi

    3 kilometers ≈ 1.86 miles

19. 290 pounds to kilograms

Unit ratio: $\dfrac{0.45 \text{ kg}}{1 \text{ lb}}$

Convert:

$290 \text{ lb} \cdot \dfrac{0.45 \text{ kg}}{1 \text{ lb}} = 290 \cdot 0.45 \text{ kg}$

$\qquad\qquad\qquad = 130.5 \text{ kg}$

290 pounds $\approx$ 130.5 kilograms

20. 85 fluid ounces to liters

Unit ratio: $\dfrac{0.03 \text{ liters}}{1 \text{ fl oz}}$

Convert:

$85 \text{ fl oz} \cdot \dfrac{0.03 \text{ liters}}{1 \text{ fl oz}}$

$\qquad = 85 \cdot 0.03 \text{ liters}$

$\qquad = 2.55 \text{ liters}$

85 fluid ounces $\approx$ 2.55 liters

21. 3 millenniums to years

Unit ratio: $\dfrac{1000 \text{ yr}}{1 \text{ millennium}}$

Convert:

$3 \text{ millenniums} \cdot \dfrac{1000 \text{ yr}}{1 \text{ millennium}}$

$\qquad = 3 \cdot 1000 \text{ yr}$

$\qquad = 3000 \text{ years}$

22. 365 days to minutes

Multiply by a sequence of unit ratios.

Convert:

$365 \text{ days} \cdot \dfrac{24 \text{ hr}}{1 \text{ day}} \cdot \dfrac{60 \text{ min}}{1 \text{ hr}}$

$= 365 \cancel{\text{ days}} \cdot \dfrac{24 \cancel{\text{ hr}}}{1 \cancel{\text{ day}}} \cdot \dfrac{60 \text{ min}}{1 \cancel{\text{ hr}}}$

$= 365 \cdot 24 \cdot 60 \text{ min}$

$= 525{,}600 \text{ minutes}$

23. 300° Fahrenheit to Celsius

$C = \dfrac{5}{9}(F - 32) = \dfrac{5}{9}(300 - 32)$

$\quad = \dfrac{5}{9} \cdot 268 = 148.888... \approx 149$

300° Fahrenheit $\approx$ 149° Celsius

24. 8° Celsius to Fahrenheit

$F = \dfrac{9}{5}C + 32 = \dfrac{9}{5} \cdot 8 + 32$

$\quad = 14.4 + 32 = 46.4 \approx 46$

8° Celsius $\approx$ 46° Fahrenheit

25.
```
   46 ft  11 in.
+  29 ft   7 in.
───────────────
   75 ft  18 in.
```

$= 75 \text{ ft} + 12 \text{ in.} + 6 \text{ in.}$

$= 75 \text{ ft} + 1 \text{ ft} + 6 \text{ in.}$

$= 76 \text{ ft}  6 \text{ in.}$

The total length of the slabs is 76 feet 6 inches.

26. Convert 15 pounds to kilograms

$15 \text{ lb} \cdot \dfrac{0.45 \text{ kg}}{1 \text{ lb}} = 15 \cdot 0.45 \text{ kg}$

$\qquad\qquad\qquad = 6.75 \text{ kg}$

Mark lost roughly 6.75 kilograms.

27. 300 milligrams at a rate of 50 milligrams per teaspoon is

$300 \text{ mg} \cdot \dfrac{1 \text{ tsp}}{50 \text{ mg}} = \dfrac{300}{50} \text{ tsp}$

$\qquad\qquad\qquad\quad = 6 \text{ tsp}$

The patient takes 6 teaspoons.

28. Convert 1.5 pints to liters

$1.5 \text{ pt} \cdot \dfrac{0.47 \text{ liters}}{1 \text{ pt}}$

$= 1.5 \cdot 0.47 \text{ liters}$

$= 0.705 \text{ liters}$

The patient drinks 0.705 liters of juice.

29. There are 60 minutes in an hour so divide 315 by 60

$$\begin{array}{r} 5 \ \ R \ \ 15 \\ 60{\overline{)315}} \\ \underline{300} \\ 15 \end{array}$$

so 315 minutes is 5 hours 15 minutes.

30. Convert $102°$ Fahrenheit to Celsius

$$C = \frac{5}{9}(F - 32) = \frac{5}{9}(102 - 32)$$

$$= \frac{5}{9} \cdot 70 = 38.888... \approx 38.9$$

To the nearest tenth of a degree, Henry's body temperature is $38.9°$ C.

# Chapter 7   Geometry

## Section 7.1   Lines and Angles

## Concept Check

1.   _Geometry_ is the branch of mathematics that deals with the measurements, properties, and relationships of shapes and sizes.

3.   A figure that lies entirely in a plane is called a _plane_ _figure_ .

5.   An object with length, width, and depth that resides in space is called a _solid_ .

7.   A _line_ is a straight row of points that extends forever in both directions.

9.   Lines that lie in the same plane and cross at some point in the plane are called _intersecting_ lines.

11.   A point at the end of a line segment is called an _endpoint_ .

13.   An _angle_ is a construct formed by uniting the endpoints of two rays.

15.   A _degree_ is a unit used to measure an angle.

17.   An _acute_ angle is an angle whose measure is greater than $0°$ and less than $90°$. A _right_ angle is an angle whose measure is $90°$. An _obtuse_ angle is an angle whose measure is greater than $90°$ and less than $180°$. A _straight_ angle is an angle whose measure is $180°$.

## Guide Problems

19.   A point is named using a single capital letter. The points shown are point $C$, point $T$, and point $W$.

21.   A line segment is named using its endpoints, with a bar above the letters.

a. Label the endpoints of the line segment with the letters Q and S.

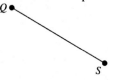

b. Name the line segment appropriately: $\overline{QS}$ or $\overline{SQ}$

*In Problems 23 through 31, identify each figure using the definitions of plane figures and according to the guidelines in Guide Problems 18 through 22.*

23.   The figure shown is a line segment because it has two endpoints. It is named $\overline{PD}$ or $\overline{DP}$.

25.   The figure shown is a ray because it begins at one point and extends forever in one direction. It is named $\overrightarrow{AI}$.

27.   The figure shown is a line segment because it has two endpoints. It is named $\overline{ZT}$ or $\overline{TZ}$.

29. The figure shown is a ray because it begins at one point and extends forever in one direction. It is named $\overrightarrow{EK}$.

31. The figure shown is a line segment because it has two endpoints. It is named $\overline{XM}$ or $\overline{MX}$.

## Guide Problems

33. An acute angle measures less than $90°$, a right angle measures $90°$, an obtuse angle measures more than $90°$ and less than $180°$, and a straight angle measures $180°$. Classify each angle as acute right, obtuse, or straight.

This angle is right because its measure is $90°$.

This angle is straight because its measure is $180°$.

This angle is obtuse because its measure is greater than $90°$ and less than $180°$.

This angle is acute because its measure is greater than $0°$ and less than $90°$.

*In Problems 35 through 41, name and classify each angle using the definitions of acute, right, obtuse, or straight angles as shown in Guide Problems 32 and 33.*

35. $\angle N$, $\angle MNP$, $\angle PNM$; this angle is right because its measure is $90°$.

37. $\angle V$; this angle is straight because its measure is $180°$.

39. $\angle C$, $\angle JCH$, $\angle HCJ$; this angle is right because its measure is $90°$.

41. $\angle F$; this angle is straight because its measure is $180°$.

## Guide Problems

43. Complementary angles are two angles, the sum of whose degree measure is $90°$. To find the complement of an angle, subtract the measure of the angle from $90°$. Find the measure of the complement of an angle measuring $25°$.
$90° - 25° = \underline{65°}$

*In Exercises 45 through 49 find the measure of the complement or supplement of each angle with the given measure using the definitions of complementary and supplementary angles.*

45. $90° - 34° = 56°$        47. $180° - 43° = 137°$        49. $180° - 40° = 140°$

51. $\angle N$ and the $72°$ angle form a right angle, so the angles are complementary.
$m\angle N = 90° - 72° = 18°$

53. $\angle WXD$ and the $64°$ angle form a right angle, so the angles are complementary.
$m\angle WXD = 90° - 64° = 26°$

55. $\angle H$, the 73° angle, and the 57° angle form a straight angle, so the angles are supplementary.

$m\angle H = 180° - 73° - 57° = 50°$

## Cumulative Skills Review

1.  Write an equation to solve for the unknown number, call it $a$.

    $10\% \cdot 25,000 = a$

    Convert the percent, 10%, to a decimal, 0.10, and multiply.

    $0.10 \cdot 25,000 = a$

    $2500 = a$

    2500 is 10% of 25,000.

3.  The rectangle is divided into 5 equal squares. 3 squares are shaded. The fraction $\dfrac{3}{5}$ represents the shaded portion.

5.  The prime factorization of 75 is $3 \cdot 5 \cdot 5 = 3 \cdot 5^2$.

7.  To determine which circle is larger, determine which diameter is longer by comparing their lengths. The LCD of $\dfrac{5}{8}$ and $\dfrac{2}{3}$ is 24. Write each fraction as an equivalent fraction with a denominator of 24.

    $\dfrac{5}{8} = \dfrac{5 \cdot 3}{8 \cdot 3} = \dfrac{15}{24}$ and $\dfrac{2}{3} = \dfrac{2 \cdot 8}{3 \cdot 8} = \dfrac{16}{24}$

    Comparing the fractions, $\dfrac{15}{24} < \dfrac{16}{24}$ so $\dfrac{5}{8} < \dfrac{2}{3}$ and thus $3\dfrac{5}{8} < 3\dfrac{2}{3}$.

    The circle with diameter of $3\dfrac{2}{3}$ is larger.

9.  Identify the parts of the percent problem.

    amount: 150

    part: 30

    base: unknown, $b$

    Substitute the values into the proportion.

    $\dfrac{150}{b} = \dfrac{30}{100}$

    Simplify the fractions in the percent proportion.

    $\dfrac{150}{b} = \dfrac{3}{10}$

    Cross multiply and set the cross products equal.

    $b \cdot 30 = 150 \cdot 10$

    $30b = 1500$

    Divide both sides of the equation by the number on the side with the unknown.

    $\dfrac{30b}{30} = \dfrac{1500}{30}, \ b = 500$

    150 is 30% of 500.

## Section 7.2    Plane and Solid Geometric Figures

### Concept Check

1.  A closed plane figure in which all sides are line segments is called a  _polygon_ .

3.  A triangle with sides of equal length and angles of equal measure is called an  _equilateral_  triangle.

5.  A triangle with all three sides of different lengths and angles of different measures is known as a  _scalene_  triangle.

7.  A  _right_  triangle is a triangle that has a right angle.

9.  A four-sided polygon is called a  _quadrilateral_ .

11. A parallelogram that has four right angles is called a  _rectangle_ .

13. A rectangle in which all sides are of equal length is called a  _square_ .

15. A  _circle_  is a plane figure that consists of all points that lie the same distance from some fixed point.

17. The length of a line segment from the center of a circle to any point of the circle is called the  _radius_  of the circle.

19. The radius is  _half_  of the diameter

21. A solid that consists of six sides known as *faces*, all of which are rectangles, is known as a  _rectangular_  solid.

23. A solid with three or more triangular-shaped faces that share a common vertex is called a  _pyramid_ .

25. A solid with two identical plane figure bases joined by line segments that are perpendicular to these bases is known as a  _cylinder_ .

### Guide Problems

27. Classify each triangle as equilateral, isosceles, or scalene.

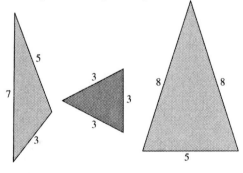

| scalene, because all sides have different lengths | equilateral or isosceles because all sides have equal length | isosceles, because two sides have equal length |

*In problems 29 through 37, classify each triangle as equilateral, isosceles, or scalene using the definitions of these triangles. Also, classify each triangle as acute, right, or obtuse using the definitions of these triangles.*

29. The triangle shown is an equilateral or isosceles triangle because all sides have equal length. The triangle is also an acute triangle because all angles are acute.

31. The triangle shown is an isosceles triangle because two sides have equal length. The triangle is also an acute triangle because all angles are acute.

33. The triangle shown is an isosceles triangle because two sides have equal length. The triangle is also an obtuse triangle because the triangle has an obtuse angle.

35. The triangle shown is an isosceles triangle because two sides have equal length. The triangle is also an obtuse triangle because the triangle has an obtuse angle.

37. The triangle shown is a scalene triangle because all sides have different lengths. The triangle is also an obtuse triangle because the triangle has an obtuse angle.

## Guide Problems

39. The sum of the measures of the three angles of a triangle is $180°$. Find the measure of the unknown angle.

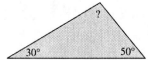

    a.   Find the sum of the two given angles.
          $50° + 30° = 80°$

    b.   Subtract the sum from $180°$.
          $180° - 80° = 100°$
          The measure of the unknown angle is $100°$.

*For problems 41 through 47, find the measure of the unknown angle of each triangle by first finding the sum of the two given angles and subtracting the sum from $180°$.*

41. $72° + 54° = 126°$, so $180° - 126° = 54°$.            43. $80° + 20° = 100°$, so $180° - 100° = 80°$.

45. $65° + 63° = 128°$, so $180° - 128° = 52°$.            47. $37° + 85° = 122°$, so $180° - 122° = 58°$.

## Guide Problems

49. Label each quadrilateral as a rectangle, square, trapezoid, or rhombus.

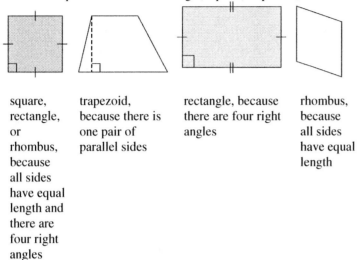

square, rectangle, or rhombus, because all sides have equal length and there are four right angles

trapezoid, because there is one pair of parallel sides

rectangle, because there are four right angles

rhombus, because all sides have equal length

*For problems 51 through 59, identify each quadrilateral as a rectangle, square, trapezoid, or rhombus using the definitions of each figure.*

51. This figure is a trapezoid because there is one pair of parallel sides.

53. This figure is a rectangle because there are four right angles.

55. This figure is a trapezoid because there is one pair of parallel sides.

57. This figure is a square, rectangle, or rhombus because all sides have equal length and there are four right angles.

59. This figure is a rhombus because all sides have equal length.

*For problems 60 through 71, identify each polygon as a triangle, quadrilateral, pentagon, hexagon, or octagon according to the number of sides of the figure.*

61. The figure is a triangle because it has 3 sides.

63. The figure is a hexagon because it has 6 sides.

65. The figure is a triangle because it has 3 sides.

67. The figure is a pentagon because it has 5 sides.

69. The figure is a triangle because it has 3 sides.

71. The figure is an octagon because it has 8 sides.

## Guide Problems

73. To find the diameter of a circle, multiply the radius by 2.
    Diameter = Radius · 2
    What is the diameter of a circle with radius of 4 centimeters?

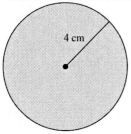

    Multiply the radius by 2.
    4 cm · 2 = 8 cm, so the diameter is 8 centimeters.

*For problems 75 through 79, find the radius or diameter using the given information. If the diameter is given divide by 2 to find the radius. If the radius is given, multiply by 2 to find the diameter.*

75.  50 mi · 2 = 100 mi               77.  7 m · 2 = 14 m               79.  11 yds · 2 = 22 yds

## Guide Problems

81. Identify each solid as a sphere, a cylinder, or a cone.

| cylinder, because the solid has two identical circular bases joined by line segments perpendicular to these bases | cone, because the solid has a circular base and a common vertex | sphere, because the solid consists of all points that lie the same distance from some fixed point |

*For problems 83 through 93, identify each solid using the definitions of the solids.*

83. The solid is a cylinder because the solid has two identical circular bases joined by line segments perpendicular to these bases.

85. The solid is a cone because the solid has a circular base and a common vertex.

87. The solid is a cube because there are six square faces.

89. The solid is a rectangular solid because there are six rectangular faces.

91. The solid is a sphere because the solid consists of all points that lie the same distance from some fixed point.

93. The solid is a pyramid because there are four triangular-shaped faces with a common vertex.

## Cumulative Skills Review

1. Angle $B$ and the given angle are supplementary angles. To find the supplement of an angle, subtract the measure of the known angle from $180°$.
$180° - 57° = 123°$, so the measure of angle $B$ is $123°$.

3. To solve $\dfrac{8}{13} = \dfrac{24}{t}$, cross multiply to find the cross products.
$13 \cdot 24 = 312$
$\quad 8 \cdot t = 8t$
Separate the cross products by an equal sign to form an equation.
$312 = 8t$
Divide both sides of the equation by the number on the side with the unknown.
$\dfrac{312}{8} = \dfrac{8t}{8}$
Simplify.
$t = 39$

5. To divide $\dfrac{21}{15} \div \dfrac{3}{7}$, rewrite the division problem as a multiplication problem using the reciprocal of the divisor, then multiply and simplify.
$$\frac{21}{15} \div \frac{3}{7} = \frac{21}{15} \cdot \frac{7}{3} = \frac{\overset{7}{\cancel{21}}}{15} \cdot \frac{7}{\underset{1}{\cancel{3}}} = \frac{7}{15} \cdot \frac{7}{1} = \frac{49}{15} \text{ or } 3\frac{4}{15}$$

7. To convert a percent to a decimal, drop the percent sign and multiply by $0.01$.
$37\% = 37 \cdot 0.01 = 0.37$

9. Write a ratio of total donations to number of days and simplify, rounding to the nearest cent.
$\dfrac{\$2893.10}{24 \text{ days}} \approx \$120.55 \text{ per day}$

## Section 7.3    Perimeter and Circumference

## Concept Check

1. The distance around a polygon is called the __perimeter__ of the polygon.

3. To find the perimeter of a rectangle, we find the sum of all sides or use the formula $P = $ __$2l + 2w$__ .

5. The distance around a circle is called the __circumference__ .

7. If $d$ represents the diameter of a circle, then the circumference $C$ is found using the formula $C = $ __$\pi d$__ .

## Guide Problems

9.   What is the perimeter of the triangle?

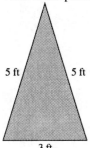

Add the lengths of all sides. 5 ft + 5 ft + 3 ft = <u>13 ft</u>

11.  What is the perimeter of the rectangle?

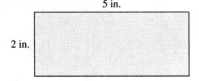

Use the formula $P = 2l + 2w$, where $l = 5$ inches and $w = 2$ inches.
$$P = (2\cdot5 \text{ in.}) + (2\cdot2 \text{ in.}) = 10 \text{ in.} + 4 \text{ in.} = 14 \text{ in.}$$

*For problems 13 through 21, find the perimeter of each polygon. Add all sides or use a perimeter formula.*

13.  $P = 4s = 4\cdot5 \text{ mi} = 20 \text{ mi}$               15.  $P = 6 \text{ yd} + 11 \text{ yd} + 7.5 \text{ yd} + 10.5 \text{ yd} = 35 \text{ yd}$

17.  $P = (2\cdot12 \text{ cm}) + (2\cdot7 \text{ cm}) = 24 \text{ cm} + 14 \text{ cm} = 38 \text{ cm}$

19.  $P = 5 \text{ m} + 9 \text{ m} + 3 \text{ m} + 13 \text{ m} + 8 \text{ m} = 38 \text{ m}$

21.  $P = 2 \text{ cm} + 3 \text{ cm} + 3 \text{ cm} + 2 \text{ cm} + 3 \text{ cm} + 3 \text{ cm} = 16 \text{ cm}$

## Guide Problems

23.  What is the circumference of a circle with the radius 6 miles? Use 3.14 for π. Round to the nearest tenth of a mile.

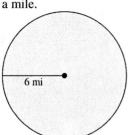

$C = 2\pi r$
$\approx (2)(3.14)(6 \text{ mi})$
$\approx \underline{37.7 \text{ miles}}$

*For problems 25 through 31, find the circumference of each circle. Use 3.14 for π. Round to the nearest hundredth. Use the formula $C = \pi d$ if the diameter is given or the formula $C = 2\pi r$ if the radius is given.*

25. The radius is 2.5 cm.
$$C = 2\pi r$$
$$\approx (2)(3.14)(2.5 \text{ cm})$$
$$\approx 15.7 \text{ cm}$$

27. The radius is 3.5 cm.
$$C = 2\pi r$$
$$\approx (2)(3.14)(3.5 \text{ cm})$$
$$\approx 21.98 \text{ cm}$$

29. The radius is 6.5 cm.
$$C = 2\pi r$$
$$\approx (2)(3.14)(6.5 \text{ cm})$$
$$\approx 40.82 \text{ cm}$$

31. The diameter is 20 mi.
$$C = \pi d$$
$$\approx (3.14)(20 \text{ mi})$$
$$\approx 62.8 \text{ mi}$$

*For problems 33 through 36, solve each problem by finding the perimeter using the methods practiced in this section.*

33. a. *P* = 32 ft + 32 ft + 32 ft = 96 ft

    b. Multiply the perimeter by the cost per foot.
    $$96 \text{ ft} \cdot \frac{\$23}{\text{foot}} = \$2208$$
    Brenda and Jack will spend $2208 to decorate the roof.

35. Find the perimeter of the course.
    *P* = 1 mi + 4.5 mi + 4.5 mi = 10 mi
    Denise will run 10 miles.

## Cumulative Skills Review

1.  Write the decimals so that the decimal points are vertically aligned. Then subtract as with whole numbers.

    $$\begin{array}{r} 10\,3\,5\,.\,6 \\ -\;\;1\,5\,.\,8 \\ \hline 10\,1\,9\,.\,8 \end{array}$$

3.  Because 40% = 0.4, we can convert either form to a fraction.
    $$0.4 = \frac{4}{10} = \frac{2}{5}$$

5.  Write the decimals so that the decimal points are vertically aligned. Then subtract as with whole numbers.

    $$\begin{array}{r} 1\,5\,.\,0\,0\,3 \\ -\,3\,.\,7\,7\,4 \\ \hline 1\,1\,.\,2\,2\,9 \end{array}$$

7.  To find the supplement of an angle, subtract the measure of the known angle from 180°.
    $180° - 116° = 64°$, so the measure of the unknown angle is 64°.

9.  The figure is a trapezoid because there is one pair of parallel sides.

## Section 7.4    Area

## Concept Check

1.   _Area_  is the measure associated with the interior of a closed plane figure.

3.   The formula for the area of a rectangle is _A = lw_ .

5.   The formula for the area of a parallelogram is _A = bh_ .

7.   The formula for the area of a trapezoid is   $A = \dfrac{1}{2}(a+b)h$  .

## Guide Problems

9.   Find the area of the rectangle.

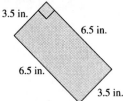

3.5 in.

6.5 in.

6.5 in.

3.5 in.

$A = lw$

$A = (6.5 \text{ in.})(3.5 \text{ in.})$

$A = \underline{22.75 \text{ in.}^2}$

*For problems 11 through 27, use the formulas for the area of a rectangle and the area of a square to find the area of each figure. The formula for the area of a rectangle is A = lw and the formula for the area of a square is A = s².*

11.  The figure is a rectangle with $l$ = 17 yd and $w$ = 5.6 yd.

$A = lw$

$A = (17 \text{ yd})(5.6 \text{ yd})$

$A = 95.2 \text{ yd}^2$

13.  The figure is a square with $s$ = 6 cm.

$A = s^2$

$A = (6 \text{ cm})^2$

$A = 36 \text{ cm}^2$

15.  The figure is a rectangle with $l$ = 31.7 in. and $w$ = 15.1 in.

$A = lw$

$A = (31.7 \text{ in.})(15.1 \text{ in.})$

$A = 478.67 \text{ in.}^2$

17. The figure is a square with $s = 14$ m.

$A = s^2$

$A = (14 \text{ m})^2$

$A = 196 \text{ m}^2$

19. The figure is a rectangle with $l = 41$ mi and $w = 18$ mi.

$A = lw$

$A = (41 \text{ mi})(18 \text{ mi})$

$A = 738 \text{ mi}^2$

21. The figure is a square with $s = 20$ ft.

$A = s^2$

$A = (20 \text{ ft})^2$

$A = 400 \text{ ft}^2$

23. The figure is a rectangle with $l = 150$ mm and $w = 100$ mm.

$A = lw$

$A = (150 \text{ mm})(100 \text{ mm})$

$A = 15,000 \text{ mm}^2$

25. The figure is a square with $s = 20$ ft.

$A = s^2$

$A = (20 \text{ ft})^2$

$A = 400 \text{ ft}^2$

27. The figure is a square with $s = 10$ m.

$A = s^2$

$A = (10 \text{ m})^2$

$A = 100 \text{ m}^2$

## Guide Problems

29. Find the area of the parallelogram.

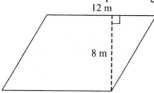

$A = bh$

$A = (12 \text{ m})(\underline{8 \text{ m}})$

$A = \underline{96 \text{ m}^2}$

31. Find the area of the trapezoid.

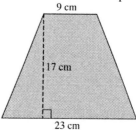

$A = \dfrac{1}{2}(a+b)h$

$A = \dfrac{1}{2}(23 \text{ cm} + \underline{9 \text{ cm}})\underline{17 \text{ cm}}$

$A = \underline{272 \text{ cm}^2}$

*For problems 33 through 47 use the formulas for the area of a triangle, the area of a parallelogram, the area of a trapezoid, and the area of a circle to find the area of each figure. Use 3.14 for π.*

33. The figure is a triangle with $b = 30$ mm and $h = 23$ mm.

$$A = \frac{1}{2}bh$$

$$A = \frac{1}{2}(30 \text{ mm})(23 \text{ mm})$$

$$A = 345 \text{ mm}^2$$

35. The figure is a triangle with $b = 8.5$ mi and $h = 4$ mi.

$$A = \frac{1}{2}bh$$

$$A = \frac{1}{2}(8.5 \text{ mi})(4 \text{ mi})$$

$$A = 17 \text{ mi}^2$$

37. The figure is a parallelogram with $b = 50$ in. and $h = 45$ in.

$$A = bh$$

$$A = (50 \text{ in.})(45 \text{ in.})$$

$$A = 2250 \text{ in.}^2$$

39. The figure is a parallelogram with $b = 15$ ft and $h = 11.5$ ft.

$$A = bh$$

$$A = (15 \text{ ft})(11.5 \text{ ft})$$

$$A = 172.5 \text{ ft}^2$$

41. The figure is a trapezoid with $a = 11$ m, $b = 6$ m, and $h = 5$ m.

$$A = \frac{1}{2}(a+b)h$$

$$A = \frac{1}{2}(11 \text{ m} + 6 \text{ m})5 \text{ m}$$

$$A = 42.5 \text{ m}^2$$

43. The figure is a trapezoid with $a = 10$ yd, $b = 6$ yd, and $h = 7.5$ yd.

$$A = \frac{1}{2}(a+b)h$$

$$A = \frac{1}{2}(10 \text{ yd} + 6 \text{ yd})7.5 \text{ yd}$$

$$A = 60 \text{ yd}^2$$

45. The figure is a circle with $d = 88$ cm, so $r = 44$ cm.

$$A = \pi r^2$$

$$A \approx (3.14)(44 \text{ cm})^2$$

$$A = 6079.04 \text{ cm}^2$$

47. The figure is a circle with $r = 2$ ft.

$$A = \pi r^2$$

$$A \approx (3.14)(2 \text{ ft})^2$$

$$A = 12.56 \text{ ft}^2$$

*For problems 49 through 57, apply your knowledge of the area formulas to solve each problem.*

49. Use the formula for the area of a square with $s = 7.5$ ft.

$$A = s^2$$

$$A = (7.5 \text{ ft})^2$$

$$A = 56.25 \text{ ft}^2$$

51. Use the formula for the area of a square with $s = 70$ mm.

$$A = s^2$$

$$A = (70 \text{ mm})^2$$

$$A = 4900 \text{ mm}^2$$

53. Use the formula for the area of a circle. Recall that the radius of a circle is half the diameter.

Dime:                                   Quarter:

$$r = \frac{1}{2}d \qquad\qquad\qquad r = \frac{1}{2}d$$

$$r = \frac{1}{2}(18 \text{ mm}) \qquad\qquad r = \frac{1}{2}(25 \text{ mm})$$

$$r = 9 \text{ mm} \qquad\qquad\qquad r = 12.5 \text{ mm}$$

$$A = \pi r^2 \qquad\qquad\qquad A = \pi r^2$$

$$A \approx (3.14)(9 \text{ mm})^2 \qquad\quad A \approx (3.14)(12.5 \text{ mm})^2$$

$$A \approx 254.3 \text{ mm}^2 \qquad\qquad A \approx 490.6 \text{ mm}^2$$

Subtract the area of the dime from the area of the quarter.
490.6 mm$^2$ − 254.3 mm$^2$ = 236.3 mm$^2$
The quarter is 236.3 mm$^2$ larger than the dime.

55. Use the formula for the area of a parallelogram with $b = 250$ yd and $h = 62.5$ yd.

$$A = bh$$

$$A = (250 \text{ yd})(62.5 \text{ yd})$$

$$A = 15{,}625 \text{ yd}^2$$

57. Let $w$ represent the area of window space for the wall. The area of the wall can be found using the formula for the area of a rectangle.

$$A = lw$$

$$A = (10 \text{ ft})(15 \text{ ft})$$

$$A = 150 \text{ ft}^2$$

Use the ratio given to find the value of $w$.

$$\frac{w}{150 \text{ ft}^2} = \frac{5}{12}$$

$$12 \cdot w = (150 \text{ ft}^2) \cdot 5$$

$$12w = 750 \text{ ft}^2$$

$$\frac{12w}{12} = \frac{750 \text{ ft}^2}{12}$$

$$w = 62.5 \text{ ft}^2$$

The area of window space should be 62.5 ft$^2$.

## Cumulative Skills Review

1. The triangle shown is an equilateral or isosceles triangle because all sides have equal length. The triangle is also an acute triangle because all angles are acute.

3. Identify the parts of the percent problem.
   Amount: unknown, $a$
   Percent: 82%
   Base: 200
   Write the problem as a percent equation.
   $a = 82\% \cdot 200$
   Convert the percent to a decimal and solve the equation.
   $a = 0.82 \cdot 200$
   $a = 164$

5. The change amount is $132 - 86 = 46$.
   $$\text{Percent change} = \frac{\text{Change amount}}{\text{Original amount}} = \frac{46}{86} \approx 0.53$$
   The percent change is approximately 53%.

7. Find the sum of the two given angles.
   $74° + 32° = 106°$
   Subtract the sum from $180°$.
   $180° - 106° = 74°$
   The measure of the unknown angle is $74°$.

9. $P = 13$ yd + 13 yd + 13 yd = 39 yd

## Section 7.5  Square Roots and the Pythagorean Theorem

## Concept Check

1. The __square__ of a number is the number times itself.

3. The symbol $\sqrt{\phantom{x}}$ is called a __radical__ __sign__ .

5. A whole number or fraction that is the square of another whole number or fraction is called a __perfect__ __square__ .

7. The two sides that meet to form the right angle of a right triangle are called __legs__ .

## Guide Problems

9. Find the principal square root of 81.
   $\left(\underline{\ 9\ }\right)^2 = 81$
   $\sqrt{81} = \underline{\ 9\ }$

*For problems 11 through 33, find each principal square root by identifying the number that must be squared to obtain the radicand.*

11. $1^2 = 1$ so $\sqrt{1} = 1$.   13. $11^2 = 121$ so $\sqrt{121} = 11$.   15. $21^2 = 441$ so $\sqrt{441} = 21$.

17. $19^2 = 361$ so $\sqrt{361} = 19$.   19. $9^2 = 81$ so $\sqrt{81} = 9$.   21. $23^2 = 529$ so $\sqrt{529} = 23$.

23. $17^2 = 289$ so $\sqrt{289} = 17$.   25. $15^2 = 225$ so $\sqrt{225} = 15$.   27. $13^2 = 169$ so $\sqrt{169} = 13$.

29. $7^2 = 49$ so $\sqrt{49} = 7$.   31. $5^2 = 25$ so $\sqrt{25} = 5$.   33. $3^2 = 9$ so $\sqrt{9} = 3$.

## Guide Problems

35. Approximate the principal square root of 17 using Appendix D or a calculator. Round to the nearest hundredth.

$$\sqrt{17} \approx \underline{4.12}$$

*For problems 37 through 51, approximate each principal square root using Appendix D or a calculator. Round to the nearest hundredth.*

37. $\sqrt{61} \approx 7.81$   39. $\sqrt{311} \approx 17.64$   41. $\sqrt{204} \approx 14.28$   43. $\sqrt{66} \approx 8.12$

45. $\sqrt{960} \approx 30.98$   47. $\sqrt{145} \approx 12.04$   49. $\sqrt{467} \approx 21.61$   51. $\sqrt{252} \approx 15.87$

## Guide Problems

53. Find the unknown length in the right triangle. Round to the nearest hundredth of a foot.

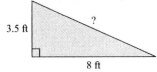

$$c = \sqrt{a^2 + b^2}$$
$$c = \sqrt{(\underline{8\ \text{ft}})^2 + (3.5\ \text{ft})^2}$$
$$c = \sqrt{\underline{64\ \text{ft}^2} + 12.25\ \text{ft}^2}$$
$$c = \sqrt{\underline{76.25\ \text{ft}^2}}$$
$$c \approx \underline{8.73\ \text{ft}}$$

*For problems 55 through 71, find the unknown length in each triangle using the Pythagorean Theorem in an appropriate form. Round to the nearest hundredth, if necessary.*

55. $b = \sqrt{c^2 - a^2}$
$b = \sqrt{(50\ \text{m})^2 - (23\ \text{m})^2}$
$b = \sqrt{2500\ \text{m}^2 - 529\ \text{m}^2}$
$b = \sqrt{1971\ \text{m}^2}$
$b \approx 44.40\ \text{m}$

57. $b = \sqrt{c^2 - a^2}$
$b = \sqrt{(35\ \text{cm})^2 - (18\ \text{cm})^2}$
$b = \sqrt{1225\ \text{cm}^2 - 324\ \text{cm}^2}$
$b = \sqrt{901\ \text{cm}^2}$
$b \approx 30.02\ \text{cm}$

59. $b = \sqrt{c^2 - a^2}$
$b = \sqrt{(9\ \text{yd})^2 - (5\ \text{yd})^2}$
$b = \sqrt{81\ \text{yd}^2 - 25\ \text{yd}^2}$
$b = \sqrt{56\ \text{yd}^2}$
$b \approx 7.48\ \text{yd}$

61. $c = \sqrt{a^2 + b^2}$

$c = \sqrt{(11 \text{ in.})^2 + (13 \text{ in.})^2}$

$c = \sqrt{121 \text{ in.}^2 + 169 \text{ in.}^2}$

$c = \sqrt{290 \text{ in.}^2}$

$c \approx 17.03 \text{ in.}$

63. $b = \sqrt{c^2 - a^2}$

$b = \sqrt{(22 \text{ mi})^2 - (12 \text{ mi})^2}$

$b = \sqrt{484 \text{ mi}^2 - 144 \text{ mi}^2}$

$b = \sqrt{340 \text{ mi}^2}$

$b \approx 18.44 \text{ mi}$

65. $b = \sqrt{c^2 - a^2}$

$b = \sqrt{(32.5 \text{ mm})^2 - (25 \text{ mm})^2}$

$b = \sqrt{1056.25 \text{ mm}^2 - 625 \text{ mm}^2}$

$b = \sqrt{431.25 \text{ mm}^2}$

$b \approx 20.77 \text{ mm}$

67. $c = \sqrt{a^2 + b^2}$

$c = \sqrt{(7.5 \text{ ft})^2 + (15 \text{ ft})^2}$

$c = \sqrt{56.25 \text{ ft}^2 + 225 \text{ ft}^2}$

$c = \sqrt{281.25 \text{ ft}^2}$

$c \approx 16.77 \text{ ft}$

69. $b = \sqrt{c^2 - a^2}$

$b = \sqrt{(19 \text{ yd})^2 - (11 \text{ yd})^2}$

$b = \sqrt{361 \text{ yd}^2 - 121 \text{ yd}^2}$

$b = \sqrt{240 \text{ yd}^2}$

$b \approx 15.49 \text{ yd}$

71. $c = \sqrt{a^2 + b^2}$

$c = \sqrt{(19 \text{ in.})^2 + (32 \text{ in.})^2}$

$c = \sqrt{361 \text{ in.}^2 + 1024 \text{ in.}^2}$

$c = \sqrt{1385 \text{ in.}^2}$

$c \approx 37.22 \text{ in.}$

*For problems 73 through 75, apply your knowledge of square roots and the Pythagorean Theorem to solve each problem.*

73. Use the Pythagorean Theorem to find the length of the hypotenuse of the sail.

$c = \sqrt{a^2 + b^2}$

$c = \sqrt{(8.6 \text{ ft})^2 + (56.5 \text{ ft})^2}$

$c = \sqrt{73.96 \text{ ft}^2 + 3192.25 \text{ ft}^2}$

$c = \sqrt{3266.21 \text{ ft}^2}$

$c \approx 57.15 \text{ ft}$

The hypotenuse of the sail is approximately 57.15 ft.

75. Use the Pythagorean Theorem to find the distance from third base to first base.

$c = \sqrt{a^2 + b^2}$

$c = \sqrt{(90 \text{ ft})^2 + (90 \text{ ft})^2}$

$c = \sqrt{8100 \text{ ft}^2 + 8100 \text{ ft}^2}$

$c = \sqrt{16,200 \text{ ft}^2}$

$c \approx 127.3 \text{ ft}$

The throw from third base to first base is approximately 127.3 ft.

## Cumulative Skills Review

1. The figure is a triangle with $b = 8.5$ mi and $h = 4$ mi.

$A = \frac{1}{2}bh$

$A = \frac{1}{2}(8.5 \text{ mi})(4 \text{ mi})$

$A = 17 \text{ mi}^2$

3.  Write an appropriate unit ratio.

$$\text{Unit ratio} = \frac{\text{New units}}{\text{Original units}} = \frac{1 \text{ t}}{2000 \text{ lb}}$$

Multiply the original measure by the unit fraction.

$$212{,}000 \text{ lb} \cdot \frac{1 \text{ t}}{2000 \text{ lb}} = 106 \text{ t}$$

5.  Identify the parts of the percent problem.
    Amount: 595
    Part: 17
    Base: unknown, $b$
    Substitute the values of the amount, base, and part into the proportion.

$$\frac{595}{b} = \frac{17}{100}$$

Cross multiply and set the products equal.

$$b \cdot 17 = 595 \cdot 100$$

$$17b = 59{,}500$$

Divide both sides of the equation by the number on the side with the unknown.

$$\frac{17b}{17} = \frac{59{,}500}{17}$$

$$b = 3500$$

7.  To find the radius of a circle, divide the diameter by 2.

$$\frac{57 \text{ in.}}{2} = 28.5 \text{ in.}$$

9.  Cross multiply and set the products equal.

$$v \cdot 90 = 30 \cdot 30$$

$$90v = 900$$

Divide both sides of the equation by the number on the side with the unknown.

$$\frac{90v}{90} = \frac{900}{90}$$

$$v = 10$$

## Section 7.6    Volume

## Concept Check

1.  The measure of the amount of interior space of a solid is called _volume_.

3.  If $l$ represents the length of a rectangular solid, $w$ the width, and $h$ the height, then the formula for the volume is _$V = lwh$_.

5.  The formula for the volume of a cylinder is _$V = \pi r^2 h$_.

7.  The formula for the volume of a pyramid is _$V = \dfrac{1}{3} Bh$_.

## Guide Problems

9.   Find the volume of the rectangular solid.

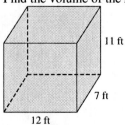

        11 ft

        7 ft

  12 ft

$V = lwh$

$V = (12 \text{ ft})(\underline{7 \text{ ft}})(\underline{11 \text{ ft}})$

$V = \underline{924 \text{ ft}^3}$

*For problems 11 through 23, find the volume of each rectangular solid. Use the formula V = lwh for a rectangular solid or V = s³ for a cube.*

11.   The figure is a rectangular solid with $l = 9$ m, $w = 3$ m, and $h = 5$ m.

    $V = lwh$

    $V = (9 \text{ m})(3 \text{ m})(5 \text{ m})$

    $V = 135 \text{ m}^3$

13.   The figure is a rectangular solid with $l = 7$ mm, $w = 5$ mm, and $h = 25$ mm.

    $V = lwh$

    $V = (7 \text{ mm})(5 \text{ mm})(25 \text{ mm})$

    $V = 875 \text{ mm}^3$

15.   The figure is a rectangular solid with $l = 26$ cm, $w = 20$ cm, and $h = 15$ cm.

    $V = lwh$

    $V = (26 \text{ cm})(20 \text{ cm})(15 \text{ cm})$

    $V = 7800 \text{ cm}^3$

17.   The figure is a rectangular solid with $l = 15$ in., $w = 11$ in., and $h = 31$ in.

    $V = lwh$

    $V = (15 \text{ in.})(11 \text{ in.})(31 \text{ in.})$

    $V = 5115 \text{ in.}^3$

19.   The figure is a rectangular solid with $l = 5.5$ m, $w = 3$ m, and $h = 9$ m.

    $V = lwh$

    $V = (5.5 \text{ m})(3 \text{ m})(9 \text{ m})$

    $V = 148.5 \text{ m}^3$

21.   The figure is a rectangular solid with $l = 26$ mi, $w = 5$ mi, and $h = 3.5$ mi.

    $V = lwh$

    $V = (26 \text{ mi})(5 \text{ mi})(3.5 \text{ mi})$

    $V = 455 \text{ mi}^3$

23. The figure is a rectangular solid with $l = 4.7$ m, $w = 1.9$ m, and $h = 1.6$ m.

$V = lwh$

$V = (4.7 \text{ m})(1.9 \text{ m})(1.6 \text{ m})$

$V = 14.288 \text{ m}^3$

## Guide Problems

25. Find the volume of the cylinder. Use 3.14 for $\pi$. Round to the nearest hundredth.

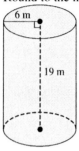

6 m

19 m

$V = \pi r^2 h$

$V \approx (3.14)\underline{(6 \text{ m})}^2 \underline{(19 \text{ m})}$

$= \underline{2147.76 \text{ m}^2}$

27. Find the volume of the cone. Use 3.14 for $\pi$. Round to the nearest hundredth.

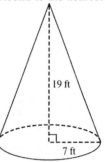

19 ft

7 ft

$V = \frac{1}{3}\pi r^2 h$

$V \approx \frac{1}{3}(3.14)\underline{(7 \text{ ft})}^2 \underline{(19 \text{ ft})}$

$\approx \underline{974.45 \text{ ft}^3}$

*For problems 29 through 59, find the volume of each solid using the appropriate formula. Use 3.14 for $\pi$. Round to the nearest hundredth, if necessary.*

29. The figure is a cylinder with $r = 10$ cm and $h = 15$ cm.

$V = \pi r^2 h$

$V \approx (3.14)(10 \text{ cm})^2 (15 \text{ cm})$

$= (3.14)(100 \text{ cm}^2)(15 \text{ cm})$

$= 4710 \text{ cm}^3$

31. The figure is a cone with $r = 7$ m and $h = 22$ m.

$V = \frac{1}{3}\pi r^2 h$

$V \approx \frac{1}{3}(3.14)(7 \text{ m})^2 (22 \text{ m})$

$= \frac{1}{3}(3.14)(49 \text{ m}^2)(22 \text{ m})$

$\approx 1128.31 \text{ m}^3$

33. The figure is a pyramid with $B = (9 \text{ m})(8 \text{ m}) = 72 \text{ m}^2$ and $h = 11$ m.

$$V = \frac{1}{3}Bh$$

$$V = \frac{1}{3}\left(72 \text{ m}^2\right)(11 \text{ m})$$

$$= 264 \text{ m}^3$$

35. The figure is a sphere with $r = 9$ m.

$$V = \frac{4}{3}\pi r^3$$

$$V \approx \frac{4}{3}(3.14)(9 \text{ m})^3$$

$$= \frac{4}{3}(3.14)\left(729 \text{ m}^3\right)$$

$$= 3052.08 \text{ m}^3$$

37. The figure is a cylinder with $r = 3$ yd and $h = 12$ yd.

$$V = \pi r^2 h$$

$$V \approx (3.14)(3 \text{ yd})^2 (12 \text{ yd})$$

$$= (3.14)\left(9 \text{ yd}^2\right)(12 \text{ yd})$$

$$= 339.12 \text{ yd}^3$$

39. The figure is a cone with $r = 9$ cm and $h = 17$ cm.

$$V = \frac{1}{3}\pi r^2 h$$

$$V \approx \frac{1}{3}(3.14)(9 \text{ cm})^2 (17 \text{ cm})$$

$$= \frac{1}{3}(3.14)\left(81 \text{ cm}^2\right)(17 \text{ cm})$$

$$= 1441.26 \text{ cm}^3$$

41. The figure is a pyramid with $B = (9 \text{ yd})(7 \text{ yd}) = 63 \text{ yd}^2$ and $h = 17$ yd.

$$V = \frac{1}{3}Bh$$

$$V = \frac{1}{3}\left(63 \text{ yd}^2\right)(17 \text{ yd})$$

$$= 357 \text{ yd}^3$$

43. The figure is a sphere with $r = 40$ cm.

$$V = \frac{4}{3}\pi r^3$$

$$V \approx \frac{4}{3}(3.14)(40 \text{ cm})^3$$

$$= \frac{4}{3}(3.14)\left(64,000 \text{ cm}^3\right)$$

$$\approx 267,946.67 \text{ cm}^3$$

45. The figure is a cylinder with $r = 2$ in. and $h = 8$ in.

$$V = \pi r^2 h$$

$$V \approx (3.14)(2 \text{ in.})^2 (8 \text{ in.})$$

$$= (3.14)\left(4 \text{ in.}^2\right)(8 \text{ in.})$$

$$= 100.48 \text{ in.}^3$$

47. The figure is a cone with $r = 5$ yd and $h = 15$ yd.

$$V = \frac{1}{3}\pi r^2 h$$

$$V \approx \frac{1}{3}(3.14)(5 \text{ yd})^2 (15 \text{ yd})$$

$$= \frac{1}{3}(3.14)(25 \text{ yd}^2)(15 \text{ yd})$$

$$= 392.5 \text{ yd}^3$$

49. The figure is a pyramid with $B = (5 \text{ in.})(8 \text{ in.}) = 40 \text{ in.}^2$ and $h = 17$ in.

$$V = \frac{1}{3}Bh$$

$$V = \frac{1}{3}(40 \text{ in.}^2)(17 \text{ in.})$$

$$\approx 226.67 \text{ in.}^3$$

51. The figure is a sphere with $r = 5$ yd.

$$V = \frac{4}{3}\pi r^3$$

$$V \approx \frac{4}{3}(3.14)(5 \text{ yd})^3$$

$$= \frac{4}{3}(3.14)(125 \text{ yd}^3)$$

$$\approx 523.33 \text{ yd}^3$$

53. The figure is a cylinder with $r = 3$ m and $h = 5$ m.

$$V = \pi r^2 h$$

$$V \approx (3.14)(3 \text{ m})^2 (5 \text{ m})$$

$$= (3.14)(9 \text{ m}^2)(5 \text{ m})$$

$$= 141.3 \text{ m}^3$$

55. The figure is a cone with $r = 8$ in. and $h = 12$ in.

$$V = \frac{1}{3}\pi r^2 h$$

$$V \approx \frac{1}{3}(3.14)(8 \text{ in.})^2 (12 \text{ in.})$$

$$= \frac{1}{3}(3.14)(64 \text{ in.}^2)(12 \text{ in.})$$

$$= 803.84 \text{ in.}^3$$

57. The figure is a pyramid with $B = (7.5 \text{ ft})(9.5 \text{ ft}) = 71.25 \text{ ft}^2$ and $h = 13$ ft.

$$V = \frac{1}{3}Bh$$

$$V = \frac{1}{3}(71.25 \text{ ft}^2)(13 \text{ ft})$$

$$= 308.75 \text{ ft}^3$$

59. The figure is a sphere with $r = 6.5$ ft.

$$V = \frac{4}{3}\pi r^3$$

$$V \approx \frac{4}{3}(3.14)(6.5 \text{ ft})^3$$

$$= \frac{4}{3}(3.14)(274.625 \text{ ft}^3)$$

$$\approx 1149.76 \text{ ft}^3$$

*For problems 61 through 63, apply your knowledge of the volume formulas to solve each problem.*

61. a. The tank is a cylinder with $d = 7$ ft and $h = 28$ ft. So $r = \dfrac{7 \text{ ft}}{2} = 3.5$ ft.

$$V = \pi r^2 h$$

$$V \approx (3.14)(3.5 \text{ ft})^2 (28 \text{ ft})$$

$$= (3.14)(12.25 \text{ ft}^2)(28 \text{ ft})$$

$$\approx 1077 \text{ ft}^3$$

b. Write an appropriate unit ratio.

$$\text{Unit ratio} = \frac{\text{New units}}{\text{Original units}} = \frac{7.5 \text{ gallons}}{1 \text{ ft}^3}$$

Multiply the original measure by the unit fraction.

$$1077 \text{ ft}^3 \cdot \frac{7.5 \text{ gallons}}{1 \text{ ft}^3} \approx 8078 \text{ gallons}$$

63. The volcano resembles a pyramid with $B = (50 \text{ ft})^2 = 2500 \text{ ft}^2$ and $h = 250$ ft.

$$V = \frac{1}{3}Bh$$

$$V = \frac{1}{3}(2500 \text{ ft}^2)(250 \text{ ft})$$

$$\approx 208,333 \text{ ft}^3$$

## Cumulative Skills Review

1. $\dfrac{2}{15} + 0.42 = \dfrac{2}{15} + \dfrac{42}{100} = \dfrac{2 \cdot 20}{15 \cdot 20} + \dfrac{42 \cdot 3}{100 \cdot 3} = \dfrac{40}{300} + \dfrac{126}{300} = \dfrac{166}{300} = \dfrac{\overset{83}{\cancel{166}}}{\underset{150}{\cancel{300}}} = \dfrac{83}{150}$

3. $P = 8 \text{ mi} + 4 \text{ mi} + 9 \text{ mi} + 5 \text{ mi} + 6 \text{ mi} + 7 \text{ mi} = 39 \text{ mi}$

5. The figure shown is a line because it extends forever in either direction. It is named $\overleftrightarrow{WY}$ or $\overleftrightarrow{YW}$.

7. $b = \sqrt{c^2 - a^2}$

$b = \sqrt{(15 \text{ in.})^2 - (7 \text{ in.})^2}$

$b = \sqrt{225 \text{ in.}^2 - 49 \text{ in.}^2}$

$b = \sqrt{176 \text{ in.}^2}$

$b \approx 13.27 \text{ in.}$

9. Find the sum of the two given angles.
$79° + 20° = 99°$
Subtract the sum from $180°$.
$180° - 99° = 81°$
The measure of the unknown angle is $81°$.

## Numerical Facts of Life

1. Find the area of each triangular region, where $b = 45$ ft and $h = 54$ ft.

$A = \dfrac{1}{2}(45 \text{ ft})(54 \text{ ft}) = 1215 \text{ ft}^2$

3. If energy efficiency is increased by 22%, only $100\% - 22\% = 78\%$ as much energy is required.
$(1,000,000 \text{ BTUs per hour})(0.78) = 780,000 \text{ BTUs per hour}$

5. $(1884 \text{ ft}^3)\left(\dfrac{7.481 \text{ gallons}}{1 \text{ ft}^3}\right) \approx 14,000 \text{ gallons}$

## Chapter 7 Review Exercises

1. The figure shown consists of point $C$, point $B$, and point $S$.

2. The figure shown is a line because it extends forever in either direction. It is named $\overleftrightarrow{BD}$ or $\overleftrightarrow{DB}$.

3. The figure shown is a line segment because it has two endpoints. It is named $\overline{GK}$ or $\overline{KG}$.

4. The figure shown is a ray because it begins at one point and extends forever in one direction. It is named $\overrightarrow{CN}$.

5. $\angle K$, $\angle DKX$, $\angle XKD$; this angle is right because its measure is $90°$.

6. $\angle W$; this angle is straight because its measure is $180°$.

7. $\angle L$, $\angle SLE$, $\angle ELS$; this angle is obtuse because its measure is greater than $90°$ and less than $180°$.

8. $\angle H$, $\angle OHM$, $\angle MHO$; this angle is acute because its measure is greater than $0°$ and less than $90°$.

9. Complementary angles are two angles, the sum of whose degree measure is $90°$. To find the complement of an angle, subtract the measure of the angle from $90°$.
$90° - 25° = 65°$

10. Supplementary angles are two angles, the sum of whose degree measures is $180°$. To find the supplement of an angle, subtract the measure of the angle from $180°$.
$180° - 130° = 50°$

11. Supplementary angles are two angles, the sum of whose degree measures is 180°. To find the supplement of an angle, subtract the measure of the angle from 180°.
$180° - 18° = 162°$

12. Complementary angles are two angles, the sum of whose degree measure is 90°. To find the complement of an angle, subtract the measure of the angle from 90°.
$90° - 47° = 43°$

13. Supplementary angles are two angles, the sum of whose degree measures is 180°. To find the supplement of an angle, subtract the measure of the angle from 180°.
$180° - 120° = 60°$ so $\angle LMO = 60°$.

14. Complementary angles are two angles, the sum of whose degree measure is 90°. To find the complement of an angle, subtract the measure of the angle from 90°.
$90° - 28° = 62°$ so $\angle PDC = 62°$.

15. Complementary angles are two angles, the sum of whose degree measure is 90°. To find the complement of an angle, subtract the measure of the angle from 90°.
$90° - 17° = 73°$ so $\angle OKH = 73°$.

16. Supplementary angles are two angles, the sum of whose degree measures is 180°. To find the supplement of an angle, subtract the measure of the angle from 180°.
$180° - 25° = 155°$ so $\angle AXZ = 155°$.

17. The first triangle shown is an isosceles triangle because two sides have equal length. The triangle is also an acute triangle because all angles are acute.
The second triangle shown is a scalene triangle because all sides have different lengths. The triangle is also an obtuse triangle because the triangle has an obtuse angle.
The third triangle shown is a scalene triangle because all sides have different lengths. The triangle is also an obtuse triangle because the triangle has an obtuse angle.
The fourth triangle shown is an equilateral or isosceles triangle because all sides have equal length. The triangle is also an acute triangle because all angles are acute.

18. The first triangle shown is a scalene triangle because all sides have different lengths. The triangle is also an obtuse triangle because the triangle has an obtuse angle.
The second triangle shown is an isosceles triangle because two sides have equal length. The triangle is also an acute triangle because all angles are acute.
The third triangle shown is an equilateral or isosceles triangle because all sides have equal length. The triangle is also an acute triangle because all angles are acute.
The fourth triangle shown is an isosceles triangle because two sides have equal length. The triangle is also an acute triangle because all angles are acute.

19. The first triangle shown is a scalene triangle because all sides have different lengths. The triangle is also a right triangle because the triangle has a right angle.
The second triangle shown is a scalene triangle because all sides have different lengths. The triangle is also an obtuse triangle because the triangle has an obtuse angle.
The third triangle shown is an equilateral or isosceles triangle because all sides have equal length. The triangle is also an acute triangle because all angles are acute.
The fourth triangle shown is an equilateral or isosceles triangle because all sides have equal length. The triangle is also an acute triangle because all angles are acute.

20. The first triangle shown is an isosceles triangle because two sides have equal length. The triangle is also an acute triangle because all angles are acute.

   The second triangle shown is a scalene triangle because all sides have different lengths. The triangle is also a right triangle because the triangle has a right angle.

   The third triangle shown is an isosceles triangle because two sides have equal length. The triangle is also an obtuse triangle because the triangle has an obtuse angle.

   The fourth triangle shown is a scalene triangle because all sides have different lengths. The triangle is also a right triangle because the triangle has a right angle.

*For problems 21 through 24, find the measure of the unknown angle of each triangle by first finding the sum of the two given angles and subtracting the sum from 180°.*

21. $20° + 128° = 148°$, so $180° - 148° = 32°$.      22. $62° + 60° = 122°$, so $180° - 122° = 58°$.

23. $118° + 15° = 133°$, so $180° - 133° = 47°$.      24. $28° + 90° = 118°$, so $180° - 118° = 62°$.

25. The first figure is a square, rectangle, or rhombus because all sides have equal length and there are four right angles.

   The second figure is a trapezoid because there is one pair of parallel sides.

   The third figure is a rectangle because there are four right angles.

   The fourth figure is a rhombus because all sides have equal length.

26. The first figure is a trapezoid because there is one pair of parallel sides.

   The second figure is a rectangle because there are four right angles.

   The third figure is a rhombus because all sides have equal length.

   The fourth figure is a square, rectangle, or rhombus because all sides have equal length and there are four right angles.

27. The first figure is a hexagon because it has 6 sides.

   The second figure is an octagon because it has 8 sides.

   The third figure is a quadrilateral because it has 4 sides.

   The fourth figure is a pentagon because it has 5 sides.

28. The first figure is an octagon because it has 8 sides.

   The second figure is a hexagon because it has 6 sides.

   The third figure is a quadrilateral because it has 4 sides.

   The fourth figure is a pentagon because it has 5 sides.

*For problems 29 through 34, find the radius or diameter using the given information. If the diameter is given divide by 2 to find the radius. If the radius is given, multiply by 2 to find the diameter.*

29. $\dfrac{56\ m}{2} = 28\ m$      30. $\dfrac{15\ in.}{2} = 7.5\ in.$      31. $25\ ft \cdot 2 = 50\ ft$

32. $49\ yd \cdot 2 = 98\ yd$      33. $\dfrac{9\ mi}{2} = 4.5\ mi$      34. $30\ cm \cdot 2 = 60\ cm$

35. The first solid is a cube because there are six square faces.

   The second solid is a rectangular solid because there are six rectangular faces.

   The third solid is a pyramid because there are four triangular-shaped faces with a common vertex.

36. The first solid is a cone because the solid has a circular base and a common vertex.
    The second solid is a sphere because the solid consists of all points that lie the same distance from some
    fixed point.
    The third solid is a cylinder because the solid has two identical circular bases joined by line segments
    perpendicular to these bases.

37. $P = 16$ cm $+ 12$ cm $+ 9.5$ cm $= 37.5$ cm          38.  $P = 22$ in $+ 7.5$ in. $+ 23$ in. $+ 8$ in. $= 60.5$ in.

39. $P = 3$ m $+ 4.5$ m $+ 3$ m $+ 4.5$ m $= 15$ m          40.  $P = 3$ mi $+ 3.5$ mi $+ 1.5$ mi $+ 2$ mi $+ 2$ mi $= 12$ mi

*For problems 24 through 32, find the circumference of each circle. Use 3.14 for π. Use the formula C = πd if
the diameter is given or the formula C = 2πr if the radius is given.*

41.  The radius is 50 ft.          42.  The radius is 15 m.          43.  The diameter is 18 mm.
     $C = 2\pi r$                        $C = 2\pi r$                        $C = \pi d$

     $\approx (2)(3.14)(50 \text{ ft})$    $\approx (2)(3.14)(15 \text{ m})$     $\approx (3.14)(18 \text{ mm})$

     $= 314$ ft                         $= 94.2$ m                         $= 56.52$ mm

44.  The radius is 8 ft.          45.  The radius is 1080 mi.
     $C = 2\pi r$                        $C = 2\pi r$

     $\approx (2)(3.14)(8 \text{ ft})$     $\approx (2)(3.14)(1080 \text{ mi})$

     $= 50.24$ ft                       $= 6782.4$ mi

46.  The diameter is 26 in.
     $C = \pi d$

     $\approx (3.14)(26 \text{ in.})$

     $= 81.64$ in.
     The circumference is 81.64 inches, and after 100 revolutions the bicycle has traveled

     $81.64 \text{ in.} \cdot 100 \text{ revolutions} = 8164 \text{ in.} = 8164 \text{ in.} \cdot \dfrac{1 \text{ ft}}{12 \text{ in.}} \approx 680 \text{ ft.}$

47.  The figure is a square with $s = 2$ ft.
     $A = s^2$

     $A = (2 \text{ ft})^2$

     $A = 4 \text{ ft}^2$

48.  The figure is a triangle with $b = 30$ mm and $h = 23$ mm.

     $A = \dfrac{1}{2}bh$

     $A = \dfrac{1}{2}(30 \text{ mm})(23 \text{ mm})$

     $A = 345 \text{ mm}^2$

49.  The figure is a parallelogram with $b = 13$ yd and $h = 5.5$ yd.
     $A = bh$

     $A = (13 \text{ yd})(5.5 \text{ yd})$

     $A = 71.5 \text{ yd}^2$

50. The figure is a rectangle with $l = 41$ mi and $w = 18$ mi.
    $$A = lw$$
    $$A = (41 \text{ mi})(18 \text{ mi})$$
    $$A = 738 \text{ mi}^2$$

51. The figure is a circle with $d = 20$ m, so $r = 10$ m.
    $$A = \pi r^2$$
    $$A \approx (3.14)(10 \text{ m})^2$$
    $$A = 314 \text{ m}^2$$

52. The figure is a circle with $r = 44$ cm.
    $$A = \pi r^2$$
    $$A \approx (3.14)(44 \text{ cm})^2$$
    $$A = 6079.04 \text{ cm}^2$$

53. The top is a rectangle with $l = 58$ cm and $w = 32$ cm.
    $$A = lw$$
    $$A = (58 \text{ cm})(32 \text{ cm})$$
    $$A = 1856 \text{ cm}^2$$

54. The runway is a rectangle with $l = 8$ yd and $w = 3.5$ yd.
    $$A = lw$$
    $$A = (8 \text{ yd})(3.5 \text{ yd})$$
    $$A = 28 \text{ yd}^2$$

55. The garden is a circle with $d = 6$ ft, so $r = 3$ ft.
    $$A = \pi r^2$$
    $$A \approx (3.14)(3 \text{ ft})^2$$
    $$A = 28.26 \text{ ft}^2$$

56. The floor space is a circle with $r = 20$ m.
    $$A = \pi r^2$$
    $$A \approx (3.14)(20 \text{ m})^2$$
    $$A = 1256 \text{ m}^2$$

57. $6^2 = 36$ so $\sqrt{36} = 6$.

58. $8^2 = 64$ so $\sqrt{64} = 8$.

59. $\sqrt{17} \approx 4.12$

60. $\sqrt{58} = 7.62$

*For problems 61 through 64, find the unknown length in each triangle using the Pythagorean Theorem in an appropriate form. Round to the nearest hundredth, if necessary.*

61. $c = \sqrt{a^2 + b^2}$
    $$c = \sqrt{(3 \text{ m})^2 + (4 \text{ m})^2}$$
    $$c = \sqrt{9 \text{ m}^2 + 16 \text{ m}^2}$$
    $$c = \sqrt{25 \text{ m}^2}$$
    $$c = 5 \text{ m}$$

62. $c = \sqrt{a^2 + b^2}$
    $$c = \sqrt{(12 \text{ ft})^2 + (6.5 \text{ ft})^2}$$
    $$c = \sqrt{144 \text{ ft}^2 + 42.25 \text{ ft}^2}$$
    $$c = \sqrt{186.25 \text{ ft}^2}$$
    $$c = 13.65 \text{ ft}$$

63. $b = \sqrt{c^2 - a^2}$
    $$b = \sqrt{(20 \text{ in.})^2 - (12 \text{ in.})^2}$$
    $$b = \sqrt{400 \text{ in.}^2 - 144 \text{ in.}^2}$$
    $$b = \sqrt{256 \text{ in.}^2}$$
    $$b = 16 \text{ in.}$$

64. $b = \sqrt{c^2 - a^2}$
    $$b = \sqrt{(4 \text{ cm})^2 - (2.5 \text{ cm})^2}$$
    $$b = \sqrt{16 \text{ cm}^2 - 6.25 \text{ cm}^2}$$
    $$b = \sqrt{9.75 \text{ cm}^2}$$
    $$b \approx 3.12 \text{ cm}$$

65. Find the hypotenuse of the triangle formed by leaning the ladder against the wall.

$$c = \sqrt{a^2 + b^2}$$
$$c = \sqrt{(3\text{ m})^2 + (2\text{ m})^2}$$
$$c = \sqrt{9\text{ m}^2 + 4\text{ m}^2}$$
$$c = \sqrt{13\text{ m}^2}$$
$$c \approx 3.6\text{ m}$$

Yes, Pete can use his ladder because the hypotenuse of the right triangle, which corresponds to the length of the ladder, is about 3.6 m.

66. Find the hypotenuse of the triangle formed by the diagonal.

$$c = \sqrt{a^2 + b^2}$$
$$c = \sqrt{(12\text{ ft})^2 + (27.5\text{ ft})^2}$$
$$c = \sqrt{144\text{ ft}^2 + 756.25\text{ ft}^2}$$
$$c = \sqrt{900.25\text{ ft}^2}$$
$$c \approx 30\text{ ft}$$

The diagonal is approximately 30 ft.

67. The figure is a rectangular solid with $l = 13$ in., $w = 6$ in., and $h = 18$ in.

$$V = lwh$$
$$V = (13\text{ in.})(6\text{ in.})(18\text{ in.})$$
$$V = 1404\text{ in.}^3$$

68. The figure is a rectangular solid with $l = 50$ cm, $w = 22$ cm, and $h = 27$ cm.

$$V = lwh$$
$$V = (50\text{ cm})(22\text{ cm})(27\text{ cm})$$
$$V = 29,700\text{ cm}^3$$

69. The figure is a cylinder with $r = 23$ cm and $h = 50$ cm.

$$V = \pi r^2 h$$
$$V \approx (3.14)(23\text{ cm})^2 (50\text{ cm})$$
$$= (3.14)(529\text{ cm}^2)(50\text{ cm})$$
$$= 83,053\text{ cm}^3$$

70. The figure is a cylinder with $r = 3$ ft and $h = 10$ ft.

$$V = \pi r^2 h$$
$$V \approx (3.14)(3\text{ ft})^2 (10\text{ ft})$$
$$= (3.14)(9\text{ ft}^2)(10\text{ ft})$$
$$= 282.6\text{ ft}^3$$

71. The figure is a cone with $r = 7$ m and $h = 22$ m.

$$V = \frac{1}{3}\pi r^2 h$$
$$V \approx \frac{1}{3}(3.14)(7\text{ m})^2 (22\text{ m})$$
$$= \frac{1}{3}(3.14)(49\text{ m}^2)(22\text{ m})$$
$$\approx 1128.31\text{ m}^3$$

72. The figure is a cone with $r = 4$ mi and $h = 10$ mi.

$$V = \frac{1}{3}\pi r^2 h$$

$$V \approx \frac{1}{3}(3.14)(4 \text{ mi})^2 (10 \text{ mi})$$

$$= \frac{1}{3}(3.14)(16 \text{ mi}^2)(10 \text{ mi})$$

$$\approx 167.47 \text{ mi}^3$$

73. The figure is a sphere with $r = 8$ ft.

$$V = \frac{4}{3}\pi r^3$$

$$V \approx \frac{4}{3}(3.14)(8 \text{ ft})^3$$

$$= \frac{4}{3}(3.14)(512 \text{ ft}^3)$$

$$\approx 2143.57 \text{ ft}^3$$

74. The figure is a sphere with $r = 7$ in.

$$V = \frac{4}{3}\pi r^3$$

$$V \approx \frac{4}{3}(3.14)(7 \text{ in.})^3$$

$$= \frac{4}{3}(3.14)(343 \text{ in.}^3)$$

$$\approx 1436.03 \text{ in.}^3$$

75. The box is a rectangular solid with $l = 5.5$ in., $w = 4$ in., and $h = 5$ in.

$$V = lwh$$

$$V = (5.5 \text{ in.})(4 \text{ in.})(5 \text{ in.})$$

$$V = 110 \text{ in.}^3$$

76. The ball is a sphere with $r = 3$ in.

$$V = \frac{4}{3}\pi r^3$$

$$V \approx \frac{4}{3}(3.14)(3 \text{ in.})^3$$

$$= \frac{4}{3}(3.14)(27 \text{ in.}^3)$$

$$\approx 113.04 \text{ in.}^3$$

77. The pyramid has dimensions $B = (95 \text{ m})(80 \text{ m}) = 7600 \text{ m}^2$ and $h = 146$ m.

$$V = \frac{1}{3}Bh$$

$$V = \frac{1}{3}(7600 \text{ m}^2)(146 \text{ m})$$

$$\approx 369{,}867 \text{ m}^3$$

## Chapter 7 Assessment Test

1. The figure shown is a line because it extends forever in either direction. It is named $\overleftrightarrow{LR}$ or $\overleftrightarrow{RL}$.

2. The figure shown is a line segment because it has two endpoints. It is named $\overline{PD}$ or $\overline{DP}$.

3. $\angle Y$, $\angle PYH$, $\angle HYP$; this angle is right because its measure is $90°$.

4.  $\angle X$; this angle is straight because its measure is $180°$.

5.  $\angle R$, $\angle ARJ$, $\angle JRA$; this angle is acute because its measure is greater than $0°$ and less than $90°$.

6.  $\angle X$, $\angle VXF$, $\angle FXV$; this angle is obtuse because its measure is greater than $90°$ and less than $180°$.

7.  Complementary angles are two angles, the sum of whose degree measure is $90°$. To find the complement of an angle, subtract the measure of the angle from $90°$.
    $90° - 38° = 52°$

8.  Supplementary angles are two angles, the sum of whose degree measures is $180°$. To find the supplement of an angle, subtract the measure of the angle from $180°$.
    $180° - 52° = 128°$

9.  Supplementary angles are two angles, the sum of whose degree measures is $180°$. To find the supplement of an angle, subtract the measure of the angle from $180°$.
    $180° - 130° = 50°$ so $\angle RPO = 50°$.

10. Complementary angles are two angles, the sum of whose degree measure is $90°$. To find the complement of an angle, subtract the measure of the angle from $90°$.
    $90° - 30° = 60°$ so $\angle TMK = 60°$.

11. The triangle shown is an isosceles triangle because two sides have equal length.

12. The triangle shown is an acute triangle because all angles are acute.

13. $72° + 90° = 162°$, so $180° - 162° = 18°$.

14. $79° + 20° = 99°$, so $180° - 99° = 81°$.

15. The first figure is a trapezoid because there is one pair of parallel sides.
    The second figure is a square, rectangle, or rhombus because all sides have equal length and there are four right angles.
    The third figure is a rhombus because all sides have equal length.
    The fourth figure is a rectangle because there are four right angles.

16. The first figure is a hexagon because it has 6 sides.
    The second figure is a pentagon because it has 5 sides.
    The third figure is a quadrilateral because it has 4 sides.
    The fourth figure is an octagon because it has 8 sides.

17. $\dfrac{20 \text{ in.}}{2} = 10 \text{ in.}$                    18.  $8 \text{ m} \cdot 2 = 16 \text{ m}$

19. The first solid is a cube because there are six square faces.
    The second solid is a rectangular solid because there are six rectangular faces.
    The third solid is a pyramid because there are four triangular-shaped faces with a common vertex.

20. The first solid is a cone because the solid has a circular base and a common vertex.
    The second solid is a sphere because the solid consists of all points that lie the same distance from some fixed point.
    The third solid is a cylinder because the solid has two identical circular bases joined by line segments perpendicular to these bases.

21. $P = 10$ yd $+ 14$ yd $+ 10$ yd $= 34$ yd

$$A = \frac{1}{2}bh$$

$$A = \frac{1}{2}(14 \text{ yd})(9 \text{ yd})$$

$$A = 63 \text{ yd}^2$$

22. $P = (2 \cdot 3.5 \text{ in.}) + (2 \cdot 6.5 \text{ in.}) = 7 \text{ in.} + 13 \text{ in.} = 20 \text{ in.}$

$$A = lw$$

$$A = (6.5 \text{ in.})(3.5 \text{ in.})$$

$$A = 22.75 \text{ in.}^2$$

23. $P = 15$ ft $+ 11.7$ ft $+ 15$ ft $+ 11.7$ ft $= 53.4$ ft

$$A = bh$$

$$A = (15 \text{ ft})(11.5 \text{ ft})$$

$$A = 172.5 \text{ ft}^2$$

24. $P = 9$ cm $+ 7.6$ cm $+ 13$ cm $+ 7.1$ cm $= 36.7$ cm

$$A = \frac{1}{2}(a+b)h$$

$$A = \frac{1}{2}(13 \text{ cm} + 9 \text{ cm})7 \text{ cm}$$

$$A = 77 \text{ cm}^2$$

25. If $d = 50$ ft, then $r = 25$ ft.

$$C = \pi d$$

$$\approx (3.14)(50 \text{ ft})$$

$$= 157 \text{ ft}$$

$$A = \pi r^2$$

$$A \approx (3.14)(25 \text{ ft})^2$$

$$A = 1962.5 \text{ ft}^2$$

26. $C = 2\pi r$

$$C \approx (2)(3.14)(9 \text{ m})$$

$$= 56.52 \text{ m}$$

$$A = \pi r^2$$

$$A \approx (3.14)(9 \text{ m})^2$$

$$A = 254.34 \text{ m}^2$$

27. If $d = 70$ mi, then $r = 35$ mi.

$$A = \pi r^2$$

$$A \approx (3.14)(35 \text{ mi})^2$$

$$A = 3846.5 \text{ mi}^2$$

28. The tile is a rectangle with $l = 115$ cm and $w = 23$ cm.

$$A = lw$$

$$A = (115 \text{ cm})(23 \text{ cm})$$

$$A = 2645 \text{ cm}^2$$

29. $12^2 = 144$ so $\sqrt{144} = 12$.

30. $\sqrt{55} \approx 7.42$

31. $c = \sqrt{a^2 + b^2}$

$$c = \sqrt{(5 \text{ yd})^2 + (9 \text{ yd})^2}$$

$$c = \sqrt{25 \text{ yd}^2 + 81 \text{ yd}^2}$$

$$c = \sqrt{106 \text{ yd}^2}$$

$$c \approx 10.30 \text{ yd}$$

32. $b = \sqrt{c^2 - a^2}$

$$b = \sqrt{(62 \text{ m})^2 - (45 \text{ m})^2}$$

$$b = \sqrt{3844 \text{ m}^2 - 2025 \text{ m}^2}$$

$$b = \sqrt{1819 \text{ m}^2}$$

$$b \approx 42.65 \text{ m}$$

33. The figure is a rectangular solid with $l = 6$ mi, $w = 4$ mi, and $h = 12$ mi.

$V = lwh$

$V = (6 \text{ mi})(4 \text{ mi})(12 \text{ mi})$

$V = 288 \text{ mi}^3$

34. The figure is a cylinder with $r = 20$ cm and $h = 60$ cm.

$V = \pi r^2 h$

$V \approx (3.14)(20 \text{ cm})^2 (60 \text{ cm})$

$= (3.14)(400 \text{ cm}^2)(60 \text{ cm})$

$= 75{,}360 \text{ cm}^3$

35. The figure is a pyramid with $B = (9 \text{ m})(8 \text{ m}) = 72 \text{ m}^2$ and $h = 11$ m.

$V = \dfrac{1}{3} Bh$

$V = \dfrac{1}{3}(72 \text{ m}^2)(11 \text{ m})$

$= 264 \text{ m}^3$

36. The figure is a cone with $r = 3$ ft and $h = 12$ ft.

$V = \dfrac{1}{3}\pi r^2 h$

$V \approx \dfrac{1}{3}(3.14)(3 \text{ ft})^2 (12 \text{ ft})$

$= \dfrac{1}{3}(3.14)(9 \text{ ft}^2)(12 \text{ ft})$

$= 113.04 \text{ ft}^3$

37. The figure is a sphere with $r = 21$ mi.

$V = \dfrac{4}{3}\pi r^3$

$V \approx \dfrac{4}{3}(3.14)(21 \text{ mi})^3$

$= \dfrac{4}{3}(3.14)(9261 \text{ mi}^3)$

$= 38{,}772.72 \text{ mi}^3$

38. The figure is a sphere with $r = 9.5$ in.

$V = \dfrac{4}{3}\pi r^3$

$V \approx \dfrac{4}{3}(3.14)(9.5 \text{ in.})^3$

$= \dfrac{4}{3}(3.14)(857.375 \text{ in.}^3)$

$\approx 3589.54 \text{ in.}^3$

39. Each leg resembles a pyramid with $B = (5 \text{ in.})(8 \text{ in.}) = 40 \text{ in.}^2$ and $h = 17$ in.

$V = \dfrac{1}{3} Bh$

$V = \dfrac{1}{3}(40 \text{ in.}^2)(17 \text{ in.})$

$\approx 226.67 \text{ in.}^3$

40. The volcano resembles a cone with $r = 6.5$ km and $h = 4.17$ km.

$$V = \frac{1}{3}\pi r^2 h$$

$$V \approx \frac{1}{3}(3.14)(6.5 \text{ km})^2 (4.17 \text{ km})$$

$$= \frac{1}{3}(3.14)(42.25 \text{ km}^2)(4.17 \text{ km})$$

$$\approx 184.4 \text{ km}^3$$

# Chapter 8    Statistics and Data Presentation

## Section 8.1    Data Presentation – Tables and Graphs

## Concept Check

1. The science of collecting, interpreting, and presenting numerical data is known as __statistics__ .

3. A _line_ graph is a "picture" of selected data changing over a period of time.

5. When constructing a line graph, the *x*-axis represents the _time_ variable and the *y*-axis represents the __numerical value__ variable.

7. A _circle_ graph or _pie_ chart is a circle divided into sections or segments that represent the component parts of a whole.

## Guide Problems

9. a. What are the titles of the columns?
There are three columns. The titles are read off the top of each column. The titles are: Occupation, Percentage Increase in Jobs from 2006, Median Salary.

b. What do the rows represent?
The first entry in each row falls under the title of occupation. The rows represent the different types of occupations.

c. What are the sources of this table's data?
The source of a table's data typically appears below the table. This table's source is the U.S. Department of Labor, Bureau of Labor Statistics.

11. Since Cunard is a cruise line, locate Cunard under the Line column. There is only one entry for Cunard. The entry in the Ship column in the row containing Cunard is the *Queen Mary 2* so Cunard owns the *Queen Mary 2*.

13. For each ship, find the ship name under the Ship title and read across the row to find the weight in the Gross Tons column. The *Adventure of the Seas* weighs 138,000 tons while the *Caribbean Princess* weighs 116,000 tons. The difference is $138,000 - 116,000 = 22,000$ tons.

15. Any cruise line with more than one ship would be listed more than once in the Line column. The cruise lines that are listed more than once are Royal Caribbean (three ships) and Princess (two ships.)

17. First note the automobile company corresponds to the make under Make and Model. For example, the Honda company manufactures the models Civic and Insight while Toyota manufactures the Camry, Prius, and Highlander. Honda and Toyota are the only companies listed with more than one hybrid model.

19. The lowest price in the Base Price column is $15,810 corresponding to the Honda Civic. The Civic is the least expensive hybrid listed.

21. The various highway mileages listed are 36, 33, 25, 40, 60, 31, 40, and 57. The smallest of these values is 25 in the row for the Lexus GS450H. Thus the Lexus GS450H gates the lowest mileage on the highway, 25 miles per gallon.

23. a. Reading the title of the graph, the line graph represents widget sales from the year 2000 to the year 2007.

b. Based on the graph title, the *y*-axis represents widget sales in billions of dollars.

c. The *x*-axis represents time, in years, from 2000 to 2007.

d. First, locate the year 2006 on the *x*-axis. Then move up the corresponding vertical line until you hit the line graph. Move to the left to the *y*-axis to see that sales were approximately $1.0 billion dollars, rounded.

e. To determine when sales reached $0.8 billion dollars, first locate the value $0.8 on the sales or *y*-axis. Move horizontally from $0.8 until you hit the line graph. Then move vertically down to locate the year on the time or *x*-axis. The corresponding year is 2005.

f. From d., sales in 2006 were approximately $1.0 billion dollars. The year 2004 corresponds to a point on the graph at a height of $0.6 billion dollars. Sales in 2006 were $1.0 - $0.6 = $0.4 billion dollars greater.

25. a. The table gives time data in units of years so label the *x*-axis "Years"

b. The numerical value data in the table are the number of Euros per U.S. dollar so label the *y*-axis "Euros per Dollar"

c. and d. Plot the points and connect the dots to get the line graph below.

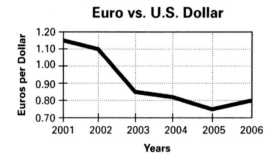

27. Locate 2003 on the *x*-axis and move vertically to the graph. From that point on the graph, move horizontally to the *y*-axis and read the 2003 sales figure. Sales in 2003 totalled 3 million units.

29. The line graph crosses the 2002 line between the 2 and 3 million unit marks on the *y*-axis. The point is closer to 2 than 3 so $2.8 million is a reasonable approximation of 2002 sales.

31. Moving vertically from the 2003 mark on the time axis to the home equity loan graph, we see that roughly 30% of homeowners applied for such a loan in 2003.

33. In 2002, the graph for home equity loans is above that for home equity lines of credit. In 2003, the loan graph falls below the lines of credit graph so 2003 was the first such year.

35. The line segment joining the years 2003 and 2004 is basically flat at 30% indicating the percentages of home equity loans in 2003 and 2004 were the same.

37. The time variable is in days of the week so label the *x*-axis "Day". The numerical value variable represents the stock price per share so label the *y*-axis "Stock Price". Then plot the data and connect the dots. The completed line graph is shown here.

**Marshall Corporation**

Per Share Stock Price

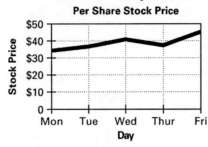

## Guide Problems

39. a. What does this bar graph illustrate?
From the title and the labels on the axes, this bar graph shows the salaries of U.S. senators from 2000-2006.

  c. What does the *y*-axis measure?
  Numerical values, salaries.

  e. How much more did senators make in 2005 than in 2004?
  The bars for 2004 and 2005 are at heights of $158,100 and $162,100 respectively. Senators made $162,100 − $158,100 = $4000 more in 2005 than in 2004.

  g. In what year did senators have the greatest pay raise over the previous year? How much was that pay raise?
  By computing the numerical difference from one year to the next or by examining the differences in the heights of neighboring bars, one can see the biggest raise occurred in 2002. The raise was $150,000 − $145,100 = $4900.

  b. What does the *x*-axis measure?
  The *x*-axis measures Time, 2000-2006.

  d. How much did senators make in 2004?
  The bar for 2004 is labeled with a salary of $158,100.

  f. In what year did senators make $154,700?
  The bar at a height of $154,700 is over the year 2003.

41. a. and b. The *x*-axis is labeled with the time unit Hours and the *y*-axis with Percent. Bars are drawn over each hour range to the height specified by the numerical value data in the table. The resulting bar graph is shown below.

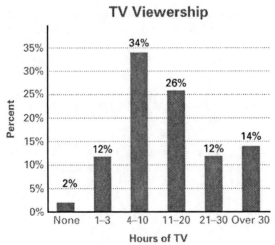

43. The bar over 1984 reaches a height of 19.9 so the number of passengers in 1984 was 19.9 million.

45. The bars over 1994 and 1974 are at heights of 21.8 and 18.7 respectively so there were $21.8 - 18.7 = 3.1$ million more passengers in 1994 than in 1974.

47. The blue bar represents sugar. For Powerade, this bar is at a height of 38 so a serving of Powerade contains 38 grams of sugar.

49. The blue bar for Coca-Cola is at a height of 72 so a serving contains 72 grams of sugar. From Exercise 47, a serving of Powerade contains 38 grams for a difference of $72 - 38 = 34$ grams.

51. From the bar graph, a serving of Coca-Cola contains 250 calories and a serving of orange juice contains 300 calories. Three servings of Coca-Cola and two servings of orange juice per week would contain $3 \cdot 250 + 2 \cdot 300 = 1350$ calories. Over six weeks you would consume $6 \cdot 1350 = 8100$ calories.

53. Label the x-axis with the Program/Network data and label the y-axis Amount Paid (in millions of dollars.) Over each program, draw a bar to the indicated dollar amounts. The resulting bar graph is shown below.

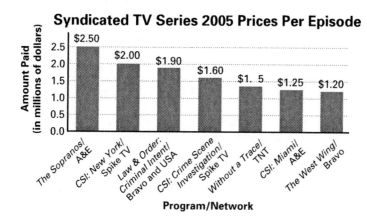

## Guide Problems

55. a. What does this circle graph illustrate?
    From the title, the circle graph represents proved oil reserves.

    b. What is the source of this information?
    The source is given below the graph and is the BP Statistical Review of World Energy 2006.

    c. What region had the greatest oil reserves? What percentage did this represent?
    The largest segment of the circle graph belongs to the Middle East and represents 61.8% of the reserves.

    d. What percent of oil reserves were from Africa?
    The segment for Africa represents 9.5% of reserves.

    e. What region had the lowest oil reserves? What percentage did this represent?
    The smallest segment of the graph belongs to Asia Pacific and represents 3.4% of the reserves.

    f. How much more (in percent) are the oil reserves Europe and Eurasia compared with the oil reserves of North America?
    Europe and Eurasia have 11.7% of the oil reserves compared to 4.9% in North America or $11.7\% - 4.9\% = 6.8\%$ more than North America.

57. The segment labeled "Two" corresponds to 24% of those surveyed.

59. Three or four dogs would be represented by the segment labeled "Three" (5%) and the segment "Four" (3%) for a total of 8%.

61. The largest segment corresponds to the top complaint which is Identity Theft.

63. The two segments representing 8% of the graph are "Shop-at-home/catalog sales" and "Foreign money offers."

65. The total number of degrees granted is
$$315 + 180 + 225 + 135 + 45 = 900.$$
Next convert degrees granted to percentages of this total and then multiply by $360°$ to determine the degrees of each segment.

| DEGREE | DEGREES GRANTED | PERCENT | DEGREES |
|---|---|---|---|
| Education | 315 | $\frac{315}{900} = 0.35 = 35\%$ | $35\% \cdot 360 = 0.35 \cdot 360 = 126$ |
| Business Administration | 180 | $\frac{180}{900} = 0.20 = 20\%$ | $20\% \cdot 360 = 0.2 \cdot 360 = 72$ |
| Health | 225 | $\frac{225}{900} = 0.25 = 25\%$ | $25\% \cdot 360 = 0.25 \cdot 360 = 90$ |
| Engineering | 135 | $\frac{135}{900} = 0.15 = 15\%$ | $15\% \cdot 360 = 0.15 \cdot 360 = 54$ |
| Computer Science | 45 | $\frac{45}{900} = 0.05 = 5\%$ | $5\% \cdot 360 = 0.05 \cdot 360 = 18$ |

After using a protractor to create the segments having the degrees computed above and adding the labels, the completed circle graph is shown below.

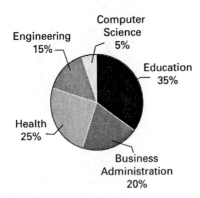

**Penta College —
Degrees Granted, 2008**

## Cumulative Skills Review

1. The denominator are 3, 5 and $6 = 2 \cdot 3$. The least common denominator is the LCM of the three denominators or $2 \cdot 3 \cdot 5 = 30$.

3. The two given angles total $90° + 45° = 135°$ so the unknown angle measures
$$180° - 135° = 45°.$$

5.  Estimate: $500 - 100 = 400$.

Actual:
$$
\begin{array}{r}
5 \not{4} \overset{13}{\not{4}} . \overset{10}{\not{1}} \overset{12}{\not{2}} 3 \\
-\ 1\ 2\ 4\ .\ 3\ 8\ 0 \\
\hline
4\ 1\ 9\ .\ 7\ 4\ 3
\end{array}
$$

7.  Solve the proportion:
$$\frac{h}{7.5} = \frac{24}{4}$$
Compute and equate cross products:
$$4h = 7.5 \cdot 24 = 180$$
So $h = \dfrac{180}{4} = 45$.

9.  $375 \text{ feet} = 375 \text{ feet} \cdot \dfrac{1 \text{ yard}}{3 \text{ feet}}$

$\qquad = 375 \ \cancel{\text{feet}} \cdot \dfrac{1 \text{ yard}}{3 \ \cancel{\text{feet}}}$

$\qquad = \dfrac{375}{3} \text{ yards} = 125 \text{ yards}$

## Section 8.2    Mean, Median, Mode, and Range

## Concept Check

1.  A number that is computed from, and describes, numerical data about a particular situation is known as a
    statistic .

3.  A  set  is a collection of numbers or objects considered as a whole.

5.  The term  arithmetic    mean  corresponds to the generally accepted meaning of the word *average*.

7.  A student's GPA is an example of a  weighted  mean.

9.  The median is a more useful measure of central tendency than the mean when one or more of the values of
    the set are significantly  larger  or  smaller  than the others.

11. The value or values in a set that occur most often is known as the mode .

13. The difference between the largest and smallest value in a set is called the range .

## Guide Problems

15. Consider this set: 34, 37, 13, 14, 53, 18, 46, 9
    a. How many values are in the set?   8

    b. Find the sum of the values?
       $34 + 37 + 13 + 14 + 53 + 18 + 46 + 9 = 224$

    c. Divide the sum of the values by the number of
       values.
       $$\frac{224}{8} = 28$$

    d. The result is called the  mean .
       The mean of a set is the sum of the values in
       the set divided by the number of values.

*In Problems 17 through 23, a mean is computed by first determining the number of values in the set, then computing the sum of the values in the set, and lastly dividing the sum by the number of values. If units are present, simply add and divide as usual. If necessary, round the mean computation to the nearest tenth.*

17. Number of values: 8
Sum of values: $15+3+6+4+7+11+13+5 = 64$

Mean: $\dfrac{64}{8} = 8$

19. Number of values: 5
Sum of values: $22 \text{ lbs} + 82 \text{ lbs} + 28 \text{ lbs}$
$+ 81 \text{ lbs} + 15 \text{ lbs} = 228 \text{ lbs}$

Mean: $\dfrac{228 \text{ lbs}}{5} = 45.6 \text{ lbs}$

21. Number of values: 8
Sum of values:
$32 \text{ ft} + 40 \text{ ft} + 46 \text{ ft} + 43 \text{ ft}$
$+ 40 \text{ ft} + 82 \text{ ft} + 73 \text{ ft} + 14 \text{ ft} = 370 \text{ ft}$

Mean: $\dfrac{370 \text{ ft}}{8} = 46.25 \text{ ft}$ or 46.3 ft rounded.

23. Number of values: 5
Sum of values: $21 \text{ in.} + 24 \text{ in.} + 93 \text{ in.}$
$+ 49 \text{ in.} + 85 \text{ in.} = 272 \text{ in.}$

Mean: $\dfrac{272 \text{ in.}}{5} = 54.4 \text{ in.}$

*In Problems 25 and 27, a GPA is computed. The grade values are the usual values:* $A = 4$, $B = 3$, $C = 2$, $D = 1$, *and* $F = 0$. *The quality points for a given course are computed by multiplying the credits by the grade value. The quality points are then added and divided by the total number of credits to find the GPA. The GPA is expressed as a decimal rounded to the hundredths place.*

25.

| COURSE | CREDITS | GRADE VALUE | QUALITY POINTS |
|--------|---------|-------------|----------------|
| Spanish | 3 | C = 2 | $3 \cdot 2 = 6$ |
| English | 3 | B = 3 | $3 \cdot 3 = 9$ |
| Algebra | 5 | C = 2 | $5 \cdot 2 = 10$ |
| Earth Science | 3 | A = 4 | $3 \cdot 4 = 12$ |
| **Total** | 14 | | 37 |

$$\text{GPA} = \frac{37}{14} = 2.642... \approx 2.64$$

27.

| COURSE | CREDITS | GRADE VALUE | QUALITY POINTS |
|--------|---------|-------------|----------------|
| Social Science | 3 | B = 3 | $3 \cdot 3 = 9$ |
| Chemistry | 5 | B = 3 | $5 \cdot 3 = 15$ |
| Chemistry Lab | 1 | C = 2 | $1 \cdot 2 = 2$ |
| Finite Math | 4 | A = 4 | $4 \cdot 4 = 16$ |
| English | 3 | C = 2 | $3 \cdot 2 = 6$ |
| **Total** | 16 | | 48 |

$$\text{GPA} = \frac{48}{16} = 3.0$$

## Guide Problems

29. Consider this set: 54, 11, 24, 36, 75, 45, 27.
a. List the values in numerical order.
11, 24, 27, 36, 45, 54, 75

b. Because there is an *odd* number of values, identify the middle value. 36

c. The result is called the  median .

*In problems 31 through 37, the median of a set is computed. First, the values in the set are listed in increasing order. If the number of values is odd, the middle value is identified as the median. If the number of values is even, the two middle values are identified and their mean is the median of the set.*

31. Values in order: 45, 52, 62, 76, 77
    Number of values: odd
    Median: 62

33. Values in order: $12, $19, $30, $32,
    $34, $60, $63, $90
    Number of values: even
    Median: $\dfrac{\$32 + \$34}{2} = \dfrac{\$66}{2} = \$33$

35. Values in order:
    133 mi, 449 mi, 632 mi, 658 mi, 904 mi
    Number of values: odd
    Median: 632 mi

37. Values in order:
    19 acres, 40 acres, 42 acres, 47 acres,
    52 acres, 60 acres, 92 acres, 93 acres
    Number of values: even
    Median: $\dfrac{47 \text{ acres} + 52 \text{ acres}}{2} = \dfrac{99 \text{ acres}}{2}$
    $= 49.5 \text{ acres}$

## Guide Problems

39. Consider this set: 4, 1, 4, 3, 6, 5, 4, 7, 5, 12, 9, 14, 6.

    a. Which value or values appear most often?
       4. The value 4 appears three times, more than
       any other value.

    b. The result is called the _mode_.

*In problems 41 through 47, a mode is determined. Analyze the data set to find the value or values that occur most often in the set. All such values are modes.*

41. Set: 32  70  57  93  70  99
    70 occurs most often
    Mode: 70

43. Set: 403 ft  310 ft  549 ft
    534 ft  804 ft  603 ft
    Each value appears only once in the set.
    Mode: no mode

45. Set: 50 cats  22 cats  53 cats  48 cats
    53 cats  53 cats  48 cats  12 cats
    53 cats appears most often
    Mode: 53 cats

47. Set: $934  $267  $860  $267  $860  $675  $259
    $267 and $860 both occur most often, twice each.
    Mode: $267, $860

## Guide Problems

49. Consider this set:
    24, 19, 5, 43, 61, 30
    a. Which value is the smallest?  5

    b. Which value is the largest?  61

    c. Calculate the difference between the largest and
       smallest values.  $61 - 5 = 56$

    d. This number is called the _range_.

*In Problems 51 through 57, the range of a set is computed. First, determine the smallest value in the set, then the largest value. Subtract the smallest value from the largest to find the range.*

51. Smallest value: 13
    Largest value:  84
    Range: $84 - 13 = 71$

53. Smallest value: 23 pens
    Largest value:  74 pens
    Range: $74 \text{ pens} - 23 \text{ pens} = 51 \text{ pens}$

55. Smallest value: 11 lbs
    Largest value:  97 lbs
    Range: $97 \text{ lbs} - 11 \text{ lbs} = 86 \text{ lbs}$

57. Smallest value: 24%
    Largest value:  86%
    Range: $86\% - 24\% = 62\%$

59. The data in ascending order is $165,000   $230,000   $290,000   $322,000.

Mean: $\dfrac{\$165,000 + \$230,000 + \$290,000 + \$322,000}{4} = \dfrac{\$1,007,000}{4} = \$251,750$

Median: $\dfrac{\$230,000 + \$290,000}{2} = \$260,000$

Range: $\$322,000 - \$165,000 = \$157,000$ miles.

61. Total credits: $4 + 3 + 3 + 4 + 2 = 16$         Total quality points: $4 \cdot 4 + 3 \cdot 3 + 3 \cdot 4 + 4 \cdot 4 + 2 \cdot 3 = 59$

GPA: $\dfrac{59}{16} = 3.6875 \approx 3.69$

63. The data in ascending order is $7300   $7300   $12,100   $15,500   $18,000.

Mean: $\dfrac{\$7300 + \$7300 + \$12,100 + \$15,500 + \$18,000}{5} = \dfrac{\$60,200}{5} = \$12,040$

Median: $12,100

Mode: $7300

Range: $\$18,000 - \$7300 = \$10,700$

65. The data in ascending order is
     22   25   25   28   30   30   34   45   48   55
The sum of the values is 342.

Mean: $\dfrac{342}{10} = 34.2$

Median: $\dfrac{30 + 30}{2} = \dfrac{60}{2} = 30$

Mode: 25, 30 (both values occur twice)

Range: $55 - 22 = 33$

67. The snowfall data in ascending order is

| 6 | 6 | 7 | 7 | 7 | 8 | 8 | 9 | 9 | 9 |
| 9 | 9 | 10 | 10 | 10 | 11 | 11 | 13 | 13 | 19 |

The sum of the values is 191.

Mean: $\dfrac{191}{20} = 9.55 \approx 9.6$ inches.

Median: $\dfrac{9 + 9}{2} = \dfrac{18}{2} = 9$

Mode: 9 inches

Range: $19 - 6 = 13$ inches

## Cumulative Skills Review

1.   $\dfrac{1}{5} = 0.2 = 0.2 \cdot 100\% = 20\%$

3.   $C = 2\pi r$
     $\quad = 2 \cdot 3.14 \cdot 6$
     $\quad = 12 \cdot 3.14 = 37.68$

5.   To convert to a decimal, divide by 100.

   $89\% = \dfrac{89}{100} = 0.89$

7.   There are $80 - 12 = 68$ inches remaining.
   The percent remaining is

   $\dfrac{68}{80} = 0.85 = 0.85 \cdot 100 = 85\%.$

9.   $823.12 \cdot 100 = 82,312$
     (To multiply by 100, move the decimal point two places to the right.)

## Chapter 8 Numerical Facts Of Life

1.   The table indicates that from 1986 to 2006, the cost of college tuition increased by 289.5%. A college
     credit that cost $180 in 1986 would cost
         $\$180.00 + \$180.00(289.5\%) = \$180.00 + \$180.00 \cdot 2.895 = \$180.00 + \$521.10 = \$701.10$
     in 2006.

3.   An office visit to a doctor falls under the category of Doctor Services. From the table, such services showed
     an increase in cost of 137.3% from 1986 to 2006. A $40 dollar office visit in 1986 would cost
         $\$40.00 + \$40.00(137.3\%) = \$40.00 + \$40.00 \cdot 1.373 = \$40.00 + \$54.92 = \$94.92$
     in 2006.

## Chapter 8 Review Exercises

1.   Locate sodium under the Amount Per Serving heading. The indicated amount is 190 mg.

2.   The information at the top of the table indicates the values are for a 3/4 cup serving. The top of the right most column accounts for 1/2 cup milk. Under skim milk, we see the calories total 130,

3.   Below the calories information, the table gives daily values as a percentage. Find the row for potassium and look under the heading for skim milk, to find that a serving of cereal with skim milk will provide 8% of the daily potassium value.

4.   Locate Vitamin C in the Daily Value portion of the table. The percentage on the left corresponds to cereal without milk. One serving without milk will provide 10% of the daily value.

5.   Under Amount Per Serving, the total Cholesterol is given as 0 mg.

6.   From the table, a single serving contains 5 g of sugars. Three servings per week would amount to $3 \cdot 5 = 15$ g per week. Over six weeks, the amount of sugar would be $6 \cdot 15 = 90$ grams.

7.   Locate Indonesia under the Country column. Move right to the Estimated Population column to find the value 336. Note under the table title that the values are in units of millions. The estimated population of Indonesia in 2050 is 336 million.

8.   Locate Brazil under the Country column. Move right to the Population 2005 column to find the value 186. Note under the table title that the values are in units of millions. The population of Brazil in 2005 was 186 million.

9.   Looking down the Estimated Population 2050 column, the largest value is 9084. However, this value does not correspond to a country but represents the world population. The largest value for a country is 1601 million corresponding to India.

10.  For each country, compute the increase in population from 2005 to 2050.

| China | $1418 - 1302 = 116$ |
| India | $1601 - 1080 = 521$ |
| United States | $420 - 296 = 124$ |
| Indonesia | $336 - 242 = 94$ |
| Brazil | $228 - 186 = 42$ |

The smallest growth belongs to Brazil.

11. From the computations in Exercise 10, the U.S. population will grow by 124 million.

12. The difference in the world total population from 2005 to 2050 is $9084 - 6449 = 2635$ million.

13. Looking at the Viewership graph on the right, the lowest point is along the horizontal line through '03 on the time axis. The lowest viewership was in 2003.

14. Using the Ad Prices graph on the left, the costs of 30-second ads in 1994 and 2005 were $643,500 and $1.6 million respectively. This is an increase of $1,600,000 - $643,500 = $956,500$.

15. Start at the 40 mark on the $y$-axis of the Viewership graph. Scan horizontally until you find a point where the graph is just above the 40 mark and note the year on the time axis. In 2002, the viewership was just over 40 million.

16. Other than the period from 2001 to 2002, ad prices have steadily increased. Viewership, on the other hand, has steadily decreased.

17. Of the three line graphs, Social Security is represented by the uppermost line graph. This graph crosses over the 700 billion mark in the year 2010.

18. Locate the '15 label on the time axis. Scan vertically to the Medicaid line graph (the lowest graph) to find the graph labeled $392. Since the indicated values are in billions, Medicaid is projected to be $392 billion in 2016.

19. Moving vertically from the '15 mark to the social security line graph, we see the graph crosses at roughly the $900 value. Social security is projected to be at $900 billion in 2015.

20. Scan vertically from the '13 mark on the time axis. The Medicaid curve is around the $300 mark but the Medicare curve is at about $700. So Medicare is predicted to reach $700 billion in 2013.

21. Label the time axis with "Academic Year": 2000-2001, 2001-2002, etc. Label the $y$-axis "Loan Rate" with percentage values. Then plot the values from the table with a dot and connect the dots. The completed line graph is shown below.

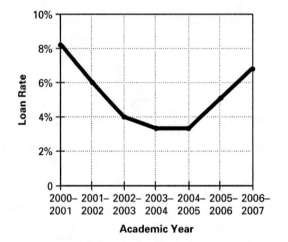

22. Label the time axis with "Year": 2000, 2001, etc. Label the $y$-axis with numbers of prints 0, 10, 20, 30, and 40. Plot the points corresponding to the Traditional column of values and connect these dots with solid lines. Label this curve Traditional. Similarly, plot the points corresponding to the Digital column of values. Connect these points with dashed lines and label the graph Digital. The completed multiple line graph is shown below.

**Photo Prints Made at Labs**
(in billions)

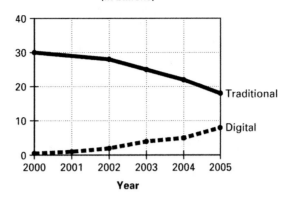

23. The bars represent the increase in home price so the shortest bar would represent the lowest increase. This occurred in 1996.

24. Start at the 8 percent mark on the *y*-axis. Scan horizontally until a bar just touches the 8 percent mark. This bar corresponds to 2003. There was an 8 percent increase in home prices in 2003.

25. Scan the tops of the bars to locate two bars that are at the same height. Home prices rose at the same rate in 2001 and 2002.

26. The top of the bar over '06 lies halfway between the 4 and 6 marks on the *y*-axis. So in 2006, home prices increased by 5%.

27. Of the two bars next to Italy, the orange bar represents 2003 according to the key. This bar indicates 714 million gallons of wine were consumed in Italy in 2003.

28. Looking only at the ends of the orange bars, for 2003, the greatest value is seen to be 774 million in France.

29. From the bar graph, Italy will consume 730 million gallons of wine in 2008 compared to 714 million consumed in 2003, an increase of $730 - 714 = 16$ million gallons.

30. Looking only at the green bars, for 2008, the largest value, 740, is associated with the United States. The United States is projected to consume the most wine in 2008.

31. Label the time axis with "Year" and use the values from the Year column: 2004, 2005, ...,2009. Label the *y*-axis "Subscribers" and mark values 20, 40, 60, 80. Indicate in the graph title or on the *y*-axis that the values are in millions. Over each year construct a bar to the height indicated by the numerical value of subscribers for that year. The resulting bar graph is shown below.

**Estimated U.S. Residential Broadband Subscribers**
(millions)

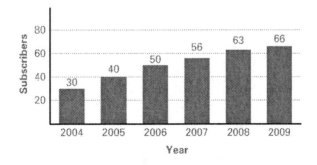

32. Label the time axis with "Position" and use the names from the Position column such as Quarterback and Defensive End. Label the *y*-axis "Salary" and mark values 1, 2, 3, and so on, representing millions of dollars. Indicate in the graph title or on the *y*-axis that the values are in millions. Over each position

construct two bar, one to the height indicated by the salary for starters and one to the height indicated by the salary for all players at that position. Use a dark color for the starters and a light color for all players. Create a key to indicate which color represents which category. The resulting bar graph is shown below.

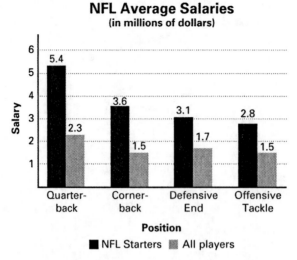

33. The largest segment belongs to SBC/AT&T so SBC/AT&T has the greatest market share.

34. The segment on the graph labeled Verizon shows a share of 15%.

35. Note that the segment for MCI is larger than that for Sprint so MCI has the greater market share. Note also that Sprint's segment is labeled 6% while MCI's segment is labeled 8%.

36. It is clear visually that if the segments for Verizon, MCI and Sprint were joined as one segment, that segment would still be smaller than that of SBC/AT&T. This can be confirmed by adding their respective shares
$$15\% + 8\% + 6\% = 29\%$$
which is less than the 37% held by SBC/AT&T.

37. From the circle graph on the left interest payments were at 7% in 1969. From the graph on the right, interest payments will be at 10% in 2009, an increase of 3%.

38. It is clear visually that the segment for Benefits grows the most in the circle graph for 2009 when compared to that for 1969.

39. National defense spending will change from 43% to 16%, a decrease of $43\% - 16\% = 27\%$.

40. Benefits and interest payments both increase while national defense and all else decrease. The decrease in all else is predicted to be 2%.

41. Since the data is already in percent form, the only calculation is to determine the number of degrees in each segment. These calculation are shown in the table below.

| REGION | PERCENT | DEGREES |
|--------|---------|---------|
| Latin America | 53% | $53\% \cdot 360 = 0.53 \cdot 360 \approx 191$ |
| Europe | 14% | $14\% \cdot 360 = 0.14 \cdot 360 \approx 50$ |
| Asia | 25% | $25\% \cdot 360 = 0.25 \cdot 360 = 90$ |
| Other | 8% | $0.08\% \cdot 360 = 0.08 \cdot 360 \approx 29$ |

Use a protractor to measure segments having the calculated angles and then label with the region and corresponding percentages to generate the circle graph shown below.

## U.S. Foreign-Born Population

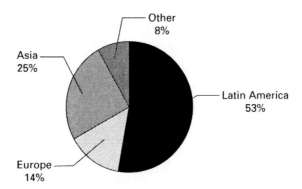

42. a. and b. The total spending is $3.9 + $3.4 + $7.6 + $2.9 = $17.8 billion dollars. Convert each dollar value in the spending column to a percent of this total. Then multiply by 360 to find the number of degrees in the corresponding segment of the circle graph. These calculation are shown in the table below.

| CATEGORY | MILLIONS OF VIEWERS | PERCENT | DEGREES |
|---|---|---|---|
| Electronics and computers | $3.9 | $\dfrac{\$3.9}{\$17.8} \approx 22\%$ | $22\% \cdot 360 = 0.22 \cdot 360 \approx 79$ |
| Shoes | $3.4 | $\dfrac{\$3.4}{\$17.8} \approx 19\%$ | $19\% \cdot 360 = 0.19 \cdot 360 \approx 68$ |
| Clothing and accessories | $7.6 | $\dfrac{\$7.6}{\$17.8} \approx 43\%$ | $43\% \cdot 360 = 0.43 \cdot 360 \approx 155$ |
| School supplies | $2.9 | $\dfrac{\$2.9}{\$17.8} \approx 16\%$ | $16\% \cdot 360 = 0.16 \cdot 360 \approx 58$ |

Use a protractor to measure segments having the calculated angles and then label with the category and corresponding percentages to generate the circle graph shown below.

## Back-to-School Spending

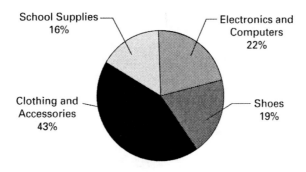

43. Values in order: 14, 21, 35, 58, 60
Number of values: odd
Median: 35
Mean: $\dfrac{35+21+58+14+60}{5}=\dfrac{188}{5}=37.6$

44. Values in order: 44, 53, 71, 85, 98, 99
Number of values: even
Median: $\dfrac{71+85}{2}=78$
Mean: $\dfrac{85+71+98+44+53+99}{6}=\dfrac{450}{6}=75$

45. Values in order: 27, 64, 77, 90
Number of values: even
Median: $\dfrac{64+77}{2}=70.5$
Mean: $\dfrac{27+90+64+77}{4}=\dfrac{258}{4}=64.5$

46. Values in order: 34, 52, 64
Number of values: odd
Median: 52
Mean: $\dfrac{64+34+52}{3}=\dfrac{150}{3}=50$

47. Values in order: 11, 17, 20, 28, 62
Number of values: odd
Median: 20
Mean: $\dfrac{62+11+28+17+20}{5}=\dfrac{138}{5}=27.6$

48. Values in order: 10, 12, 30, 84
Number of values: even
Median: $\dfrac{12+30}{2}=21$
Mean: $\dfrac{84+12+10+30}{5}=\dfrac{136}{4}=34$

49. Set: 56, 63, 48, 85, 48, 54, 77, 79
48 occurs most often
Mode: 48
Smallest value:  48
Largest value:   85
Range: $85-48=37$

50. Set: 18, 49, 19, 50, 68, 68
68 occurs most often
Mode: 68
Smallest value:  18
Largest value:   68
Range: $68-18=50$

51. Set: 66, 36, 13, 28
No value occurs more than once
Mode: none
Smallest value:  13
Largest value:   66
Range: $66-13=53$

52. Set: 51, 35, 35, 79, 72, 51, 60
35 and 51 occur most often, twice each
Modes: 35 and 51
Smallest value:  35
Largest value:   79
Range: $79-35=44$

53. Set: 70, 43, 37, 80, 43
43 occurs most often
Mode: 43
Smallest value:  37
Largest value:   80
Range: $80-37=43$

54. Set: 14, 14, 15, 14, 73, 15
14 occurs most often
Mode: 14
Smallest value:  14
Largest value:   73
Range: $73-14=59$

55. Set: 22, 43, 24, 16, 24, 8, 16, 9, 12
16 and 24 occur most often, twice each
Modes: 16 and 24
Smallest value:   8
Largest value:   43
Range: $43-8=35$

56. Set: 135, 180, 240, 160, 210, 201
No value occurs more than once
Mode: none
Smallest value:  135
Largest value:   240
Range: $240-135=105$

# Chapter 8 Assessment Test

1.  Look in the row corresponding to the academic year 2000-2001 and find the value under the column heading of Private Colleges. The value $22,240 is the average cost for a private college in 2000-2001.

2.  Look in the row corresponding to the academic year 1995-1996 and find the value under the column heading of Public Colleges. The value $6743 is the average cost for a public college in 1995-1996.

3.  From the table, the average costs in 2005-2006 for private and public colleges were $29,026 and $12,127 respectively. The costs for a private college were $29,026 - \$12,127 = \$16,899$ more than that for a public college.

4.  Locate Sept. on the time axis. Move vertically until you reach the blue line graph, corresponding to cell phones with cameras. The point on the graph lies about halfway between the 40 and 60 marks on the Sales axis so about 50 phones were sold with cameras in September.

5.  Locate Nov. on the time axis. Move vertically until you reach the red line graph, corresponding to cell phones without cameras. The point on the graph lies about halfway between the 60 and 80 marks on the Sales axis so about 70 phones were sold without cameras in November.

6.  The lowest point on the red line graph (without cameras) is above July on the time axis. The fewest phones without cameras were sold in July.

7.  Under the heading of the business travel, the year 2006 shows a yellow bar (representing major carriers) labeled 81%. Thus 81% of business travelers choose major carriers in 2006.

8.  Under the heading of personal travelers, the year 2004 shows a green bar (representing low-cost carriers) labeled 41%. Thus 41% of personal travelers chose low-cost carriers in 2004.

9.  Under business travel, the low cost carrier (green bar) labeled 23% corresponds to the year 2004.

10. Looking at the personal travel portion of the bar graph and the major carrier bars (yellow bar), 53% chose major carriers in 2004 and 65% chose major carriers in 2006. Thus 2006 was the year 65% of personal travelers chose major carriers.

11. The largest and smallest segments of the circle graph correspond to Phillips and Samsung respectively. Thus, Phillips had the highest sales while Samsung had the lowest sales.

12. Toshiba had 18% of DVD player sales while Panasonic had 30% of the sales for combined sales of $18\% + 30\% = 48\%$.

13. Panasonic's 30% share of sales was higher than Samsung's 12% by $30\% - 12\% = 18\%$.

14. Since 40% of sales were attributed to Phillips, if 2000 DVD players were sold then $40\% \cdot 2000 = 0.40 \cdot 2000 = 800$ units were sold by Phillips.

15. The $x$-axis is labeled with the different types of food in the table. The $y$-axis is labeled with values for steps. Bars are drawn over each food to the height specified by the numerical value of steps in the table. The resulting bar graph is shown here.

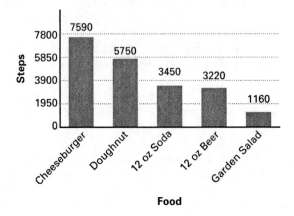

16. The total revenue from all division is

$$\$6 + \$12 + \$9 + \$3 = \$30 \text{ billion.}$$

Next convert each division's revenue to percentages of this total and then multiply by $360°$ to determine the degrees of each segment in the circle graph.

| DEGREE | DEGREES GRANTED | PERCENT | DEGREES |
|---|---|---|---|
| Electronics | $6 | $\frac{\$6}{\$30} = 0.2 = 20\%$ | $20\% \cdot 360 = 0.2 \cdot 360 = 72$ |
| Games | $12 | $\frac{\$12}{\$30} = 0.4 = 40\%$ | $40\% \cdot 360 = 0.4 \cdot 360 = 144$ |
| Pictures and Music | $9 | $\frac{\$9}{\$30} = 0.3 = 30\%$ | $30\% \cdot 360 = 0.3 \cdot 360 = 108$ |
| Other | $3 | $\frac{\$3}{\$30} = 0.1 = 10\%$ | $10\% \cdot 360 = 0.1 \cdot 360 = 36$ |

After using a protractor to create the segments having the degrees computed above and adding the labels, the completed circle graph is shown below.

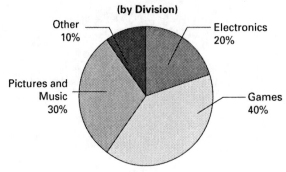

17. The mean is $\dfrac{24+49+39+46+31+51}{6} = \dfrac{240}{6} = 40$.

18. The data in ascending order is: 28 32 36 40 46 54. The number of values is even so the median is given by $\dfrac{36+40}{2} = 38$.

19. Both values 60 and 69 occur three times in the set, the most any value occurs. The modes are 60 and 69.

20. The largest value is 293 and the smallest value is 75. The range is $293 - 75 = 218$.

# Chapter 9    Signed Numbers

## Section 9.1    Introduction to Signed Numbers

### Concept Check

1.  A  <u>positive</u>  number is a number that is greater than 0.

3.  The number 0 on the number line is known as the  <u>origin</u> .

5.  Two numbers that lie the same distance from the origin on opposite sides of the origin are called  <u>opposites</u> .

7.  A number that can be written in the form $\dfrac{a}{b}$, where $a$ and $b$ are integers and $b \neq 0$ is a  <u>rational</u>  number.

### Guide Problems

9.  The opposite of a positive number is negative, so the opposite of 8 is −8.

*In Problems 11 through 17, find the opposite of the number by changing the sign of the number.*

11. −14          13. 23.5          15. $-\dfrac{2}{9}$          17. $\dfrac{5}{2}$

### Guide Problems

19. Graph the signed numbers −2, 5, 7, and 0.

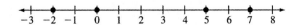

*In Problems 21 and 23, graph the given numbers on a number line.*

21. −5, −3, 1, 4

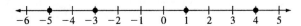

23. $-0.25, 2.4, -2\dfrac{1}{5}, 3\dfrac{1}{10}$

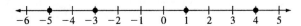

### Guide Problems

25. $\left|15\right| = 15$

*In Exercises 27 through 33, the absolute value of a positive number is the number itself, the absolute value of a negative number is the opposite of the number, and the absolute value of 0 is 0.*

27. $\left|43\right| = 43$          29. $\left|-31\right| = 31$          31. $\left|-2.7\right| = 2.7$          33. $\left|7\dfrac{2}{9}\right| = 7\dfrac{2}{9}$

## Guide Problems

35. Compare −3.2 and 8.5
    Using a number line, we see that −3.2 is to the left of 8.5, so −3.2 < 8.5. Alternatively, 8.5 > −3.2

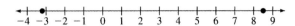

*For problems 37–47, when comparing two signed numbers, the number farthest to the left on a number line is the smaller number.*

37. 2 < 7, 7 > 2

39. −24.8 < −24.0, −24.0 > −24.8

41. −15 < 21, 21 > −15

43. −11.0 > −11.5, −11.5 < −11.0

45. $-\dfrac{7}{9} < -\dfrac{2}{3}, -\dfrac{2}{3} > -\dfrac{7}{9}$

47. $-\dfrac{6}{5} < -\dfrac{13}{12}, -\dfrac{13}{12} > -\dfrac{6}{5}$

49. 2 points

51. −6.5 sec, 2.4 sec, −4.1 sec

53. 6000°K

## Cumulative Skills Review

1.  $2\dfrac{1}{3} + 3\dfrac{2}{5}$     The LCD of $\dfrac{1}{3}$ and $\dfrac{2}{5}$ is 15.

    Rewrite the fractional parts.

    $\dfrac{1}{3} = \dfrac{1 \cdot 5}{3 \cdot 5} = \dfrac{5}{15}$     $\dfrac{2}{5} = \dfrac{2 \cdot 3}{5 \cdot 3} = \dfrac{6}{15}$

    $2\dfrac{5}{15}$

    $+ 3\dfrac{6}{15}$

    $5\dfrac{11}{15}$

3.  1822.4̲33 to the nearest tenth is 1822.4.

5.  60 minutes to 1200 seconds
    Smaller unit: seconds
    60 minutes = $60 \cdot 60 = 3600$ seconds

    $\dfrac{3600 \text{ seconds}}{1200 \text{ seconds}} = \dfrac{3}{1}$

7.  12 hg to g
    Move the decimal point 2 places to the right.
    12 hg = 1200 g

9.  Find the mode, if any, of 15, 18, 28, 15, 17, 65. The number 15 occurs twice, so 15 is the mode.

## Section 9.2    Adding Signed Numbers

## Concept Check

1.  When adding two positive numbers, the sum will be __positive__.

3.  To add signed numbers with different signs, determine the absolute value of each addend. __Subtract__ the smaller absolute value from the larger. Then, attach the sign of the addend having the larger absolute value to this difference.

## Guide Problems

5.   Add $-12.3 + (-31.5)$

   a.   Determine the absolute value of each addend.
      $|-12.3| = \underline{12.3};$       $|-31.5| = \underline{31.5}$

   b.   Add the absolute values of the addends.
   $$\begin{array}{r} 12.3 \\ + \ 31.5 \\ \hline 43.8 \end{array}$$

   c.   Attach the common sign of the addends to the sum of part b.
      $-12.3 + (-31.5) = \underline{-43.8}$

*In Problems 7 through 45, add the signed numbers using the procedure outlined above. Simplify fractions, if possible. All of the steps are shown for problems 9 through 13.*

7.   $6 + 3 = 9$

9.   $-9 + (-1)$
   $|-9| = 9; \ |-1| = 1$
   $9 + 1 = 10$
   $-9 + (-1) = -10$

11.  $-7 + (-8)$
   $|-7| = 7; \ |-8| = 8$
   $7 + 8 = 15$
   $-7 + (-8) = -15$

13.  $-43 + (-25)$
   $|-43| = 43; \ |-25| = 25$
   $43 + 25 = 68$
   $-43 + (-25) = -68$

15.  $-56 + (-89) = -145$

17.  $-512 + (-106) = -618$

19.  $-518 + (-31) = -549$

21.  $-126 + (-339) = -465$

23.  $2.51 + 3.54 = 6.05$

25.  $-6.1 + (-31.2) = -37.3$

27.  $-3.21 + (-14.5) = -17.71$

29.  $-83.31 + (-14.031) = -97.341$

*In Problems 31 through 45, find the least common denominator and then write each fraction as an equivalent fraction with the LCD before adding the signed numbers.*

31.  $-\dfrac{1}{4} + \left(-\dfrac{3}{4}\right)$

   Determine the absolute value of each addend.
   $$\left|-\dfrac{1}{4}\right| = \dfrac{1}{4}; \quad \left|-\dfrac{3}{4}\right| = \dfrac{3}{4}$$
   Add the absolute values of the addends.
   LCD = 4
   $$\dfrac{1}{4} + \dfrac{3}{4} = \dfrac{4}{4} = 1$$
   Attach the common sign of the addends to the sum.
   $$-\dfrac{1}{4} + \left(-\dfrac{3}{4}\right) = -1$$

33.  $-\dfrac{3}{7} + \left(-\dfrac{1}{3}\right)$

   Determine the absolute value of each addend.
   $$\left|-\dfrac{3}{7}\right| = \dfrac{3}{7}; \quad \left|-\dfrac{1}{3}\right| = \dfrac{1}{3}$$
   Add the absolute values of the addends.
   LCD = 21
   $$\dfrac{3}{7} + \dfrac{1}{3} = -\dfrac{3 \cdot 3}{7 \cdot 3} + \dfrac{1 \cdot 7}{3 \cdot 7}$$
   $$= \dfrac{9}{21} + \dfrac{7}{21} = \dfrac{16}{21}$$
   Attach the common sign of the addends to the sum.
   $$-\dfrac{3}{7} + \left(-\dfrac{1}{3}\right) = -\dfrac{16}{21}$$

35. $-\dfrac{1}{8}+\left(-\dfrac{5}{6}\right)$

Determine the absolute value of each addend.

$\left|-\dfrac{1}{8}\right|=\dfrac{1}{8};\quad\left|-\dfrac{5}{6}\right|=\dfrac{5}{6}$

Add the absolute values of the addends.
LCD = 24

$\dfrac{1}{8}+\dfrac{5}{6}=\dfrac{1\cdot3}{8\cdot3}+\dfrac{5\cdot4}{6\cdot4}$

$\qquad=\dfrac{3}{24}+\dfrac{20}{24}=\dfrac{23}{24}$

Attach the common sign of the addends
to the sum.

$-\dfrac{1}{8}+\left(-\dfrac{5}{6}\right)=-\dfrac{23}{24}$

37. $-\dfrac{3}{10}+\left(-\dfrac{4}{5}\right)$

Determine the absolute value of each addend.

$\left|-\dfrac{3}{10}\right|=\dfrac{3}{10};\quad\left|-\dfrac{4}{5}\right|=\dfrac{4}{5}$

Add the absolute values of the addends.
LCD = 10

$\dfrac{3}{10}+\dfrac{4}{5}=\dfrac{3}{10}+\dfrac{4\cdot2}{5\cdot2}$

$\qquad=\dfrac{3}{10}+\dfrac{8}{10}=\dfrac{11}{10}=1\dfrac{1}{10}$

Attach the common sign of the addends
to the sum.

$-\dfrac{3}{10}+\left(-\dfrac{4}{5}\right)=-1\dfrac{1}{10}$

39. $-1\dfrac{1}{3}+\left(-4\dfrac{5}{12}\right)$

Determine the absolute value of each addend.

$\left|-1\dfrac{1}{3}\right|=1\dfrac{1}{3};\quad\left|-4\dfrac{5}{12}\right|=4\dfrac{5}{12}$

Add the absolute values of the addends.
LCD = 12

$1\dfrac{1}{3}+4\dfrac{5}{12}=\dfrac{4}{3}+\dfrac{53}{12}=\dfrac{4\cdot4}{3\cdot4}+\dfrac{53}{12}$

$\qquad=\dfrac{16}{12}+\dfrac{53}{12}=\dfrac{69}{12}=5\dfrac{9}{12}=5\dfrac{3}{4}$

Attach the common sign of the addends
to the sum.

$-1\dfrac{1}{3}+\left(-4\dfrac{5}{12}\right)=-5\dfrac{3}{4}$

41. $-21\dfrac{5}{9}+\left(-33\dfrac{5}{12}\right)$

Determine the absolute value of each addend.

$\left|-21\dfrac{5}{9}\right|=21\dfrac{5}{9};\quad\left|-33\dfrac{5}{12}\right|=33\dfrac{5}{12}$

Add the absolute values of the addends.
LCD = 72

$21\dfrac{5}{9}+33\dfrac{5}{12}=21\dfrac{5\cdot8}{9\cdot8}+33\dfrac{5\cdot6}{12\cdot6}$

$\qquad=21\dfrac{40}{72}+33\dfrac{30}{72}=54\dfrac{70}{72}=54\dfrac{35}{36}$

Attach the common sign of the addends
to the sum.

$-21\dfrac{5}{9}+\left(-33\dfrac{5}{12}\right)=-54\dfrac{35}{36}$

43. $-12\dfrac{2}{3}+\left(-17\dfrac{5}{6}\right)$

Determine the absolute value of each addend.

$\left|-12\dfrac{2}{3}\right|=12\dfrac{2}{3};\quad\left|-17\dfrac{5}{6}\right|=17\dfrac{5}{6}$

Add the absolute values of the addends.
LCD = 6

$12\dfrac{2}{3}+17\dfrac{5}{6}=12\dfrac{2\cdot2}{3\cdot2}+17\dfrac{5}{6}$

$\qquad=12\dfrac{4}{6}+17\dfrac{5}{6}=29\dfrac{9}{6}=29+1\dfrac{3}{6}$

$\qquad=30\dfrac{3}{6}=30\dfrac{1}{2}$

Attach the common sign of the addends
to the sum.

$-12\dfrac{2}{3}+\left(-17\dfrac{5}{6}\right)=-30\dfrac{1}{2}$

45. $-39\dfrac{8}{9}+\left(-29\dfrac{7}{12}\right)$

Determine the absolute value of each addend.

$\left|-39\dfrac{8}{9}\right|=39\dfrac{8}{9};\quad\left|-29\dfrac{7}{12}\right|=29\dfrac{7}{12}$

Add the absolute values of the addends.
LCD = 72

$39\dfrac{8}{9}+29\dfrac{7}{12}=39\dfrac{8\cdot8}{9\cdot8}+29\dfrac{7\cdot6}{12\cdot6}$

$\qquad=39\dfrac{64}{72}+29\dfrac{42}{72}=68\dfrac{106}{72}$

$\qquad=68+1\dfrac{34}{72}=69\dfrac{17}{36}$

Attach the common sign of the addends
to the sum.

$-39\dfrac{8}{9}+\left(-29\dfrac{7}{12}\right)=-69\dfrac{17}{36}$

## Guide Problems

47. Add 15.2 + (−42.9)

    a.   Determine the absolute value of each addend.
$$|15.2| = \underline{15.2}; \qquad |-42.9| = \underline{42.9}$$

    b.   Subtract the smaller absolute value from the larger.
$$42.9 - 15.2 = 27.7$$

    c.   Attach the sign of the addend having the larger absolute value to the difference of part b.
$$15.2 + (-42.9) = -27.7$$

*In Problems 49 through 95, add the signed numbers using the procedure outlined above. Simplify fractions, if possible. All of the steps are shown for problems 57 through 67.*

49.  6 + (−3) = 3          51.  15 + (−8) = 7          53.  −10 + 14 = 4          55.  −8 + 8 = 0

57.  16 + (−27)
Determine the absolute value of each addend.
$$|16| = 16; \quad |-27| = 27$$
Subtract the smaller absolute value from the larger.
$$27 - 16 = 11$$
Attach the sign of the addend having the larger absolute value to the difference.
$$16 + (-27) = -11$$

59.  81 + (−45)
Determine the absolute value of each addend.
$$|81| = 81; \quad |-45| = 45$$
Subtract the smaller absolute value from the larger.
$$81 - 45 = 36$$
Attach the sign of the addend having the larger absolute value to the difference.
$$81 + (-45) = 36$$

61.  −139 + 76
Determine the absolute value of each addend.
$$|-139| = 139; \quad |76| = 76$$
Subtract the smaller absolute value from the larger.
$$139 - 76 = 63$$
Attach the sign of the addend having the larger absolute value to the difference.
$$-139 + 76 = -63$$

63.  170 + (−530)
Determine the absolute value of each addend.
$$|170| = 170; \quad |-530| = 530$$
Subtract the smaller absolute value from the larger.
$$530 - 170 = 360$$
Attach the sign of the addend having the larger absolute value to the difference.
$$170 + (-530) = -360$$

65.  −1216 + (3105)
$$|-1216| = 1216; \quad |3105| = 3105$$
Subtract the smaller absolute value from the larger.
$$3105 - 1216 = 1889$$
Attach the sign of the addend having the larger absolute value to the difference.
$$-1216 + (3105) = 1889$$

67.  −13,245 + 2178
$$|-13,245| = 13,245; \quad |2178| = 2178$$
Subtract the smaller absolute value from the larger.
$$13,245 - 2178 = 11,067$$
Attach the sign of the addend having the larger absolute value to the difference.
$$-13,245 + 2178 = -11,067$$

69.  −2.41 + (3.51) = 1.1          71.  −1.01 + 1.34 = 0.33          73.  11.3 + (−3.7) = 7.6

75.  −13.43 + 18.5 = 5.07          77.  198.3 + (−213.6) = −15.3          79.  −147.23 + 211.9 = 64.67

*In problems 81 through 95, find the least common denominator and then write each fraction as an equivalent fraction with the LCD before adding. All steps are shown for these problems.*

81. $-\dfrac{2}{5}+\dfrac{1}{5}$

$\left|-\dfrac{2}{5}\right|=\dfrac{2}{5};\quad \left|\dfrac{1}{5}\right|=\dfrac{1}{5}$

Subtract the smaller absolute value from the larger.

LCD = 5

$\dfrac{2}{5}-\dfrac{1}{5}=\dfrac{1}{5}$

Attach the sign of the addend having the larger absolute value to the difference.

$-\dfrac{2}{5}+\dfrac{1}{5}=-\dfrac{1}{5}$

83. $-\dfrac{2}{3}+\dfrac{5}{6}$

$\left|-\dfrac{2}{3}\right|=\dfrac{2}{3};\quad \left|\dfrac{5}{6}\right|=\dfrac{5}{6}$

LCD = 6

Subtract the smaller absolute value from the larger.

$\dfrac{5}{6}-\dfrac{2}{3}=\dfrac{5}{6}-\dfrac{2\cdot 2}{3\cdot 2}=\dfrac{5}{6}-\dfrac{4}{6}=\dfrac{1}{6}$

Attach the sign of the addend having the larger absolute value to the difference.

$-\dfrac{2}{3}+\dfrac{5}{6}=\dfrac{1}{6}$

85. $\dfrac{1}{8}+\left(-\dfrac{5}{6}\right)$

$\left|\dfrac{1}{8}\right|=\dfrac{1}{8};\quad \left|-\dfrac{5}{6}\right|=\dfrac{5}{6}$

LCD = 24

Subtract the smaller absolute value from the larger.

$\dfrac{5}{6}-\dfrac{1}{8}=\dfrac{5\cdot 4}{6\cdot 4}-\dfrac{1\cdot 3}{8\cdot 3}=\dfrac{20}{24}-\dfrac{3}{24}=\dfrac{17}{24}$

Attach the sign of the addend having the larger absolute value to the difference.

$\dfrac{1}{8}+\left(-\dfrac{5}{6}\right)=-\dfrac{17}{24}$

87. $\dfrac{5}{9}+\left(-\dfrac{1}{6}\right)$

$\left|\dfrac{5}{9}\right|=\dfrac{5}{9};\quad \left|-\dfrac{1}{6}\right|=\dfrac{1}{6}$

LCD = 18

Subtract the smaller absolute value from the larger.

$\dfrac{5}{9}-\dfrac{1}{6}=\dfrac{5\cdot 2}{9\cdot 2}-\dfrac{1\cdot 3}{6\cdot 3}=\dfrac{10}{18}-\dfrac{3}{18}=\dfrac{7}{18}$

Attach the sign of the addend having the larger absolute value to the difference.

$\dfrac{5}{9}+\left(-\dfrac{1}{6}\right)=\dfrac{7}{18}$

89. $-7\dfrac{1}{3}+2\dfrac{1}{8}$

$\left|-7\dfrac{1}{3}\right|=7\dfrac{1}{3};\quad \left|2\dfrac{1}{8}\right|=2\dfrac{1}{8}$

LCD = 24

Subtract the smaller absolute value from the larger.

$7\dfrac{1}{3}-2\dfrac{1}{8}=7\dfrac{1\cdot 8}{3\cdot 8}-2\dfrac{1\cdot 3}{8\cdot 3}$

$=7\dfrac{8}{24}-2\dfrac{3}{24}=5\dfrac{5}{24}$

Attach the sign of the addend having the larger absolute value to the difference.

$-7\dfrac{1}{3}+2\dfrac{1}{8}=-5\dfrac{5}{24}$

91. $8\dfrac{2}{3}+\left(-15\dfrac{11}{12}\right)$

$\left|8\dfrac{2}{3}\right|=8\dfrac{2}{3};\quad \left|-15\dfrac{11}{12}\right|=15\dfrac{11}{12}$

LCD = 12

Subtract the smaller absolute value from the larger.

$15\dfrac{11}{12}-8\dfrac{2}{3}=15\dfrac{11}{12}-8\dfrac{2\cdot 4}{3\cdot 4}$

$=15\dfrac{11}{12}-8\dfrac{8}{12}=7\dfrac{3}{12}=7\dfrac{1}{4}$

Attach the sign of the addend having the larger absolute value to the difference.

$8\dfrac{2}{3}+\left(-15\dfrac{11}{12}\right)=-7\dfrac{1}{4}$

93.  $-25 + 17\frac{3}{5}$

   $\left|-25\right| = 25;$   $\left|17\frac{3}{5}\right| = 17\frac{3}{5}$

   LCD = 5
   Subtract the smaller absolute value from the larger.

   $25 - 17\frac{3}{5} = 24\frac{5}{5} - 17\frac{3}{5} = 7\frac{2}{5}$

   Attach the sign of the addend having the larger absolute value to the difference.

   $-25 + 17\frac{3}{5} = -7\frac{2}{5}$

95.  $41\frac{8}{9} + \left(-53\frac{1}{6}\right)$

   $\left|41\frac{8}{9}\right| = 41\frac{8}{9};$   $\left|-53\frac{1}{6}\right| = 53\frac{1}{6}$

   LCD = 18
   Subtract the smaller absolute value from the larger.

   $53\frac{1}{6} - 41\frac{8}{9} = 53\frac{1 \cdot 3}{6 \cdot 3} - 41\frac{8 \cdot 2}{9 \cdot 2}$

   $\qquad\qquad = 53\frac{3}{18} - 41\frac{16}{18}$

   $\qquad\qquad = 52\frac{21}{18} - 41\frac{16}{18} = 11\frac{5}{18}$

   Attach the sign of the addend having the larger absolute value to the difference.

   $41\frac{8}{9} + \left(-53\frac{1}{6}\right) = -11\frac{5}{18}$

97.  The year 100 BC can be represented by $-100$. Since Julius Caesar was killed 56 years after 100 BC, this translates to $-100 + 56 = -44$. He was killed in the year 44 BC.

99.  The lowest point can be represented by $-282$ feet. Since Mt. Whitney is 14,777 feet higher, this translates to $-282 + 14,777 = 14,495$. Mount Whitney is 14,495 feet tall.

## Cumulative Skills Review

1.  $\dfrac{5}{8} - \dfrac{1}{6}$     The LCD of $\dfrac{5}{8}$ and $\dfrac{1}{6}$ is 24.
   Rewrite the fractional parts.

   $\dfrac{5}{8} = \dfrac{5 \cdot 3}{8 \cdot 3} = \dfrac{15}{24}$        $\dfrac{1}{6} = \dfrac{1 \cdot 4}{6 \cdot 4} = \dfrac{4}{24}$

   $\dfrac{5}{8} - \dfrac{1}{6} = \dfrac{15}{24} - \dfrac{4}{24} = \dfrac{11}{24}$

3.     881.42
     $\underline{+\ 78.4\ \ \ }$
       959.82

5.  $-128.5°\text{F}$

7.  Find the range by subtracting the smallest number from the largest number in the set of numbers.
   $3052 - 2545 = 507$
   The range of daily visits is 507.

9.  Graph $-5, 3, -1, 0$.

## Section 9.3     Subtracting Signed Numbers

## Concept Check

1.  The subtraction problem $a - b$ is the same as the addition problem $a + \ \underline{\ -b\ }$.

## Guide Problems

3. Subtract $8 - 15$

   a. Write the subtraction problem as an addition problem.
      $8 - 15 = 8 + \underline{(-15)}$

   b. Determine the absolute value of each addend in the addition problem of part a.
      $|8| = 8; \quad |-15| = 15$

   c. Since the addends in part a have different signs, subtract the smaller absolute value from the larger.
      $15 - 8 = 7$

   d. Attach the sign of the addend having the larger absolute value to the difference of part c.
      $8 - 15 = 8 + (-15) = -7$

*In Problems 5 through 59, subtract the signed numbers using the procedure outlined above. Simplify fractions, if possible.*

5. $13 - 6 = 7$

7. $27 - 11 = 16$

9. $5 - 11$
   Write the subtraction problem as an addition problem.
   $5 - 11 = 5 + (-11)$
   Determine the absolute value of each addend in the addition problem of part a.
   $|5| = 5; \quad |-11| = 11$
   Since the addends in the addition have different signs, subtract the smaller absolute value from the larger.
   $11 - 5 = 6$
   Attach the sign of the addend having the larger absolute value to the difference.
   $5 - 11 = 5 + (-11) = -6$

11. $15 - 32$
    Write the subtraction problem as an addition problem.
    $15 - 32 = 15 + (-32)$
    Determine the absolute value of each addend in the addition problem.
    $|15| = 15; \quad |-32| = 32$
    Since the addends in the addition have different signs, subtract the smaller absolute value from the larger.
    $32 - 15 = 17$
    Attach the sign of the addend having the larger absolute value to the difference
    $15 - 32 = 15 + (-32) = -17$

13. $13 - (-14)$
    Write the subtraction problem as an addition problem.
    $13 - (-14) = 13 + 14$
    Both addends are positive, so the sum is positive.
    $13 - (-14) = 13 + 14 = 27$

15. $48 - (-33)$
    Write the subtraction problem as an addition problem.
    $48 - (-33)) = 48 + 33$
    Both addends are positive, so the sum is positive.
    $48 - (-33)) = 48 + 33 = 81$

17. $413 - 106 = 307$ (Subtract as usual.)

19. $-27 - (-10)$
    Write the subtraction problem as an addition problem: $-27 - (-10) = -27 + 10$
    Determine the absolute value of each addend in the addition problem: $|-27| = 27; \quad |10| = 10$
    Since the addends in the addition have different signs, subtract the smaller absolute value from the larger: $27 - 10 = 17$
    Attach the sign of the addend having the larger absolute value to the difference:
    $-27 - (-10) = -27 + 10 = -17$

21.  −2145 − 1478

Write the subtraction problem as an addition problem.

−2145 − 1478 = −2145 + (−1478)

Determine the absolute value of each addend in the addition problem.

$\left|-2145\right| = 2145;$    $\left|-1478\right| = 1478$

Since the addends in the addition have the same signs, add the absolute values.

2145 + 1478 = 3623

Since both addends are negative, the sum is negative.

−2145 − 1478 = −2145 + (−1478) = −3623

23.  1039 − 908 = 131

25.  −147 −(−409)

Write the subtraction problem as an addition problem.

−147 −(−409) = −147 + 409

Determine the absolute value of each addend in the addition problem.

$\left|-147\right| = 147;$    $\left|409\right| = 409$

Since the addends in the addition have different signs, subtract the smaller absolute value from the larger.

409 − 147 = 262

Attach the sign of the addend having the larger absolute value to the difference.

−147 −(−409) = −147 + 409 = 262

27.  540 − 1329

Write the subtraction problem as an addition problem.

540 − 1329 = 540 + (−1329)

Determine the absolute value of each addend in the addition problem.

$\left|540\right| = 540;$    $\left|-1329\right| = 1329$

Since the addends in the addition have different signs, subtract the smaller absolute value from the larger.

1329 − 540 = 789

Attach the sign of the addend having the larger absolute value to the difference.

540 − 1329 = 540 + (−1329) = −789

*In problems 29 through 43, find the least common denominator and then write each fraction as an equivalent fraction with the LCD before subtracting.*

29.  $-\dfrac{1}{5} - \dfrac{3}{5}$

Write the subtraction problem as an addition problem.

$$-\frac{1}{5} - \frac{3}{5} = -\frac{1}{5} + \left(-\frac{3}{5}\right)$$

$$\left|-\frac{1}{5}\right| = \frac{1}{5};\quad \left|-\frac{3}{5}\right| = \frac{3}{5}$$

Since the addends in the addition have the same signs, add the absolute values.

LCD = 5

$$\frac{1}{5} + \frac{3}{5} = \frac{4}{5}$$

Since both addends are negative, the sum is negative.

$$-\frac{1}{5} - \frac{3}{5} = -\frac{1}{5} - \frac{3}{5} = -\frac{1}{5} + \left(-\frac{3}{5}\right) = -\frac{4}{5}$$

31.  $\dfrac{1}{6} - \dfrac{3}{8}$

Write the subtraction problem as an addition problem.

$$\frac{1}{6} - \frac{3}{8} = \frac{1}{6} + \left(-\frac{3}{8}\right)$$

$$\left|\frac{1}{6}\right| = \frac{1}{6};\quad \left|-\frac{3}{8}\right| = \frac{3}{8}$$

Since the addends in the addition have different signs, subtract the smaller absolute value from the larger.

LCD = 24

$$\frac{3}{8} - \frac{1}{6} = \frac{3\cdot3}{8\cdot3} - \frac{1\cdot4}{6\cdot4} = \frac{9}{24} - \frac{4}{24} = \frac{5}{24}$$

Attach the sign of the addend having the larger absolute value to the difference.

$$\frac{1}{6} - \frac{3}{8} = \frac{1}{6} + \left(-\frac{3}{8}\right) = -\frac{5}{24}$$

33. $-\dfrac{2}{3}-\left(-\dfrac{1}{6}\right)$

Write the subtraction problem as an addition problem.

$-\dfrac{2}{3}-\left(-\dfrac{1}{6}\right)=-\dfrac{2}{3}+\dfrac{1}{6}$

$\left|-\dfrac{2}{3}\right|=\dfrac{2}{3};\quad \left|\dfrac{1}{6}\right|=\dfrac{1}{6}$

Since the addends in the addition have different signs, subtract the smaller absolute value from the larger.

LCD = 6

$\dfrac{2}{3}-\dfrac{1}{6}=\dfrac{2\cdot 2}{3\cdot 2}-\dfrac{1}{6}=\dfrac{4}{6}-\dfrac{1}{6}=\dfrac{3}{6}=\dfrac{1}{2}$

Attach the sign of the addend having the larger absolute value to the difference.

$-\dfrac{2}{3}-\left(-\dfrac{1}{6}\right)=-\dfrac{2}{3}+\dfrac{1}{6}=-\dfrac{1}{2}$

35. $\dfrac{5}{12}-\left(-\dfrac{7}{18}\right)$

Write the subtraction problem as an addition problem.

$\dfrac{5}{12}-\left(-\dfrac{7}{18}\right)=\dfrac{5}{12}+\dfrac{7}{18}$

Both addends are positive, so add. The sum is positive.

LCD = 36

$\dfrac{5}{12}+\dfrac{7}{18}=\dfrac{5\cdot 3}{12\cdot 3}+\dfrac{7\cdot 2}{18\cdot 2}=\dfrac{15}{36}+\dfrac{14}{36}=\dfrac{29}{36}$

$\dfrac{5}{12}-\left(-\dfrac{7}{18}\right)=\dfrac{5}{12}+\dfrac{7}{18}=\dfrac{29}{36}$

37. $2\dfrac{1}{5}-6\dfrac{3}{7}$

Write the subtraction problem as an addition problem.

$2\dfrac{1}{5}-6\dfrac{3}{7}=2\dfrac{1}{5}+\left(-6\dfrac{3}{7}\right)$

$\left|2\dfrac{1}{5}\right|=2\dfrac{1}{5};\quad \left|-6\dfrac{3}{7}\right|=6\dfrac{3}{7}$

Since the addends in the addition have different signs, subtract the smaller absolute value from the larger.

LCD = 35

$6\dfrac{3}{7}-2\dfrac{1}{5}=6\dfrac{3\cdot 5}{7\cdot 5}-2\dfrac{1\cdot 7}{5\cdot 7}$

$\qquad =6\dfrac{15}{35}-2\dfrac{7}{35}=4\dfrac{8}{35}$

Attach the sign of the addend having the larger absolute value to the difference.

$2\dfrac{1}{5}-6\dfrac{3}{7}=2\dfrac{1}{5}+\left(-6\dfrac{3}{7}\right)=-4\dfrac{8}{35}$

39. $14\dfrac{2}{3}-\left(-11\dfrac{5}{9}\right)$

Write the subtraction problem as an addition problem.

$14\dfrac{2}{3}-\left(-11\dfrac{5}{9}\right)=14\dfrac{2}{3}+11\dfrac{5}{9}$

Since the addends in the addition are both positive, just add. The sign of the sum is positive.

LCD = 9

$14\dfrac{2}{3}+11\dfrac{5}{9}=14\dfrac{2\cdot 3}{3\cdot 3}+11\dfrac{5}{9}=14\dfrac{6}{9}+11\dfrac{5}{9}$

$\qquad =25\dfrac{11}{9}=25+1\dfrac{2}{9}=26\dfrac{2}{9}$

$14\dfrac{2}{3}-\left(-11\dfrac{5}{9}\right)=14\dfrac{2}{3}+11\dfrac{5}{9}=26\dfrac{2}{9}$

41. $5\dfrac{1}{8} - 7\dfrac{7}{28}$

Write the subtraction problem as an addition problem.

$$5\dfrac{1}{8} - 7\dfrac{7}{28} = 5\dfrac{1}{8} + \left(-7\dfrac{7}{28}\right)$$

$$\left|5\dfrac{1}{8}\right| = 5\dfrac{1}{8}; \quad \left|-7\dfrac{7}{28}\right| = 7\dfrac{7}{28}$$

Since the addends in the addition have different signs, subtract the smaller absolute value from the larger.

LCD = 56

$$7\dfrac{7}{28} - 5\dfrac{1}{8} = 7\dfrac{7\cdot 2}{28\cdot 2} - 5\dfrac{1\cdot 7}{8\cdot 7} = 7\dfrac{14}{56} - 5\dfrac{7}{56}$$

$$= 2\dfrac{7}{56} = 2\dfrac{1}{8}$$

Attach the sign of the addend having the larger absolute value to the difference.

$$5\dfrac{1}{8} - 7\dfrac{7}{28} = 5\dfrac{1}{8} + \left(-7\dfrac{7}{28}\right) = -2\dfrac{1}{8}$$

43. $1\dfrac{8}{9} - 4\dfrac{11}{12}$

Write the subtraction problem as an addition problem.

$$1\dfrac{8}{9} - 4\dfrac{11}{12} = 1\dfrac{8}{9} + \left(-4\dfrac{11}{12}\right)$$

$$\left|1\dfrac{8}{9}\right| = 1\dfrac{8}{9}; \quad \left|-4\dfrac{11}{12}\right| = 4\dfrac{11}{12}$$

Since the addends in the addition have different signs, subtract the smaller absolute value from the larger.

LCD = 72

$$4\dfrac{11}{12} - 1\dfrac{8}{9} = 4\dfrac{11\cdot 6}{12\cdot 6} - 1\dfrac{8\cdot 8}{9\cdot 8} = 4\dfrac{66}{72} - 1\dfrac{64}{72}$$

$$= 3\dfrac{2}{72} = 3\dfrac{1}{36}$$

Attach the sign of the addend having the larger absolute value to the difference.

$$1\dfrac{8}{9} - 4\dfrac{11}{12} = 1\dfrac{8}{9} + \left(-4\dfrac{11}{12}\right) = -3\dfrac{1}{36}$$

45. $9.78 - 4.25 = 5.53$

47. $4.32 - 6.87 = 4.32 + (-6.87) = -2.55$

49. $-7.64 - 1.32 = -7.64 + (-1.32) = -8.96$

51. $-4.18 - (-8.96) = -4.18 + 8.96 = 4.78$

53. $12.25 - 45.30 = 12.25 + (-45.30) = -33.05$

55. $-35.14 - (-51.2) = -35.14 + 51.2 = 16.06$

57. $-120.3 - 35.8 = -120.3 + (-35.8) = -156.1$

59. $143.8 - (-87.3) = 143.8 + 87.3 = 231.1$

61. The year 212 BC can be represented as $-212$. Find the difference in the years:
    $1777 - (-212) = 1777 + 212 = 1989$ years.

63. $212°F - (-143.6°F) = 212°F + 143.6°F = 355.6°F$.
    The difference between the two temperatures is 355.6°F.

## Cumulative Skills Review

1. The problem translates to: what percent of $225,000 is $70,000? Solving the corresponding percent
   equation gives $p \cdot 225,000 = 70,000$ or $p = \dfrac{70,000}{225,000} \approx 0.31 = 31.1\%$ or 31% to the nearest percent.

3. line, $\overrightarrow{FB}$ or $\overrightarrow{BF}$

5. Convert 24 square yards to square feet.

   $$\dfrac{\text{New units}}{\text{Original units}} = \dfrac{9 \text{ square feet}}{1 \text{ square yard}}$$

   $$24 \text{ sq yd} \cdot \dfrac{9 \text{ sq ft}}{1 \text{ sq yd}} = 216 \text{ square feet}$$

7. Since $-12$ is farther to the left on a number line,
   $-12 < 7$ or $7 > -12$.

9. $100\% + 35\% = 135\%$
   $a = 135\% \cdot 150 = 1.35 \cdot 150 = 202.5$

## Section 9.4 Multiplying and Dividing Signed Numbers

### Concept Check

1. The product or quotient of two numbers with different signs is always __negative__.

### Guide Problems

3. Multiply $-7 \cdot 12$

    a. Determine the absolute value of each factor.
$$|-7| = 7; \quad |12| = 12$$

    b. Multiply the absolute values
$$7 \cdot 12 = 84$$

    c. Since the factors have different signs, the product is negative. Attach a negative sign to the result of part b.
$$-7 \cdot 12 = -84$$

5. Multiply $-\dfrac{12}{25} \cdot \dfrac{5}{32}$. Simplify, if possible.

    a. Determine the absolute value of each factor.
$$\left|-\frac{12}{25}\right| = \frac{12}{25}; \quad \left|\frac{5}{32}\right| = \frac{5}{32}$$

    b. Multiply the absolute values
$$\frac{\overset{3}{\cancel{12}}}{\underset{5}{\cancel{25}}} \cdot \frac{\overset{1}{\cancel{5}}}{\underset{8}{\cancel{32}}} = \frac{3}{40}$$

    c. Since the factors have different signs, the product is negative. Attach a negative sign to the result of part b.
$$-\frac{12}{25} \cdot \frac{5}{32} = -\frac{3}{40}$$

*In Problems 7 through 65, multiply the signed numbers using the procedure outlined in the Guide problems. Simplify fractions, if possible.*

7. $6 \cdot 4 = 24$

9. $-3 \cdot 9 = -27$

11. $5 \cdot (-20) = -100$

13. $-11 \cdot (-12) = 132$

15. $128 \cdot (-4) = -512$

17. $-103 \cdot (-7) = 721$

19. $11 \cdot (-312) = -3432$

21. $-13 \cdot (-514) = 6682$

23. $\dfrac{3}{6} \cdot \left(-\dfrac{4}{11}\right) = \dfrac{\overset{1}{\cancel{3}}}{\underset{2}{\cancel{6}}} \cdot \left(-\dfrac{\overset{2}{\cancel{4}}}{11}\right) = -\dfrac{2}{11}$

25. $-\dfrac{5}{4} \cdot \left(-\dfrac{2}{7}\right) = -\dfrac{5}{\underset{2}{\cancel{4}}} \cdot \left(-\dfrac{\overset{1}{\cancel{2}}}{7}\right) = \dfrac{5}{14}$

27. $-\dfrac{4}{9} \cdot \dfrac{5}{12} = -\dfrac{\overset{1}{\cancel{4}}}{9} \cdot \dfrac{5}{\underset{3}{\cancel{12}}} = -\dfrac{5}{27}$

29. $-\dfrac{2}{5} \cdot \left(-\dfrac{5}{8}\right) = -\dfrac{\overset{1}{\cancel{2}}}{\underset{1}{\cancel{5}}} \cdot \left(-\dfrac{\overset{1}{\cancel{5}}}{\underset{4}{\cancel{8}}}\right) = \dfrac{1}{4}$

31. $\dfrac{5}{12} \cdot \dfrac{9}{10} = \dfrac{\overset{1}{\cancel{5}}}{\underset{4}{\cancel{12}}} \cdot \dfrac{\overset{3}{\cancel{9}}}{\underset{2}{\cancel{10}}} = \dfrac{3}{8}$

33. $-\dfrac{5}{21}\cdot\dfrac{14}{25}=-\dfrac{\overset{1}{\cancel{5}}}{\underset{3}{\cancel{21}}}\cdot\dfrac{\overset{2}{\cancel{14}}}{\underset{5}{\cancel{25}}}=-\dfrac{2}{15}$

35. $-7\dfrac{1}{2}\cdot 3\dfrac{1}{5}=-\dfrac{15}{2}\cdot\dfrac{16}{5}=-\dfrac{\overset{3}{\cancel{15}}}{\underset{1}{\cancel{2}}}\cdot\dfrac{\overset{8}{\cancel{16}}}{\underset{1}{\cancel{5}}}=-24$

37. $-3\dfrac{4}{5}\cdot\left(-3\dfrac{3}{4}\right)=-\dfrac{19}{5}\cdot\left(-\dfrac{15}{4}\right)=-\dfrac{19}{\underset{1}{\cancel{5}}}\cdot\left(-\dfrac{\overset{3}{\cancel{15}}}{4}\right)$

$\qquad\qquad =\dfrac{57}{4}=14\dfrac{1}{4}$

39. $\dfrac{1}{7}\cdot 2=\dfrac{2}{7}$

41. $-4\cdot\dfrac{5}{8}=-\dfrac{4}{1}\cdot\dfrac{5}{8}=-\dfrac{\overset{1}{\cancel{4}}}{1}\cdot\dfrac{5}{\underset{2}{\cancel{8}}}=-\dfrac{5}{2}=-2\dfrac{1}{2}$

43. $6.1\cdot 34=207.4$

45. $10.2\cdot(-1.8)=-18.36$

47. $-0.13\cdot 0.018=-0.00234$

49. $2.7\cdot(-0.51)=-1.377$

51. $-49\cdot 12=-588$

53. $72\cdot(-218)=-15,696$

55. $11\cdot 10\cdot(-2)=-220$

57. $24\cdot(-3)\cdot(-2)=144$

59. $2\cdot(-45)\cdot 18=-1620$

61. $-\dfrac{6}{7}\cdot\dfrac{1}{3}\cdot\dfrac{3}{14}=-\dfrac{\overset{\overset{1}{\cancel{2}}}{\cancel{6}}}{7}\cdot\dfrac{1}{\underset{1}{\cancel{3}}}\cdot\dfrac{3}{\underset{7}{\cancel{14}}}=-\dfrac{3}{49}$

63. $-3\dfrac{1}{2}\left(-3\dfrac{3}{4}\right)\left(-\dfrac{1}{5}\right)=-\dfrac{7}{2}\left(-\dfrac{15}{4}\right)\left(-\dfrac{1}{5}\right)=-\dfrac{7}{2}\left(-\dfrac{\overset{3}{\cancel{15}}}{4}\right)\left(-\dfrac{1}{\underset{1}{\cancel{5}}}\right)=-\dfrac{21}{8}=-2\dfrac{5}{8}$

65. $-2.3(-3.1)(0.2)=1.426$

## Guide Problems

67. Divide $-56\div 8$

    a.   Determine the absolute values of the dividend and divisor.
$$|-56|=56;\qquad |8|=8$$

    b.   Divide the absolute values
$$56\div 8=7$$

    c.   Since the dividend and divisor have different signs, the quotient is negative. Attach a negative sign to the result of part b.
$$-7\cdot 12=-84$$

69. Divide $-\dfrac{3}{8} \div \left(-\dfrac{9}{4}\right)$. Simplify, if possible.

   a. Determine the absolute values of the dividend and divisor.

   $$\left|-\dfrac{3}{8}\right| = \dfrac{3}{8}; \qquad \left|-\dfrac{9}{4}\right| = \dfrac{9}{4}$$

   b. Divide the absolute values:

   $$\dfrac{3}{8} \div \dfrac{9}{4} = \dfrac{3}{8} \cdot \dfrac{4}{9} = \dfrac{1}{6}$$

   c. Since the dividend and divisor have the same sign, the quotient is positive.

   $$-\dfrac{3}{8} \div \left(-\dfrac{9}{4}\right) = \dfrac{1}{6}$$

*In Problems 71 through 113, divide the signed numbers using the procedure outlined in the Guide problems. Simplify fractions, if possible.*

71. $12 \div 4 = 3$

73. $21 \div 3 = 7$

75. $49 \div (-7) = -7$

77. $-81 \div (-9) = 9$

79. $132 \div (-4) = -33$

81. $-135 \div 9 = -15$

83. $266 \div (-14) = -19$

85. $273 \div (-21) = -13$

87. $\dfrac{2}{3} \div \left(-\dfrac{2}{7}\right) = \dfrac{2}{3} \cdot \left(-\dfrac{7}{2}\right) = -\dfrac{7}{3} = -2\dfrac{1}{3}$

89. $-\dfrac{7}{14} \div \dfrac{1}{7} = -\dfrac{7}{14} \cdot \dfrac{7}{1} = -\dfrac{7}{2} = -3\dfrac{1}{2}$

91. $\dfrac{3}{4} \div \left(-\dfrac{1}{8}\right) = \dfrac{3}{4} \cdot \left(-\dfrac{8}{1}\right) = -6$

93. $-\dfrac{1}{10} \div \dfrac{6}{11} = -\dfrac{1}{10} \cdot \dfrac{11}{6} = -\dfrac{11}{60}$

95. $-\dfrac{2}{3} \div 2\dfrac{1}{3} = -\dfrac{2}{3} \div \dfrac{7}{3} = -\dfrac{2}{3} \cdot \dfrac{3}{7} = -\dfrac{2}{7}$

97. $-5\dfrac{3}{4} \div \left(-1\dfrac{1}{2}\right) = -\dfrac{23}{4} \div \left(-\dfrac{3}{2}\right) = -\dfrac{23}{4} \cdot \left(-\dfrac{2}{3}\right) = \dfrac{23}{6} = 3\dfrac{5}{6}$

99. $2\dfrac{2}{3} \div \left(-3\dfrac{1}{2}\right) = \dfrac{8}{3} \div \left(-\dfrac{7}{2}\right) = \dfrac{8}{3} \cdot \left(-\dfrac{2}{7}\right) = -\dfrac{16}{21}$

101. $-51 \div 1\dfrac{1}{2} = -51 \div \dfrac{3}{2} = -\dfrac{51}{1} \cdot \dfrac{2}{3} = -34$

103. $\dfrac{-2.2}{-50} = 0.044$

105. $\dfrac{-4.69}{-67} = 0.07$

107. $36.668 \div (-4.12) = -8.9$

109. $-1.786 \div 89.3 = -0.02$

111. $-67.95 \div (-1.5) = 45.3$

113. $643.786 \div (-45.02) = -14.3$

115. A loss of $10.6 billion can be represented by $-10.6$. $3 \cdot (-10.6) = -31.8$, so General Motors would lose $31.8 billion.

117. The bonds were purchased at $12 per bond and sold at $15 per bond, a profit of $3 per bond. For 75 bonds, you earned $3 \cdot 75 = \$225$.

## Cumulative Skills Review

1.  $\dfrac{288 \text{ students}}{24 \text{ pecan pies}} = 12$ students for every pie

3.  Convert 12.5 years to months

$$\frac{\text{New units}}{\text{Original units}} = \frac{12 \text{ months}}{1 \text{ year}}$$

$$12.5 \text{ years} \cdot \frac{12 \text{ months}}{1 \text{ year}} = 150 \text{ months}$$

5.  $V = l \times w \times h = 79 \text{ mm} \times 122 \text{ mm} \times 86 \text{ mm}$

    $= 828,868 \text{ mm}^3$

7.  $-25 + (-33) = -58$

9.  $-55 - 7 = -55 + (-7) = -62$

## Section 9.5    Signed Numbers and Order of Operations

## Concept Check

1.  When applying the rules for order of operations, we perform all operations within  grouping symbols  first.

3.  Then, we perform all  multiplications  and  divisions  as they occur from left to right.

## Guide Problems

5.  Simplify $7 + 2^4 \cdot 5$ using order of operations.

$$7 + 2^4 \cdot 5 = 7 + 16 \cdot 5$$
$$= 7 + 80$$
$$= 87$$

*In Problems 7 through 52, use the order of operations to simplify each expression. First perform all operations within grouping symbols (parentheses, brackets, curly braces, fraction bars.) Next evaluate all exponential expressions. Then perform all multiplications and divisions as they appear in reading from left to right, followed by all additions and subtractions as they appear in reading from left to right.*

7.  $6 + 5 + (-4) = 11 + (-4)$
    $= 7$

9.  $8 - 2 \cdot (-3) - 4 = 8 - (-6) - 4$
    $= 14 - 4$
    $= 10$

11.  $8 \div 4 + (-3)(-5) = 2 + 15$
     $= 17$

13.  $-18 \div 2 + (-5) \cdot 4 + (-3) = -9 + (-20) + (-3)$
     $= -29 + (-3)$
     $= -32$

15.  $-5 \cdot 2^2 \cdot 3^2 = -5 \cdot 4 \cdot 9$
     $= -20 \cdot 9$
     $= -180$

17.  $(-8)^2 + 5 - 12 = 64 + 5 - 12$
     $= 69 - 12$
     $= 57$

19. $6 + (-2)^4 \cdot 7 = 6 + 16 \cdot 7$
$\phantom{6 + (-2)^4 \cdot 7} = 6 + 112$
$\phantom{6 + (-2)^4 \cdot 7} = 118$

21. $\left(7^2 - 9\right) \div (-5) = (49 - 9) \div (-5)$
$\phantom{\left(7^2 - 9\right) \div (-5)} = 40 \div (-5)$
$\phantom{\left(7^2 - 9\right) \div (-5)} = -8$

23. $-6 - (-3)^2 - (-4) = -6 - 9 - (-4)$
$\phantom{-6 - (-3)^2 - (-4)} = -15 - (-4)$
$\phantom{-6 - (-3)^2 - (-4)} = -11$

25. $(36 + 4) \div (20 \div 4) = 40 \div 5$
$\phantom{(36 + 4) \div (20 \div 4)} = 8$

27. $-3 + \left[2 \cdot 3 + (-4)\right]^2 \div 4 = -3 + \left[6 + (-4)\right]^2 \div 4$
$\phantom{-3 + [2 \cdot 3 + (-4)]^2 \div 4} = -3 + 2^2 \div 4$
$\phantom{-3 + [2 \cdot 3 + (-4)]^2 \div 4} = -3 + 4 \div 4$
$\phantom{-3 + [2 \cdot 3 + (-4)]^2 \div 4} = -3 + 1$
$\phantom{-3 + [2 \cdot 3 + (-4)]^2 \div 4} = -2$

29. $12^2 - \left(4 \cdot 5^2 - 20\right) \div 2^3 = 12^2 - \left(4 \cdot 25 - 20\right) \div 8$
$\phantom{12^2 - (4 \cdot 5^2 - 20) \div 2^3} = 12^2 - \left(100 - 20\right) \div 8$
$\phantom{12^2 - (4 \cdot 5^2 - 20) \div 2^3} = 12^2 - 80 \div 8$
$\phantom{12^2 - (4 \cdot 5^2 - 20) \div 2^3} = 144 - 80 \div 8$
$\phantom{12^2 - (4 \cdot 5^2 - 20) \div 2^3} = 144 - 10 = 134$

31. $(8 - 4)^2 \div \left[3^2 + (-5)\right] = 4^2 \div \left[9 + (-5)\right]$
$\phantom{(8 - 4)^2 \div [3^2 + (-5)]} = 16 \div 4$
$\phantom{(8 - 4)^2 \div [3^2 + (-5)]} = 4$

33. $2 + (7 - 5)^2 \cdot \left(5^2 - 13\right) = 2 + 2^2 \cdot (25 - 13)$
$\phantom{2 + (7 - 5)^2 \cdot (5^2 - 13)} = 2 + 4 \cdot 12$
$\phantom{2 + (7 - 5)^2 \cdot (5^2 - 13)} = 2 + 48$
$\phantom{2 + (7 - 5)^2 \cdot (5^2 - 13)} = 50$

35. $-\dfrac{3}{8} - \dfrac{1}{4}\left(-\dfrac{1}{3}\right) = -\dfrac{3}{8} - \left(-\dfrac{1}{12}\right)$
$\phantom{-\dfrac{3}{8} - \dfrac{1}{4}} = -\dfrac{3}{8} + \dfrac{1}{12} = -\dfrac{3 \cdot 3}{8 \cdot 3} + \dfrac{1 \cdot 2}{12 \cdot 2}$
$\phantom{-\dfrac{3}{8} - \dfrac{1}{4}} = -\dfrac{9}{24} + \dfrac{2}{24} = -\dfrac{7}{24}$

37. $-\dfrac{5}{27} - \left(-\dfrac{2}{3}\right)^2 = -\dfrac{5}{27} - \dfrac{4}{9}$
$\phantom{-\dfrac{5}{27} - (-\dfrac{2}{3})^2} = -\dfrac{5}{27} - \dfrac{4 \cdot 3}{9 \cdot 3} = -\dfrac{5}{27} - \dfrac{12}{27}$
$\phantom{-\dfrac{5}{27} - (-\dfrac{2}{3})^2} = -\dfrac{17}{27}$

39. $2^2 + \left[-\dfrac{2}{3} \div \left(\dfrac{1}{2}\right)^2\right] \cdot \dfrac{3}{4} = 2^2 + \left[-\dfrac{2}{3} \div \dfrac{1}{4}\right] \cdot \dfrac{3}{4}$
$\phantom{2^2 + [-\frac{2}{3} \div (\frac{1}{2})^2] \cdot \frac{3}{4}} = 2^2 + \left[-\dfrac{2}{3} \cdot \dfrac{4}{1}\right] \cdot \dfrac{3}{4}$
$\phantom{2^2 + [-\frac{2}{3} \div (\frac{1}{2})^2] \cdot \frac{3}{4}} = 4 + \left(-\dfrac{8}{3}\right) \cdot \dfrac{3}{4}$
$\phantom{2^2 + [-\frac{2}{3} \div (\frac{1}{2})^2] \cdot \frac{3}{4}} = 4 + (-2) = 2$

41. $4.21(-0.3) + 4(-0.06) = -1.263 + (-0.24)$
$\phantom{4.21(-0.3) + 4(-0.06)} = -1.503$

43.  $(-0.5)^2 + 0.3(-8) = 0.25 + 0.3(-8)$
$$= 0.25 + (-2.4)$$
$$= -2.15$$

45.  $-\left(1.5 + 0.1^2\right)^2 - 4\left(-0.2\right)^2$
$$= -\left(1.5 + 0.01\right)^2 - 4\left(-0.2\right)^2$$
$$= -\left(1.51\right)^2 - 4\left(-0.2\right)^2$$
$$= -2.2801 - 4\left(0.04\right)$$
$$= -2.2801 - 0.16 = -2.4401$$

47.  $196 \div 2\left[10 + \left(4 - 2\right)^2\right] - \left(-5\right)^2 = 196 \div 2\left[10 + 2^2\right] - \left(-5\right)^2 = 196 \div 2\left[10 + 4\right] - \left(-5\right)^2$
$$= 196 \div 2\left[14\right] - \left(-5\right)^2 = 196 \div 2\left[14\right] - 25 = 98 \cdot 14 - 25 = 1372 - 25 = 1347$$

49.  $6 + \dfrac{16 - 2^3}{\left(-2\right)^2 + 4} - 2 = 6 + \dfrac{16 - 8}{4 + 4} - 2$
$$= 6 + \dfrac{8}{8} - 2$$
$$= 6 + 1 - 2$$
$$= 7 - 2 = 5$$

51.  $4^3 + \left(\dfrac{5^2 - 15}{3^2 - 4}\right)^2 - 14 = 4^3 + \left(\dfrac{25 - 15}{9 - 4}\right)^2 - 14$
$$= 4^3 + \left(\dfrac{10}{5}\right)^2 - 14$$
$$= 4^3 + 2^2 - 14$$
$$= 64 + 4 - 14 = 68 - 14 = 54$$

53.  A birdie is one stroke under par (−1), an eagle is two strokes under par (−2), and a birdie is one stroke over par (1). Represent par by 0.

   a.    $3(-1) + 3(-2) + 2(0) + 1(1)$

   b.    $3(-1) + 3(-2) + 2(0) + 1(1) = -3 + (-6) + 0 + 1$
   $$= -9 + 0 + 1 = -8$$

55.  $-16(4)^2 + 80(4) = -16(16) + 80(4) = -256 + 320 = 64$
   The object will be 64 feet above the point after 4 seconds.

## Cumulative Skills Review

1.   $(-36) \div 6 = -6$

3.   $78.322 > 78.32$ or $78.32 < 78.322$

5.   $\dfrac{24 \text{ bottles}}{4 \text{ cases}} = \dfrac{6 \text{ bottles}}{1 \text{ case}}$

7.   $24 + (-9) = 15$

9.   $V = l \times w \times h$
   $$= 7 \text{ yd} \times 4 \text{ yd} \times 6 \text{ yd}$$
   $$= 168 \text{ yd}^3$$

## Chapter 9 Numerical Facts of Life

1.   Using the chart in the text, we find that the wind chill when the actual temperature is −5°F and the wind speed is 30 mph is −33°F.

3. If the actual temperature decreased by 10 degrees, then the new temperature is –5ºF – 10ºF = –15º. If the wind speed increased by 5 mph, then the new wind speed is 30 mph + 5 mph = 35 mph. Using the chart in the text, we find that the wind chill when the actual temperature is –15ºF and the wind speed is 35 mph is –48ºF.

## Chapter 9 Review Exercises

1. The opposite of 15 is –15.

2. The opposite of 2.34 is –2.34.

3. The opposite of –47 is 47.

4. The opposite of –8.094 is 8.094.

5.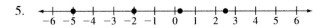

6. 

7. $|12| = 12$

8. $|-52| = 52$

9. $|-2.314| = 2.314$

10. $|7436| = 7436$

11. $2 < 7, 7 > 2$

12. $-3 > -8, -8 < -3$

13. $-2.05 < -2, -2 > -2.05$

14. $7 < 7.4, 7.4 > 7$

15. –210ºF

16. 1,000,000ºK

17. –$45.00

18. 3500 ft

19. $21 + 89 = 110$

20. $-117 + (-37) = -154$

21. $\dfrac{4}{5} + \left(-\dfrac{1}{6}\right)$

Determine the absolute value of each addend.

$\left|\dfrac{4}{5}\right| = \dfrac{4}{5};$   $\left|-\dfrac{1}{6}\right| = \dfrac{1}{6}$

Subtract the smaller absolute value from the larger.

LCD = 30

$\dfrac{4}{5} - \dfrac{1}{6} = \dfrac{4 \cdot 6}{5 \cdot 6} - \dfrac{1 \cdot 5}{6 \cdot 5}$

$= \dfrac{24}{30} - \dfrac{5}{30} = \dfrac{19}{30}$

Attach the sign of the addend having the larger absolute value to the difference.

$\dfrac{4}{5} + \left(-\dfrac{1}{6}\right) = \dfrac{19}{30}$

22. $-\dfrac{8}{9} + \dfrac{5}{12}$

Determine the absolute value of each addend.

$\left|-\dfrac{8}{9}\right| = \dfrac{8}{9};$   $\left|\dfrac{5}{12}\right| = \dfrac{5}{12}$

Subtract the smaller absolute value from the larger.

LCD = 36

$\dfrac{8}{9} - \dfrac{5}{12} = \dfrac{8 \cdot 4}{9 \cdot 4} - \dfrac{5 \cdot 3}{12 \cdot 3}$

$= \dfrac{32}{36} - \dfrac{15}{36} = \dfrac{17}{36}$

Attach the sign of the addend having the larger absolute value to the difference.

$-\dfrac{8}{9} + \dfrac{5}{12} = -\dfrac{17}{36}$

23. $-2\frac{1}{2}+1\frac{3}{7}$

Determine the absolute value of each addend.

$$\left|-2\frac{1}{2}\right|=\left|-\frac{5}{2}\right|=\frac{5}{2}; \quad \left|1\frac{3}{7}\right|=\left|\frac{10}{7}\right|=\frac{10}{7}$$

Subtract the smaller absolute value from the larger.

LCD = 14

$$\frac{5}{2}-\frac{10}{7}=\frac{5\cdot7}{2\cdot7}-\frac{10\cdot2}{7\cdot2}$$

$$=\frac{35}{14}-\frac{20}{14}=\frac{15}{14}=1\frac{1}{14}$$

Attach the sign of the addend having the larger absolute value to the difference.

$$-2\frac{1}{2}+1\frac{3}{7}=-1\frac{1}{14}$$

24. $4\frac{5}{6}+\left(-1\frac{1}{8}\right)$

Determine the absolute value of each addend.

$$\left|4\frac{5}{6}\right|=\left|\frac{29}{6}\right|=\frac{29}{6}; \quad \left|-1\frac{1}{8}\right|=\left|\frac{9}{8}\right|=\frac{9}{8}$$

Subtract the smaller absolute value from the larger.

LCD = 24

$$\frac{29}{6}-\frac{9}{8}=\frac{29\cdot4}{6\cdot4}-\frac{9\cdot3}{8\cdot3}$$

$$=\frac{116}{24}-\frac{27}{24}=\frac{89}{24}=3\frac{17}{24}$$

Attach the sign of the addend having the larger absolute value to the difference.

$$4\frac{5}{6}+\left(-1\frac{1}{8}\right)=3\frac{17}{24}$$

25. $-0.45 + 2.01$

Determine the absolute value of each addend.

$$\left|-0.45\right|=0.45; \quad \left|2.01\right|=2.01$$

Subtract the smaller absolute value from the larger.

$2.01 - 0.45 = 1.56$

Attach the sign of the addend having the larger absolute value to the difference.

$-0.45 + 2.01 = 1.56$

26. $-12.1 + (-4.8)$

The signs are the same, so add the absolute values and attach the common negative sign to the sum.

$-12.1 + (-4.8) = -16.9$

27. $54 - 82$

Write the subtraction problem as an addition problem.

$54 - 82 = 54 + (-82)$

Since the addends in the addition have different signs, subtract the smaller absolute value from the larger.

$82 - 54 = 28$

Attach the sign of the addend having the larger absolute value to the difference.

$54 - 82 = -28$

28. $2.3 - (-5.41)$

Write the subtraction problem as an addition problem.

$2.3 - (-5.41) = 2.3 + 5.41$

Since the addends in the addition have the same sign, add the absolute values and attach the common positive sign to the sum.

$2.3 - (-5.41) = 2.3 + 5.41 = 7.71$

29. $\frac{1}{4}-\frac{5}{8}$

Write the subtraction problem as an addition problem: $\frac{1}{4}-\frac{5}{8}=\frac{1}{4}+\left(-\frac{5}{8}\right)$

$$\left|\frac{1}{4}\right|=\frac{1}{4}; \quad \left|-\frac{5}{8}\right|=\frac{5}{8}$$

Since the addends in the addition have different signs, subtract the smaller absolute value from the larger.

LCD = 8

$$\frac{5}{8}-\frac{1}{4}=\frac{5}{8}-\frac{2}{8}=\frac{3}{8}$$

Attach the sign of the addend having the larger absolute value to the difference: $\frac{1}{4}-\frac{5}{8}=-\frac{3}{8}$

30. $5\dfrac{2}{9} - 7\dfrac{1}{3}$

Write the subtraction problem as an addition problem: $5\dfrac{2}{9} - 7\dfrac{1}{3} = 5\dfrac{2}{9} + \left(-7\dfrac{1}{3}\right)$

$\left|5\dfrac{2}{9}\right| = \left|\dfrac{47}{9}\right| = \dfrac{47}{9}$; $\left|-7\dfrac{1}{3}\right| = \left|-\dfrac{22}{3}\right| = \dfrac{22}{3}$

Since the addends in the addition have different signs, subtract the smaller absolute value from the larger.
LCD = 9

$\dfrac{22}{3} - \dfrac{47}{9} = \dfrac{66}{9} - \dfrac{47}{9} = \dfrac{19}{9} = 2\dfrac{1}{9}$

Attach the sign of the addend having the larger absolute value to the difference: $5\dfrac{2}{9} - 7\dfrac{1}{3} = -2\dfrac{1}{9}$

31. $-3 \cdot 5 = -15$

32. $-0.6 \cdot (-1.5) = 0.9$

33. $\dfrac{2}{3}\left(-\dfrac{7}{8}\right) = \dfrac{\overset{1}{\cancel{2}}}{3}\left(-\dfrac{7}{\underset{4}{\cancel{8}}}\right) = -\dfrac{7}{12}$

34. $\left(-1\dfrac{2}{5}\right)\left(-2\dfrac{6}{7}\right) = \left(-\dfrac{7}{5}\right)\left(-\dfrac{20}{7}\right)$

$= \left(-\dfrac{\overset{1}{\cancel{7}}}{\cancel{5}}\right)\left(-\dfrac{\overset{4}{\cancel{20}}}{\underset{1}{\cancel{7}}}\right) = 4$

35. $81 \div (-9) = -9$

36. $-1.21 \div (-0.11) = 11$

37. $-\dfrac{2}{9} \div \dfrac{1}{3} = -\dfrac{2}{9} \cdot \dfrac{3}{1} = -\dfrac{2}{\underset{3}{\cancel{9}}} \cdot \dfrac{\overset{1}{\cancel{3}}}{1} = -\dfrac{2}{3}$

38. $-2\dfrac{7}{8} \div 3\dfrac{3}{4} = -\dfrac{23}{8} \div \dfrac{15}{4} = -\dfrac{23}{8} \cdot \dfrac{4}{15}$

$= -\dfrac{23}{\underset{2}{\cancel{8}}} \cdot \dfrac{\overset{1}{\cancel{4}}}{15} = -\dfrac{23}{30}$

39. $\left[9^2 - (3-8)^2\right] \div 8 = \left[9^2 - (-5)^2\right] \div 8 = [81 - 25] \div 8 = 56 \div 8 = 7$

40. $-2 + \left[-2^2 - (21+7)\right]^2 + 10 = -2 + \left[-2^2 - 28\right]^2 + 10 = -2 + [-4 - 28]^2 + 10$

$= -2 + [-32]^2 + 10 = -2 + 1024 + 10 = 1022 + 10 = 1032$

41. $(-4)^2 + \dfrac{3^2 - 1}{2^2} - 3 = (-4)^2 + \dfrac{9-1}{4} - 3 = (-4)^2 + \dfrac{8}{4} - 3 = (-4)^2 + 2 - 3 = 16 + 2 - 3 = 16 + 2 - 3 = 18 - 3 = 15$

42. $5 - \left(\dfrac{8^2 - 1}{2^4 + 5}\right)^2 = 5 - \left(\dfrac{64-1}{16+5}\right)^2 = 5 - \left(\dfrac{63}{21}\right)^2 = 5 - 3^2 = 5 - 9 = -4$

43. a. Represent the deposits as positive numbers, and represent the withdrawals as negative numbers.
$\$250 + \$105 + \$215 + (-\$55)$

 b. $\$250 + \$105 + \$215 + (-\$55) = \$570 + (-\$55) = \$515$
Mr. Cortez has $515 in his account at the end of the month.

44.  a.   Represent 323 BC as −323, and represent 46 AD as 46. 46 AD is the later year, so the subtraction is
          46 − (−323).

     b.   46 − (−323) = 46 + 323 = 369.
          369 years separate Alexander the Great from his primary historian.

45.  188 − 226 = 188 + (−226) = −38.
     The base of the construction site is 38 feet below ground level.

46.  a.   Represent the price decrease as −$2.40.
          500(−$2.40)

     b.   500(−$2.40) = −$1200.  You lost $1200.

47.  Represent the salary increase as $175, the tax increase as −$62, and the social security deduction as −$27.
     $175 + (−$62) + (−$27) = $113 + (−$27) = $86.
     Your net raise is $86.

## Chapter 9 Assessment Test

1.   The opposite of 457 is −457.                    2.        The opposite of −5713 is 5713.

3.

4.

5.   $|350| = 350$                                   6.   $|-453| = 453$

7.   $7 > 4, 4 < 7$                                   8.   $-19.5 < -19.0, -19.0 > -19.5$

9.   −39°C                                            10.  −$3.24

11.  −25 + (−78)                                      12.  −91 + 118
     Since the addends in the addition have the same     Since the addends in the addition have different
     sign, add the absolute values and attach the        signs, subtract the smaller absolute value from the
     common positive sign to the sum.                    larger. Then, attach the sign of the addend having
     −25 + (−78) = −103                                  the larger absolute value to the difference.
                                                         −91 + 118 = 27

13.  −14.7 + 2.64                                     14.  13.05 + (−7.2)
     Since the addends in the addition have different     Since the addends in the addition have different
     signs, subtract the smaller absolute value from the  signs, subtract the smaller absolute value from the
     larger. Then, attach the sign of the addend having   larger. Then, attach the sign of the addend having
     the larger absolute value to the difference.         the larger absolute value to the difference.
     −14.7 + 2.64 = −12.06                               13.05 + (−7.2) = 5.85

15. $-\dfrac{1}{12}+\left(-\dfrac{5}{16}\right)$

Since the addends in the addition have the same signs, add the absolute values. Then, attach the common negative sign of the addends to the sum.

LCD = 48

$-\dfrac{1}{12}+\left(-\dfrac{5}{16}\right)=-\dfrac{4}{48}+\left(-\dfrac{15}{48}\right)=-\dfrac{19}{48}$

16. $2\dfrac{4}{5}+\left(-5\dfrac{1}{2}\right)$

Since the addends in the addition have different signs, subtract the smaller absolute value from the larger. Then, attach the sign of the addend having the larger absolute value to the difference.

LCD = 10

$2\dfrac{4}{5}+\left(-5\dfrac{1}{2}\right)=\dfrac{14}{5}+\left(-\dfrac{11}{2}\right)$

$=\dfrac{28}{10}+\left(-\dfrac{55}{10}\right)=-\dfrac{27}{10}=-2\dfrac{7}{10}$

17. $75-48=27$

18. $-0.47-8.9$

Write the subtraction problem as an addition problem.

$-0.47+(-8.9)=-9.37$

19. $\dfrac{5}{6}-\dfrac{4}{9}=\dfrac{15}{18}-\dfrac{8}{18}=\dfrac{7}{18}$

20. $-3\dfrac{7}{8}-1\dfrac{1}{6}=-3\dfrac{21}{24}-1\dfrac{4}{24}=-3\dfrac{21}{24}+\left(-1\dfrac{4}{24}\right)$

$=-4\dfrac{25}{24}=-5\dfrac{1}{24}$

21. $1.2\cdot(-0.5)=-0.6$

22. $\left(-\dfrac{8}{15}\right)\left(-\dfrac{1}{6}\right)=\left(-\dfrac{\cancel{8}^{4}}{15}\right)\left(-\dfrac{1}{\cancel{6}_{3}}\right)=\dfrac{4}{45}$

23. $-72\div(-0.9)=80$

24. $1\dfrac{4}{5}\div\left(-4\dfrac{1}{2}\right)=\dfrac{9}{5}\div\left(-\dfrac{9}{2}\right)=\dfrac{9}{5}\cdot\left(-\dfrac{2}{9}\right)=-\dfrac{2}{5}$

25. $4^{3}+\left(3^{3}-30\right)^{2}+(-65)=4^{3}+(27-30)^{2}+(-65)=4^{3}+(-3)^{2}+(-65)=64+9+(-65)=73+(-65)=8$

26. $100-\dfrac{(-12)^{2}+6}{(-4)^{2}-1}+(-50)=100-\dfrac{144+6}{16-1}+(-50)=100-\dfrac{150}{15}+(-50)=100-10+(-50)=90+(-50)=40$

27. Represent 1300 feet below sea level as $-1300$ feet.
$29{,}000-(-1300)=29{,}000+1300=30{,}300$ feet
The difference in altitude between the Dead Sea and Mount Everest is 30,300 feet.

28. $-\$37.21+150.00-\$60.00=\$112.79-\$60.00=\$52.72$
Steven's balance is $52.72.

29. The average change is $-3420\div12=-285$ students per year.

30. There are 2 large orders at $10 each, 2 medium orders at $8 each, 2 side dishes for each of the 4 orders at $2.50 each, a $4 delivery charge, a $6 tip, and a $3 discount.
$2(\$10.00)+2(\$8.00)+2(4)(\$2.50)+\$4.00+\$6.00-\$3.00$

$=\$20.00+\$16.00+\$20.00+\$4.00+\$6.00-\$3.00=\$63.00$
The total cost of the order is $63.00

# Chapter 10  Introduction to Algebra

## Section 10.1   Algebraic Expressions

## Concept Check

1.  A letter or some other symbol that represents a number whose value is unknown is called a __variable__ .

3.  When we __evaluate__ an expression, we replace each variable of an algebraic expression by a particular value and find the value of the expression.

5.  A term that contains a variable is known as a __variable__ term, whereas a term that is a number is called a __constant__ term.

7.  Terms with the same variable factors are called __like__ terms.

## Guide Problems

9.  Evaluate each algebraic expression when $x = 7$.

  a.  $3x$
  $3(\underline{7}) = \underline{21}$

  b.  $2 - x$
  $2 - \underline{7} = \underline{-5}$

  c.  $4(x+1)$
  $4(\underline{7}+1) = 4(\underline{8}) = \underline{32}$

*In Problems 11 through 15, evaluate each algebraic expression when x = 7.*

11.  $31 - x + 5$
$31 - 7 + 5 = 24 + 5 = 29$

13.  $5x + 4x$
$5(7) + 4(7) = 35 + 28 = 63$

15.  $2 + 3x + x^2$
$2 + 3(7) + 7^2 = 2 + 21 + 49 = 72$

*In Problems 17 through 19, evaluate each algebraic expression when y = 5.*

17.  $2y + 7$
$2(5) + 7 = 10 + 7 = 17$

19.  $-9y - 7$
$-9(5) - 7 = -45 - 7 = -52$

*In Problems 21 through 25, evaluate each algebraic expression when x = −3 and y = 2.*

21.  $3x - 2y$
$3(-3) - 2(2) = -9 - 4 = -13$

23.  $\dfrac{8x^2}{3y}$
$\dfrac{8(-3)^2}{3(2)} = \dfrac{8(9)}{3(2)} = \dfrac{72}{6} = 12$

25.  $(x+1)(y+3)$
$(-3+1)(2+3) = (-2)(5) = -10$

*In Problems 27 through 29, evaluate each algebraic expression when x = 10 and y = 15.*

27.  $3x - 4y$
$3(10) - 4(15) = 30 - 60 = -30$

29.  $6y \div x$
$6(15) \div 10 = 90 \div 10 = 9$

## Guide Problems

31. Consider $3x^4 - 2x^3 - 9x^2 + x - 1$

    a.  The variable terms are $\underline{3x^4}$, $\underline{-2x^3}$, $\underline{-9x^2}$, and $\underline{x}$.

    b.  The constant term is $\underline{-1}$.

    c.  The coefficient in the first term is $\underline{3}$.
        The coefficient in the second term is $\underline{-2}$.
        The coefficient in the third term is $\underline{-9}$.
        The coefficient in the fourth term is $\underline{1}$.

*In Problems 33 and 35, identify the variable and constant terms in each expression, along with the coefficient of each variable term.*

33. $4a^3 + 3a^2 - 9a - 5$

    Variable terms: $4a^3, 3a^2, -9a$

    Constant term: $-5$
    The coefficient in the first term is $\underline{4}$.
    The coefficient in the second term is $\underline{3}$.
    The coefficient in the third term is $\underline{-9}$.

35. $-y^4 - y^2 - 3y - 9$

    Variable terms: $-y^4, -y^2, -3y$

    Constant term: $-9$
    The coefficient in the first term is $\underline{-1}$.
    The coefficient in the second term is $\underline{-1}$.
    The coefficient in the third term is $\underline{-3}$.

*In Exercises 37 through 49, remember that like terms have the same variables and same exponents.*

37. $5t + 7s + 2t + s = (7s + s) + (5t + 2t)$
    $$= 8s + 7t$$

39. $2x + \dfrac{1}{4}y + \dfrac{3}{4}x + y = \left(2x + \dfrac{3}{4}x\right) + \left(\dfrac{1}{4}y + y\right)$
    $$= \left(2\dfrac{3}{4}\right)x + \left(1\dfrac{1}{4}\right)y$$
    $$= \dfrac{11}{4}x + \dfrac{5}{4}y$$

41. $x^2 - xy - xy + y^2 = x^2 + (-xy - xy) + y^2$
    $$= x^2 - 2xy + y^2$$

43. $2u^2v - 3uv^2 + 6u^2v - 2uv^2$
    $$= \left(2u^2v + 6u^2v\right) + \left(-3uv^2 - 2uv^2\right)$$
    $$= 8u^2v + \left(-5uv^2\right) = 8u^2v - 5uv^2$$

45. $3rst - 5r^2s + 4rst + 7rs - 3rs - 4r^2s = \left(-5r^2s - 4r^2s\right) + (7rs - 3rs) + (3rst + 4rst) = -9r^2s + 4rs + 7rst$

47. $5x^2 + 10xy + 5y^2 - 3x^2 - 6xy - 3y^2 = \left(5x^2 - 3x^2\right) + (10xy - 6xy) + \left(5y^2 - 3y^2\right) = 2x^2 + 4xy + 2y^2$

49. $\dfrac{1}{5}m^4 + \dfrac{1}{5} - 2m^2 + \dfrac{1}{10} - \dfrac{2}{15}m^4 + 4m^2 = \left(\dfrac{1}{5}m^4 - \dfrac{2}{15}m^4\right) + \left(-2m^2 + 4m^2\right) + \left(\dfrac{1}{5} + \dfrac{1}{10}\right)$
    $$= \left(\dfrac{3}{15}m^4 - \dfrac{2}{15}m^4\right) + \left(-2m^2 + 4m^2\right) + \left(\dfrac{2}{10} + \dfrac{1}{10}\right) = \dfrac{1}{15}m^4 + 2m^2 + \dfrac{3}{10}$

## Guide Problems

51. Multiply. Simplify, if possible.

   a.  $3(5x) = (3 \cdot 5)x$          b.  $-0.4(5y) = (-0.4 \cdot 5)y$       c.  $\frac{2}{9}\left(-\frac{3}{5}z\right) = \left(\frac{2}{9} \cdot -\frac{3}{5}\right)z$
   $\qquad = 15x$                      $\qquad\qquad = -2y$                      $\qquad\qquad\qquad = -\frac{2}{15}z$

*For problems 53–63, apply the associative property of multiplication and the commutative property of multiplication as necessary to multiply the expression.*

53.  $7(3x) = (7 \cdot 3)x$           55.  $(-7b)5 = (-7 \cdot 5)b$          57.  $-(8n) = -8n$
   $\qquad = 21x$                      $\qquad\quad = -35b$

59.  $0.3(8d) = (0.3 \cdot 8)d$       61.  $(0.5n)0.4 = (0.5 \cdot 0.4)n$    63.  $-\frac{4}{7}\left(-\frac{5}{8}x\right) = -\frac{4}{7}\left(-\frac{5}{8}\right)x$
   $\qquad = 2.4d$                     $\qquad\qquad = 0.2n$                     $\qquad\qquad\qquad = \frac{5}{14}x$

*For problems 65–87, use the distributive property $a(b+c) = ab + ac$  or  $(a+b)c = ac + bc$.*

65.  $5(x+2) = 5x + 10$              67.  $4(a-1) = 4a - 4$               69.  $-2(n-2) = -2n + 4$

71.  $9(3x-8) = 27x - 72$           73.  $(3n+7)3 = 9n + 21$             75.  $(6x+7y)8 = 48x + 56y$

77.  $-\frac{1}{2}(4t-6) = -2t + 3$   79.  $\frac{5}{7}\left(\frac{3}{4}x - 6\right) = \frac{5}{7}\left(\frac{3}{4}x\right) + \frac{5}{7}(-6) = \frac{15}{28}x - \frac{30}{7}$

81.  $0.1(0.4y+3) = 0.040y + 0.3$   83.  $0.3(2.4d-9) = 0.72d - 2.7$

85.  $3(y^2 - 2y + 6) = 3y^2 - 6y + 18$   87.  $-2(k^2 + 2k - 3) = -2k^2 - 4k + 6$

89.  The total of seventeen times a number and three can be represented by $17x + 3$.

91.  a.   Since Taipei 101 is taller than the Empire State Building, represent the height of the Empire State Building by $x - 421$.

   b.   If $x = 1671$ feet, then $x - 421 = 1671 - 421 = 1250$ feet. The Empire State Building is 1250 feet tall.

## Cumulative Skills Review

1.   2 days $\times \dfrac{24 \text{ hours}}{1 \text{ day}} \times \dfrac{60 \text{ minutes}}{1 \text{ hour}} = 2880$ minutes          3.   $x^2 + 12^2 = 22^2$
   $\qquad\qquad\qquad\qquad\qquad\qquad\qquad\qquad\qquad x^2 + 144 = 484$
   $\qquad\qquad\qquad\qquad\qquad\qquad\qquad\qquad\qquad\qquad x^2 = 340$
   $\qquad\qquad\qquad\qquad\qquad\qquad\qquad\qquad\qquad x = \sqrt{340} \approx 18.4 \text{ km}$

5.  $\dfrac{17}{100}$;  17 to 100;  17 : 100

7.   $99.12444 \cdot 100 = 9912.444$

9.  Graph $-3.5$, $-1$, 1, 4, and 0 on the number line.

## Section 10.2   Solving an Equation Using the Addition Property of Equality

### Concept Check

1.  An  _equation_  is a mathematical statement consisting of two expressions on either side of an equals (=) sign.

3.  The _addition_  property of equality states that adding the same value to each side of an equation preserves equality.

5.  The process of finding a solution to an equation involving a variable is know as _solving_  an equation.

### Guide Problems

7.  Determine whether 3 is a solution of the equation $3x + 9 = 18$

$$3x + 9 = 18$$

$$3\left(\underline{\ 3\ }\right) + 9 \overset{?}{=} 18$$

$$\underline{\ 9\ } + 9 \overset{?}{=} 18$$

$$\underline{18} \overset{?}{=} 18$$

3 is a solution to the equation $3x + 9 = 18$

*In Problems 7 through 35, determine whether the given value is a solution to the given equation by substituting the value for the unknown in the equation.*

9.  $s = 12$;  $s + 10 = 22$

$s + 10 = 22$

$12 + 10 \overset{?}{=} 22$

$22 = 22$

12 is a solution.

11.  $b = 32$;  $b - 8 = 22$

$b - 8 = 22$

$32 - 8 \overset{?}{=} 22$

$24 \neq 22$

32 is not a solution.

13.  $s = 18$;  $6 + s = 12$

$6 + s = 12$

$6 + 18 \overset{?}{=} 12$

$24 \neq 12$

18 is not a solution.

15.  $x = 49$; $x - 17 = 32$

$x - 17 = 32$

$49 - 17 \overset{?}{=} 32$

$32 = 32$

49 is a solution.

17.  $k = 105$; $k + 27 = 132$

$k + 27 = 132$

$105 + 27 \overset{?}{=} 132$

$132 = 132$

105 is a solution.

19.  $x = 213$;  $x - 35 = 168$

$x - 35 = 168$

$213 - 35 \overset{?}{=} 168$

$178 \neq 168$

213 is not a solution.

21. $t = \dfrac{5}{3}$; $t - \dfrac{2}{3} = 1$

$t - \dfrac{2}{3} = 1$

$\dfrac{5}{3} - \dfrac{2}{3} \overset{?}{=} 1$

$1 = 1$

$\dfrac{5}{3}$ is a solution.

23. $c = 34$; $21 - c = -13$

$21 - c = -13$

$21 - 34 \overset{?}{=} -13$

$-13 = -13$

34 is a solution.

25. $q = -20$; $7 - q = 27$

$7 - q = 27$

$7 - (-20) \overset{?}{=} 27$

$7 + 20 \overset{?}{=} 27$

$27 = 27$

$-20$ is a solution.

27. $x = -30$; $8 - x = 22$

$8 - x = 22$

$8 - (-30) \overset{?}{=} 22$

$8 + 30 \overset{?}{=} 23$

$38 \neq 27$

$-30$ is not a solution.

29. $r = 2.5$; $r - 1.3 = 3.8$

$r - 1.3 = 3.8$

$2.5 - 1.3 \overset{?}{=} 3.8$

$1.2 \neq 3.8$

2.5 is not a solution.

31. $n = 1.6$; $n + 2.3 = 3.9$

$n + 2.3 = 3.9$

$1.6 + 2.3 \overset{?}{=} 3.9$

$3.9 = 3.9$

1.6 is a solution.

33. $m = -6$; $m - 6 = 0$

$m - 6 = 0$

$-6 - 6 \overset{?}{=} 0$

$-12 \neq 0$

$-6$ is not a solution.

35. $t = 10$; $t - 5 = 5$

$t - 5 = 5$

$10 - 5 \overset{?}{=} 5$

$5 = 5$

10 is a solution.

## Guide Problems

37. Solve $x + 8 = 21$.

$\begin{aligned} x + 8 &= 21 \\ \underline{-8\phantom{x}} & \underline{-8} \\ x + 0 &= 13 \\ x &= 13 \end{aligned}$   Check:

$13 + 8 \overset{?}{=} 21$

$21 = 21$ [a]

*Use the method shown in the Guide Problem to solve problems 39 through 77. Be sure to check each solution.*

39. $\begin{aligned} x + 6 &= 14 \\ \underline{-6\phantom{x}} & \underline{-6} \\ x + 0 &= 8 \\ x &= 8 \end{aligned}$   Check:

$8 + 6 \overset{?}{=} 14$

$14 = 14$ [a]

41. $\begin{aligned} n + 18 &= 30 \\ \underline{-18\phantom{x}} & \underline{-18} \\ n + 0 &= 12 \\ n &= 12 \end{aligned}$   Check:

$12 + 18 \overset{?}{=} 30$

$30 = 30$ [a]

43. $\begin{aligned} t + 9 &= 7 \\ \underline{-9\phantom{x}} & \underline{-9} \\ t + 0 &= -2 \\ t &= -2 \end{aligned}$   Check:

$-2 + 9 \overset{?}{=} 7$

$7 = 7$ [a]

45. $\begin{aligned} 9 &= s + 5 \\ \underline{-5\phantom{x}} & \underline{-5} \\ 4 &= s + 0 \\ 4 &= s \end{aligned}$   Check:

$9 \overset{?}{=} 4 + 5$

$9 = 9$ [a]

47. $3 = k + 8$    Check:

$$\frac{-8 \qquad -8}{-5 = k + 0}$$
$$-5 = k$$

$3 \overset{?}{=} -5 + 8$
$3 = 3^a$

49. $a + 6 = 0$    Check:

$$\frac{-6 \quad -6}{a + 0 = -6}$$
$$a = -6$$

$-6 + 6 \overset{?}{=} 0$
$0 = 0^a$

51. $k - 5 = 15$    Check:

$$\frac{+5 \quad +5}{k + 0 = 20}$$
$$k = 20$$

$20 - 5 \overset{?}{=} 15$
$15 = 15^a$

53. $v - 16 = 32$    Check:

$$\frac{+16 \quad +16}{v + 0 = 48}$$
$$v = 48$$

$48 - 16 \overset{?}{=} 32$
$32 = 32^a$

55. $40 = x - 27$    Check:

$$\frac{+27 \qquad +27}{67 = x + 0}$$
$$67 = x$$

$40 \overset{?}{=} 67 - 27$
$40 = 40^a$

57. $21 = x - 34$    Check:

$$\frac{+34 \qquad +34}{55 = x + 0}$$
$$55 = x$$

$21 \overset{?}{=} 55 - 34$
$21 = 21^a$

59. $d - 28 = 28$    Check:

$$\frac{+28 \quad +28}{d + 0 = 56}$$
$$d = 56$$

$56 - 28 \overset{?}{=} 28$
$28 = 28^a$

61. $m - 13 = 0$    Check:

$$\frac{+13 \quad +13}{m + 0 = 13}$$
$$m = 13$$

$13 - 13 \overset{?}{=} 0$
$0 = 0^a$

63. $t - 9 = -9$    Check:

$$\frac{+9 \quad +9}{t + 0 = 0}$$
$$t = 0$$

$0 - 9 \overset{?}{=} -9$
$-9 = -9^a$

65. $k + 17 = 27$    Check:

$$\frac{-17 \quad -17}{k + 0 = 10}$$
$$k = 10$$

$10 + 17 \overset{?}{=} 27$
$27 = 27^a$

67. $0.8 = x - 2.7$    Check:

$$\frac{+2.7 \qquad +2.7}{3.5 = x + 0}$$
$$3.5 = x$$

$0.8 \overset{?}{=} 3.5 - 2.7$
$0.8 = 0.8^a$

69. $0 = d - 1.47$    Check:

$$\frac{+1.47 \qquad +1.47}{1.47 = d + 0}$$
$$1.47 = d$$

$0 \overset{?}{=} 1.47 - 1.47$
$0 = 0^a$

71. $b - 2.45 = 13.35$    Check:

$$\frac{+2.45 \quad +2.45}{b + 0 = 15.8}$$
$$b = 15.8$$

$15.8 - 2.45 \overset{?}{=} 13.35$
$13.35 = 13.35^a$

73. $p + \dfrac{3}{4} = \dfrac{7}{12}$    Check:

$$p + \frac{3}{4} - \frac{3}{4} = \frac{7}{12} - \frac{3}{4}$$
$$p = \frac{7}{12} - \frac{9}{12}$$
$$p = -\frac{2}{12} = -\frac{1}{6}$$

$-\dfrac{1}{6} + \dfrac{3}{4} \overset{?}{=} \dfrac{7}{12}$
$-\dfrac{2}{12} + \dfrac{9}{12} \overset{?}{=} \dfrac{7}{12}$
$\dfrac{7}{12} = \dfrac{7}{12}^a$

75. $\dfrac{7}{8} = \dfrac{1}{4} + z$     Check:

$\dfrac{7}{8} - \dfrac{1}{4} = \dfrac{1}{4} + z - \dfrac{1}{4}$    $\dfrac{7}{8} \overset{?}{=} \dfrac{1}{4} + \dfrac{5}{8}$

$\dfrac{7}{8} - \dfrac{2}{8} = z$    $\dfrac{7}{8} \overset{?}{=} \dfrac{2}{8} + \dfrac{5}{8}$

$\dfrac{5}{8} = z$    $\dfrac{7}{8} = \dfrac{7}{8}$ₐ

77. $a + 1\dfrac{1}{2} = 6$     Check:

$a + 1\dfrac{1}{2} - 1\dfrac{1}{2} = 6 - 1\dfrac{1}{2}$    $4\dfrac{1}{2} + 1\dfrac{1}{2} \overset{?}{=} 6$

$a = 5\dfrac{2}{2} - 1\dfrac{1}{2}$    $5\dfrac{2}{2} \overset{?}{=} 6$

$a = 4\dfrac{1}{2}$    $6 = 6$ₐ

*For problems 79–81, use the formula R − C = P, where R represents a company's revenue, C represents a company's cost, and P represents a company's profit.*

79.  $R - C = P$, $C = \$18{,}788{,}000{,}000$, $P = \$3{,}796{,}000{,}000$

$R - 18{,}788{,}000{,}000 = \quad 3{,}796{,}000{,}000$

$\underline{+\,18{,}788{,}000{,}000 \quad +\,18{,}788{,}000{,}000}$

$R + 0 \qquad\quad = \quad 22{,}584{,}000{,}000$

$R = 22{,}584{,}000{,}000$

Verizon's revenue for the quarter was $22,584,000,000.

81.  $R - C = P$, $C = \$4{,}490{,}000{,}000$, $P = \$770{,}000{,}000$

$R - 4{,}490{,}000{,}000 = \quad 770{,}000{,}000$

$\underline{+\,4{,}490{,}000{,}000 \quad +\,4{,}490{,}000{,}000}$

$R + 0 \qquad\quad = \quad 5{,}260{,}000{,}000$

$R = 5{,}260{,}000{,}000$

Apple's revenue for the quarter was $5,260,000,000.

## Cumulative Skills Review

1.  $\$21{,}125.34 - (\$302.45 + \$18.99 + \$57.76) = \$21{,}125.34 - \$379.20 = \$20{,}746.14$

3.  $32 \text{ weeks} \times \dfrac{7 \text{ days}}{1 \text{ week}} = 224 \text{ days}$     5.   14 is what percent of 98 translates to $\dfrac{14}{98} = \dfrac{p}{100}$ .

7.  $-8(q + 15) = -8q - 120$     9.   $d = 2r = 2 \times 13 = 26 \text{ units}$

## Section 10.3   Solving an Equation Using the Multiplication Property of Equality

## Concept Check

1.  The __multiplication__ property of equality states that multiplying each side of an equation by the same value preserves equality.

## Guide Problems

3.  Solve $9x = 45$.

$$\frac{9x}{9} = \frac{45}{9} \quad \text{Check:}$$
$$x = \underline{5} \qquad 9 \cdot 5 \overset{?}{=} 45$$
$$45 = 45^{\text{ a}}$$

5.  Solve $\dfrac{2}{7}x = \dfrac{1}{3}$.

$$\frac{\cancel{7}}{\cancel{2}} \cdot \frac{\cancel{2}}{\cancel{7}} x = \frac{1}{3} \cdot \frac{7}{\cancel{2}} \quad \text{Check:}$$
$$x = \frac{7}{6} \qquad \frac{2}{7} \cdot \frac{7}{6} \overset{?}{=} \frac{1}{3}$$
$$\frac{1}{3} = \frac{1}{3} \text{ a}$$

*Use the method shown in the Guide Problem to solve problems 7 through 41. Be sure to check each solution.*

7.  $3x = 12$   Check:

$$\frac{3x}{3} = \frac{12}{3} \qquad 3 \cdot 4 \overset{?}{=} 12$$
$$x = 4 \qquad 12 = 12 \text{ a}$$

9.  $9t = -45$   Check:

$$\frac{9t}{9} = \frac{-45}{9} \qquad 9 \cdot (-5) \overset{?}{=} -45$$
$$t = -5 \qquad -45 = -45^{\text{a}}$$

11.   $-r = -3$   Check:

$$(-1)(-r) = (-3)(-1) \qquad -(3) \overset{?}{=} -3$$
$$r = 3 \qquad\qquad -3 = -3^{\text{a}}$$

13.   $-p = 5$   Check:

$$(-1)(-p) = (5)(-1) \qquad -(-5) \overset{?}{=} 5$$
$$p = -5 \qquad\qquad 5 = 5^{\text{a}}$$

15.  $72 = 6a$   Check:

$$\frac{72}{6} = \frac{6a}{6} \qquad 72 \overset{?}{=} 6 \cdot 12$$
$$12 = a \qquad 72 = 72 \text{ a}$$

17.  $50 = -2m$   Check:

$$\frac{50}{-2} = \frac{-2m}{-2} \qquad 50 \overset{?}{=} -2 \cdot (-25)$$
$$-25 = m \qquad 50 = 50 \text{ a}$$

19.  $\dfrac{1}{2}x = 3$   Check:

$$\frac{2}{1} \cdot \frac{1}{2}x = 3 \cdot 2 \qquad \frac{1}{2} \cdot 6 \overset{?}{=} 3$$
$$x = 6 \qquad\qquad 3 = 3^{\text{a}}$$

21.   $-\dfrac{b}{2} = 7$   Check:

$$\left(-\frac{2}{1}\right) \cdot \left(-\frac{b}{2}\right) = 7 \cdot (-2) \qquad -\frac{-14}{2} \overset{?}{=} 7$$
$$b = -14 \qquad -(-7) \overset{?}{=} 7$$
$$7 = 7^{\text{a}}$$

23.   $\dfrac{1}{2}x = 23$   Check:

$$\frac{2}{1} \cdot \frac{1}{2}x = 23 \cdot 2 \qquad \frac{1}{2} \cdot 46 \overset{?}{=} 23$$
$$x = 46 \qquad\qquad 23 = 23^{\text{a}}$$

25.   $\dfrac{1}{2}r = \dfrac{3}{10}$   Check:

$$\frac{2}{1} \cdot \frac{1}{2}r = \frac{3}{10} \cdot \frac{2}{1} \qquad \frac{1}{2} \cdot \frac{3}{5} \overset{?}{=} \frac{3}{10}$$
$$r = \frac{3}{5} \qquad\qquad \frac{3}{10} = \frac{3}{10} \text{ a}$$

27.    $-5m = \dfrac{1}{4}$   Check:

$$\left(-\frac{1}{5}\right)(-5m) = \frac{1}{4}\left(-\frac{1}{5}\right) \qquad -5\left(-\frac{1}{20}\right) \overset{?}{=} \frac{1}{4}$$
$$m = -\frac{1}{20} \qquad\qquad \frac{1}{4} = \frac{1}{4} \text{ a}$$

29.    $-\dfrac{5}{8}z = -12$   Check:

$$\left(-\frac{8}{5}\right) \cdot \left(-\frac{5}{8}\right)z = -12 \cdot \left(-\frac{8}{5}\right) \qquad -\frac{5}{8} \cdot \left(19\frac{1}{5}\right) \overset{?}{=} -12$$
$$z = \frac{96}{5} = 19\frac{1}{5} \qquad -\frac{5}{8} \cdot \frac{96}{5} \overset{?}{=} -12$$
$$-12 = -12^{\text{a}}$$

31.           $-\dfrac{3}{8}x = 9$           Check:

$$\left(-\dfrac{8}{3}\right)\cdot\left(-\dfrac{3}{8}\right)x = 9\cdot\left(-\dfrac{8}{3}\right)\qquad -\dfrac{3}{8}\cdot(-24)\overset{?}{=}9$$

$$x = -24 \qquad\qquad\qquad\qquad\qquad\qquad 9 = 9^{a}$$

33.           $\dfrac{1}{9} = \dfrac{2}{7}b$           Check:

$$\dfrac{7}{2}\cdot\dfrac{1}{9} = \dfrac{2}{7}b\cdot\dfrac{7}{2}\qquad \dfrac{1}{9}\overset{?}{=}\dfrac{2}{7}\cdot\dfrac{7}{18}$$

$$\dfrac{7}{18} = b \qquad\qquad\qquad \dfrac{1}{9} = \dfrac{1}{9}{}_{a}$$

35.  $-2.5v = 10$       Check:

$$\dfrac{-2.5v}{-2.5} = \dfrac{10}{-2.5}\qquad -2.5\cdot(-4)\overset{?}{=}10$$

$$v = -4 \qquad\qquad\qquad 10 = 10^{\,a}$$

37.  $-3.3c = 24.75$     Check:

$$\dfrac{-3.3c}{-3.3} = \dfrac{24.75}{-3.3}\qquad -3.3\cdot(-7.5)\overset{?}{=}24.75$$

$$c = -7.5 \qquad\qquad\qquad 24.75 = 24.75^{\,a}$$

39.  $6.3x = 88.2$       Check:

$$\dfrac{6.3x}{6.3} = \dfrac{88.2}{6.3}\qquad 6.3\cdot 14\overset{?}{=}88.2$$

$$x = 14 \qquad\qquad\qquad 88.2 = 88.2^{a}$$

41.  $1.6r = 5.44$       Check:

$$\dfrac{1.6r}{1.6} = \dfrac{5.44}{1.6}\qquad 1.6\cdot 3.4\overset{?}{=}5.44$$

$$r = 3.4 \qquad\qquad\qquad 5.44 = 5.44^{a}$$

*For problems 43–45, use the formula d = rt, where d represents distance, r represents rate, and t represents time.*

43.  $d = rt,\ d = 270$ miles, $r = 60$ mph

$$270 = 60t$$

$$\dfrac{270}{60} = \dfrac{60t}{60}$$

$$\dfrac{27}{6} = 4\dfrac{1}{2} = t$$

It will take $4\dfrac{1}{2}$ hours (or 4 hours 30 minutes) to make the trip.

45.  $d = rt,\ d = 4000$ miles, $t = 7$ hours 30 minutes $= 7\dfrac{1}{2}$ hours

$$4000 = \left(7\dfrac{1}{2}\right)r$$

$$4000 = \dfrac{15}{2}r$$

$$\dfrac{2}{15}\cdot 4000 = \dfrac{15}{2}r\cdot\dfrac{2}{15}$$

$$\dfrac{1600}{3} = 533\dfrac{1}{3} = r$$

The average speed of the plane is $533\dfrac{1}{3}$ miles per hour.

## Cumulative Skills Review

1.  $89.9\% = 0.899$

3.  segment, $\overline{PD}$ or $\overline{DP}$

5.  $11.75$ km $= 11{,}750$ m

7.  $-20 - 7w + 82;\ w = 4$

$$-20 - 7(4) + 82 = -20 - 28 + 82 = -48 + 82 = 34$$

9.  $t = 9; \ 17 - t = 8$

$$17 - t = 8$$
$$17 - 9 \overset{?}{=} 8$$
$$8 = 8$$

9 is a solution.

## Section 10.4   Solving an Equation Using the Addition and Multiplication Properties

### Concept Check

1.  To solve the equation $2x + 1 = 7$, first __subtract__ 1 from each side of the equation. Then __divide__ each side of the equation by 2.

### Guide Problems

3.  Solve $3x + 14 = 35$.

   a.   Apply the addition property to isolate the variable term on one side of the equation.

$$3x + 14 = \ 35$$
$$\underline{\phantom{3x+} -14 \ \ -14}$$
$$3x + \ 0 = \ 21$$
$$3x = 21$$

   b.   Apply the multiplication property to solve for the variable.

$$\frac{3x}{3} = \frac{21}{3}$$
$$x = 7$$

   c.   Check the solution.

$$3(7) + 14 \overset{?}{=} 35$$
$$21 + 14 \overset{?}{=} 35$$
$$35 = 35^{a}$$

5.  Solve $4x = 2x + 36$.

   a.   Apply the addition property to isolate the variable term on one side of the equation.

$$4x = \ 2x + 36$$
$$\underline{-2x \ \ \ -2x \phantom{+36}}$$
$$2x = \ 0 \ + 36$$
$$2x = 36$$

   b.   Apply the multiplication property to solve for the variable.

$$\frac{2x}{2} = \frac{36}{2}$$
$$x = 18$$

   c.   Check the solution.

$$4(18) \overset{?}{=} 2(18) + 36$$
$$72 \overset{?}{=} 36 + 36$$
$$72 = 72^{a}$$

*Use the method shown in the Guide Problem to solve problems 7 through 65. Be sure to check each solution. Note that we drop the zeros starting with problem 47. Starting with problem 63, we use the horizontal format in order to save space.*

7.  $2x + 6 = \ 12$      Check:

$$\underline{\phantom{2x} -6 \ \ -6}$$
$$2x + 0 = \ \ 6$$
$$2x = 6$$
$$\frac{2x}{2} = \frac{6}{2}$$
$$x = 3$$

$$2 \cdot (3) + 6 \overset{?}{=} 12$$
$$6 + 6 \overset{?}{=} 12$$
$$12 = 12 \ ^{a}$$

9.  $3n + 6 = \ 24$      Check:

$$\underline{\phantom{3n} -6 \ \ -6}$$
$$3n + 0 = \ \ 18$$
$$3n = 18$$
$$\frac{3n}{3} = \frac{18}{3}$$
$$n = 6$$

$$3 \cdot (6) + 6 \overset{?}{=} 24$$
$$18 + 6 \overset{?}{=} 24$$
$$24 = 24 \ ^{a}$$

11. $5x + 7 = 7$     Check:

$$\frac{-7 \quad -7}{5x + 0 = \quad 0}$$

$5x = 0$

$$\frac{5x}{5} = \frac{0}{5}$$

$x = 0$

$5 \cdot (0) + 7 \overset{?}{=} 7$

$0 + 7 \overset{?}{=} 7$

$7 = 7$ ᵃ

13. $3r - 6 = 21$     Check:

$$\frac{+6 \quad +6}{3r + 0 = \quad 27}$$

$3r = 27$

$$\frac{3r}{3} = \frac{27}{3}$$

$r = 9$

$3 \cdot (5) + 6 \overset{?}{=} 21$

$15 + 6 \overset{?}{=} 21$

$21 = 21$ ᵃ

15. $3q + 4 = -14$     Check:

$$\frac{-4 \quad -4}{3q + 0 = \quad -18}$$

$3q = -18$

$$\frac{3q}{3} = \frac{-18}{3}$$

$q = -6$

$3 \cdot (-6) + 4 \overset{?}{=} -14$

$-18 + 4 \overset{?}{=} -14$

$-14 = -14$ᵃ

17. $-3p + 6 = 15$     Check:

$$\frac{-6 \quad -6}{-3p + 0 = \quad 9}$$

$-3p = 9$

$$\frac{-3p}{-3} = \frac{9}{-3}$$

$p = -3$

$-3 \cdot (-3) + 6 \overset{?}{=} 15$

$9 + 6 \overset{?}{=} 15$

$15 = 15$ ᵃ

19. $-7m - 12 = \quad 51$     Check:

$$\frac{+12 \quad +12}{-7m + \ 0 = \quad 63}$$

$-7m = 63$

$$\frac{-7m}{-7} = \frac{63}{-7}$$

$m = -9$

$-7 \cdot (-9) - 12 \overset{?}{=} 51$

$63 - 12 \overset{?}{=} 51$

$51 = 51$ ᵃ

21. $\frac{x}{3} + 5 = 11$     Check:

$$\frac{-5 \quad -5}{\frac{x}{3} + 0 = \quad 6}$$

$\frac{x}{3} = 6$

$3 \cdot \frac{x}{3} = 6 \cdot 3$

$x = 18$

$\frac{18}{3} + 5 \overset{?}{=} 11$

$6 + 5 \overset{?}{=} 11$

$11 = 11$ ᵃ

23. $\frac{y}{2} + 15 = -42$     Check:

$$\frac{-15 \quad -15}{\frac{y}{2} + 0 \ = \ -57}$$

$\frac{y}{2} = -57$

$2 \cdot \frac{y}{2} = -57 \cdot 2$

$y = -114$

$\frac{-114}{2} + 15 \overset{?}{=} -42$

$-57 + 15 \overset{?}{=} -42$

$-42 = -42$ ᵃ

25. $\frac{z}{2} - 7 = -6$     Check:

$$\frac{+7 \quad +7}{\frac{z}{2} + 0 = \quad 1}$$

$\frac{z}{2} = 1$

$2 \cdot \frac{z}{2} = 1 \cdot 2$

$z = 2$

$\frac{2}{2} - 7 \overset{?}{=} -6$

$1 - 7 \overset{?}{=} -6$

$-6 = -6$ ᵃ

27. $-6 + \dfrac{x}{3} = 5$  Check:

$\underline{+6} \qquad \underline{+6}$  $-6 + \dfrac{33}{3} \overset{?}{=} 5$

$0 + \dfrac{x}{3} = 11$  $-6 + 11 \overset{?}{=} 5$

$\dfrac{x}{3} = 11$  $5 = 5^{\text{ a}}$

$3 \cdot \dfrac{x}{3} = 11 \cdot 3$

$x = 33$

29. $4x + 9 = 3$  Check:

$\dfrac{-9 \quad -9}{4x + 0 = -6}$  $4 \cdot \left(-\dfrac{3}{2}\right) + 9 \overset{?}{=} 3$

$4x = -6$  $-6 + 9 \overset{?}{=} 3$

$\dfrac{4x}{4} = \dfrac{-6}{4}$  $3 = 3^{\text{a}}$

$x = -\dfrac{3}{2}$

31. $\dfrac{1}{4} - \dfrac{x}{16} = \dfrac{1}{8}$  Check:

$\underline{-\dfrac{1}{4}} \qquad \underline{-\dfrac{1}{4}}$  $\dfrac{1}{4} - \dfrac{2}{16} \overset{?}{=} \dfrac{1}{8}$

$0 - \dfrac{x}{16} = \dfrac{1}{8} - \dfrac{2}{8}$  $\dfrac{1}{4} - \dfrac{1}{8} \overset{?}{=} \dfrac{1}{8}$

$-\dfrac{x}{16} = -\dfrac{1}{8}$  $\dfrac{2}{8} - \dfrac{1}{8} \overset{?}{=} \dfrac{1}{8}$

$-16 \cdot \left(-\dfrac{x}{16}\right) = \left(-\dfrac{1}{8}\right) \cdot (-16)$  $\dfrac{1}{8} = \dfrac{1}{8}{}^{\text{ a}}$

$x = 2$

33. $\dfrac{1}{3}x + \dfrac{2}{3} = \dfrac{3}{4}$  Check:

$\dfrac{1}{3}x + \dfrac{2}{3} - \dfrac{2}{3} = \dfrac{3}{4} - \dfrac{2}{3}$  $\dfrac{1}{3}\left(\dfrac{1}{4}\right) + \dfrac{2}{3} \overset{?}{=} \dfrac{3}{4}$

$\dfrac{1}{3}x + 0 = \dfrac{9}{12} - \dfrac{8}{12}$  $\dfrac{1}{12} + \dfrac{2}{3} \overset{?}{=} \dfrac{3}{4}$

$\dfrac{1}{3}x = \dfrac{1}{12}$  $\dfrac{1}{12} + \dfrac{8}{12} \overset{?}{=} \dfrac{3}{4}$

$3 \cdot \left(\dfrac{1}{3}x\right) = \left(\dfrac{1}{12}\right) \cdot 3$  $\dfrac{9}{12} \overset{?}{=} \dfrac{3}{4}$

$x = \dfrac{3}{12} = \dfrac{1}{4}$  $\dfrac{3}{4} = \dfrac{3}{4}{}^{\text{ a}}$

35. $6 - 5x = 24$  Check:

$\dfrac{-6 \qquad -6}{0 - 5x = 18}$  $6 - 5 \cdot \left(-\dfrac{18}{5}\right) \overset{?}{=} 24$

$-5x = 18$  $6 + 18 \overset{?}{=} 24$

$\dfrac{-5x}{-5} = \dfrac{18}{-5}$  $24 = 24^{\text{a}}$

$x = -\dfrac{18}{5}$

37. $3.3m - 2.1 = 7.8$  Check:

$\dfrac{+2.1 \quad +2.1}{3.3m + 0 \;= 9.9}$  $3.3 \cdot 3 - 2.1 \overset{?}{=} 7.8$

$3.3m = 9.9$  $9.9 - 2.1 \overset{?}{=} 7.8$

$\dfrac{3.3m}{3.3} = \dfrac{9.9}{3.3}$  $7.8 = 7.8^{\text{a}}$

$m = 3$

39. $3x = 2x + 6$  Check:

$\dfrac{-2x \quad -2x}{x = 0 + 6}$  $3 \cdot 6 \overset{?}{=} 2 \cdot 6 + 6$

$x = 6$  $18 \overset{?}{=} 12 + 6$

$18 = 18^{\text{ a}}$

41. $12t = 3t + 72$  Check:

$\dfrac{- 3t \quad - 3t}{9t = 0 + 72}$  $12 \cdot 8 \overset{?}{=} 3 \cdot 8 + 72$

$9t = 72$  $96 \overset{?}{=} 24 + 72$

$\dfrac{9t}{9} = \dfrac{72}{9}$  $96 = 96^{\text{ a}}$

$t = 8$

**43.**
$$9a - 3 = 3a$$

$$\underline{-9a \qquad -9a}$$
$$0 - 3 = -6a$$
$$-3 = -6a$$
$$\frac{-3}{-6} = \frac{-6a}{-6}$$
$$\frac{1}{2} = a$$

Check:
$$9 \cdot \left(\frac{1}{2}\right) - 3 \overset{?}{=} 3 \cdot \left(\frac{1}{2}\right)$$
$$\frac{9}{2} - 3 \overset{?}{=} \frac{3}{2}$$
$$4\frac{1}{2} - 3 \overset{?}{=} \frac{3}{2}$$
$$1\frac{1}{2} \overset{?}{=} \frac{3}{2}$$
$$\frac{3}{2} = \frac{3}{2}\,^a$$

**45.**
$$5w = -7w + 8$$

$$\underline{+7w \qquad +7w}$$
$$12w = 0 + 8$$
$$12w = 8$$
$$\frac{12w}{12} = \frac{8}{12}$$
$$w = \frac{2}{3}$$

Check:
$$5 \cdot \left(\frac{2}{3}\right) \overset{?}{=} -7 \cdot \left(\frac{2}{3}\right) + 8$$
$$\frac{10}{3} \overset{?}{=} -\frac{14}{3} + 8$$
$$\frac{10}{3} \overset{?}{=} -\frac{14}{3} + \frac{24}{3}$$
$$\frac{10}{3} = \frac{10}{3}\,^a$$

**47.**
$$2d - 2 = 6d + 6$$

$$\underline{-6d \qquad -6d}$$
$$-4d - 2 = \qquad 6$$
$$\underline{+2 \qquad +2}$$
$$-4d = \qquad 8$$
$$\frac{-4d}{-4} = \frac{8}{-4}$$
$$d = -2$$

Check:
$$2 \cdot (-2) - 2 \overset{?}{=} 6 \cdot (-2) + 6$$
$$-4 - 2 \overset{?}{=} -12 + 6$$
$$-6 = -6\,^a$$

**49.**
$$6 + 2u = 3u - 12$$

$$\underline{-2u \qquad -2u}$$
$$6 = u - 12$$
$$\underline{+12 \qquad +12}$$
$$18 = u$$

Check:
$$6 + 2 \cdot 18 \overset{?}{=} 3 \cdot 18 - 12$$
$$6 + 36 \overset{?}{=} 54 - 12$$
$$42 = 42\,^a$$

**51.**
$$6m - 1 = m - 26$$

$$\underline{-m \qquad -m}$$
$$5m - 1 = \qquad -26$$
$$\underline{+1 \qquad +1}$$
$$\frac{5m}{5} = \frac{-25}{5}$$
$$m = -5$$

Check:
$$6(-5) - 1 \overset{?}{=} -5 - 26$$
$$-30 - 1 \overset{?}{=} -31$$
$$-31 = -31\,^a$$

**53.**
$$2p - 17 = 3p - 3$$

$$\underline{-2p \qquad -2p}$$
$$-17 = \qquad p - 3$$
$$\underline{+3 \qquad +3}$$
$$-14 = p$$

Check:
$$2(-14) - 17 \overset{?}{=} 3(-14) - 3$$
$$-28 - 17 \overset{?}{=} -42 - 3$$
$$-45 = -45\,^a$$

**55.**
$$8.2q + 4.1 = 3.4q - 5.5$$

$$\underline{-3.4q \qquad -3.4q}$$
$$4.8q + 4.1 = \qquad -5.5$$
$$\underline{-4.1 \qquad -4.1}$$
$$\frac{4.8q}{4.8} = \frac{-9.6}{4.8}$$
$$m = -2$$

Check:
$$8.2(-2) + 4.1 \overset{?}{=} 3.4(-2) - 5.5$$
$$-16.4 + 4.1 \overset{?}{=} -6.8 - 5.5$$
$$-12.3 = -12.3\,^a$$

**57.**
$$0.8x - 3.7 = 0.9x + 0.8$$

$$\underline{-0.8x \qquad -0.8x}$$
$$-3.7 = \qquad 0.1x + 0.8$$
$$\underline{-0.8 \qquad -0.8}$$
$$\frac{-4.5}{0.1} = \frac{0.1x}{0.1}$$
$$-45 = x$$

Check:
$$0.8(-45) - 3.7 \overset{?}{=} 0.9(-45) + 0.8$$
$$-36 - 3.7 \overset{?}{=} -40.5 + 0.8$$
$$-39.7 = -39.7\,^a$$

**59.**

$$3c + 4 = c + 5 \quad \text{Check:}$$

$$\underline{-c} \qquad \underline{-c}$$

$$2c + 4 = 5$$

$$\underline{-4} \qquad \underline{-4}$$

$$\frac{2c}{2} = \frac{1}{2}$$

$$c = \frac{1}{2}$$

$$3\left(\frac{1}{2}\right) + 4 \stackrel{?}{=} \frac{1}{2} + 5$$

$$\frac{3}{2} + 4 \stackrel{?}{=} 5\frac{1}{2}$$

$$1\frac{1}{2} + 4 \stackrel{?}{=} 5\frac{1}{2}$$

$$5\frac{1}{2} = 5\frac{1}{2}\text{a}$$

**61.**

$$5q - 1 = 2q + 6 \quad \text{Check:}$$

$$\underline{-2q} \qquad \underline{-2q}$$

$$3q - 1 = 6$$

$$\underline{+1} \qquad \underline{+1}$$

$$3q = 7$$

$$\frac{3q}{3} = \frac{7}{3}$$

$$q = \frac{7}{3}$$

$$5 \cdot \left(\frac{7}{3}\right) - 1 \stackrel{?}{=} 2 \cdot \left(\frac{7}{3}\right) + 6$$

$$\frac{35}{3} - 1 \stackrel{?}{=} \frac{14}{3} + 6$$

$$\frac{35}{3} - \frac{3}{3} \stackrel{?}{=} \frac{14}{3} + \frac{18}{3}$$

$$\frac{32}{3} = \frac{32}{3}\text{a}$$

**63.**

$$-\frac{1}{2}a - 2 = a + 4$$

$$-\frac{1}{2}a - 2 + \frac{1}{2}a = a + 4 + \frac{1}{2}a$$

$$-2 = \frac{3}{2}a + 4$$

$$-2 - 4 = \frac{3}{2}a + 4 - 4$$

$$-6 = \frac{3}{2}a$$

$$\left(\frac{2}{3}\right) \cdot -6 = \left(\frac{2}{3}\right) \cdot \frac{3}{2}a$$

$$-4 = a$$

Check:

$$-\frac{1}{2} \cdot (-4) - 2 \stackrel{?}{=} -4 + 4$$

$$2 - 2 \stackrel{?}{=} 0$$

$$0 = 0\text{a}$$

**65.**

$$\frac{1}{3}m - \frac{1}{2} = -\frac{2}{3}m + \frac{1}{4}$$

$$\frac{1}{3}m - \frac{1}{2} + \frac{2}{3}m = -\frac{2}{3}m + \frac{1}{4} + \frac{2}{3}m$$

$$m - \frac{1}{2} = \frac{1}{4}$$

$$m - \frac{1}{2} + \frac{1}{2} = \frac{1}{4} + \frac{1}{2}$$

$$m = \frac{1}{4} + \frac{2}{4}$$

$$m = \frac{3}{4}$$

Check:

$$\frac{1}{3} \cdot \left(\frac{3}{4}\right) - \frac{1}{2} \stackrel{?}{=} -\frac{2}{3}\left(\frac{3}{4}\right) + \frac{1}{4}$$

$$\frac{1}{4} - \frac{1}{2} \stackrel{?}{=} -\frac{1}{2} + \frac{1}{4}$$

$$\frac{1}{4} - \frac{2}{4} \stackrel{?}{=} -\frac{2}{4} + \frac{1}{4}$$

$$-\frac{1}{4} = -\frac{1}{4}\text{a}$$

## Guide Problems

**67.** Solve the equation $3(x + 2) = 2(x + 6)$.

a. Apply the distribute property.

$$3(x + 2) = 2(x + 6)$$

$$3x + 6 = 2x + 12$$

b. Use the addition property to get the variable term on one side of the equation.

$$3x + 6 = 2x + 12$$

$$\underline{-2x} \qquad \underline{-2x}$$

$$x + 6 = 0x + 12$$

$$x + 6 = 12$$

c.  Use the addition property to isolate the variable term on one side of the equation.

$$x + 6 = 12$$
$$\underline{-6 \quad -6}$$
$$x + 0 = 6$$
$$x = 6$$

d.  Check the solution

$$3(6+2) \overset{?}{=} 2(6+6)$$
$$3(8) \overset{?}{=} 2(12)$$
$$24 = 24^{a}$$

*Use the method shown in the Guide Problem to solve problems 69 through 83. Be sure to check each solution.*

69.  $3(2x - 5) = 21$

$$6x - 15 = 21$$
$$6x - 15 + 15 = 21 + 15$$
$$6x = 36$$
$$\frac{6x}{6} = \frac{36}{6}$$
$$x = 6$$

Check:

$$3(2 \cdot 6 - 5) \overset{?}{=} 21$$
$$3(12 - 5) \overset{?}{=} 21$$
$$3(7) \overset{?}{=} 21$$
$$21 = 21^{\,a}$$

71.  $6(5r + 2) = 36$

$$30r + 12 = 36$$
$$30r + 12 - 12 = 36 - 12$$
$$30r = 24$$
$$\frac{30r}{30} = \frac{24}{30}$$
$$r = \frac{4}{5}$$

Check:

$$6\left(5 \cdot \frac{4}{5} + 2\right) \overset{?}{=} 36$$
$$6(4 + 2) \overset{?}{=} 36$$
$$6(6) \overset{?}{=} 36$$
$$36 = 36^{\,a}$$

73.  $2(3s - 5) = 8s$

$$6s - 10 = 8s$$
$$6s - 10 - 6s = 8s - 6s$$
$$-10 = 2s$$
$$\frac{-10}{2} = \frac{2s}{2}$$
$$-5 = s$$

Check:

$$2(3 \cdot (-5) - 5) \overset{?}{=} 8(-5)$$
$$2(-15 - 5) \overset{?}{=} -40$$
$$2(-20) \overset{?}{=} -40$$
$$-40 = -40^{\,a}$$

75.  $12x - 3(x - 5) = -12$

$$12x - 3x + 15 = -12$$
$$9x + 15 = -12$$
$$9x + 15 - 15 = -12 - 15$$
$$9x = -27$$
$$\frac{9x}{9} = \frac{-27}{9}$$
$$x = -3$$

Check:

$$12(-3) - 3(-3 - 5) \overset{?}{=} -12$$
$$12(-3) - 3(-8) \overset{?}{=} -12$$
$$-36 + 24 \overset{?}{=} -12$$
$$-12 = -12^{\,a}$$

77.  $5(x + 8) = 7(x - 4)$

$$5x + 40 = 7x - 28$$
$$5x + 40 - 5x = 7x - 28 - 5x$$
$$40 = 2x - 28$$
$$40 + 28 = 2x - 28 + 28$$
$$68 = 2x$$
$$\frac{68}{2} = \frac{2x}{2}$$
$$34 = x$$

Check:

$$5(34 + 8) \overset{?}{=} 7(34 - 4)$$
$$5(42) \overset{?}{=} 7(30)$$
$$210 = 210^{\,a}$$

79. $4(2a-1)=2(3a-2)$ Check:

$$8a-4=6a-4$$

$$8a-4-6a=6a-4-6a$$

$$2a-4=-4$$

$$2a-4+4=-4+4$$

$$2a=0$$

$$\frac{2a}{2}=\frac{0}{2}$$

$$a=0$$

$$4(2\cdot0-1)\overset{?}{=}2(3\cdot0-2)$$

$$4(0-1)\overset{?}{=}2(0-2)$$

$$4(-1)\overset{?}{=}2(-2)$$

$$-4=-4 \text{ }^{a}$$

81. $\frac{1}{3}(4b+9)=\frac{2}{3}b-7$ Check:

$$\frac{4}{3}b+3=\frac{2}{3}b-7$$

$$\frac{4}{3}b+3-\frac{2}{3}b=\frac{2}{3}b-7-\frac{2}{3}b$$

$$-\frac{2}{3}b+3=-7$$

$$-\frac{2}{3}b+3-3=-7-3$$

$$\frac{2}{3}b=-10$$

$$\frac{3}{2}\cdot\frac{2}{3}b=-10\cdot\frac{3}{2}$$

$$b=-15$$

$$\frac{1}{3}(4\cdot(-15)+9)\overset{?}{=}\frac{2}{3}(-15)-7$$

$$\frac{1}{3}(-60+9)\overset{?}{=}-10-7$$

$$\frac{1}{3}(-51)\overset{?}{=}-17$$

$$-17=-17 \text{ }^{a}$$

83. $\frac{1}{3}(x-2)=\frac{4}{3}-x$ Check:

$$\frac{1}{3}x-\frac{2}{3}=\frac{4}{3}-x$$

$$\frac{1}{3}x-\frac{2}{3}+x=\frac{4}{3}-x+x$$

$$\frac{4}{3}x-\frac{2}{3}=\frac{4}{3}$$

$$\frac{4}{3}x-\frac{2}{3}+\frac{2}{3}=\frac{4}{3}+\frac{2}{3}$$

$$\frac{4}{3}x=\frac{6}{3}$$

$$\frac{3}{4}\cdot\frac{4}{3}x=2\cdot\frac{3}{4}$$

$$x=\frac{3}{2}$$

$$\frac{1}{3}\left(\frac{3}{2}-2\right)\overset{?}{=}\frac{4}{3}-\frac{3}{2}$$

$$\frac{1}{3}\left(\frac{3}{2}-\frac{4}{2}\right)\overset{?}{=}\frac{4}{3}-\frac{3}{2}$$

$$\frac{1}{3}\left(-\frac{1}{2}\right)\overset{?}{=}\frac{4}{3}-\frac{3}{2}$$

$$-\frac{1}{6}\overset{?}{=}\frac{8}{6}-\frac{9}{6}$$

$$-\frac{1}{6}=-\frac{1}{6} \text{ }^{a}$$

*For problems 85 – 87, use the formula P = 2l + 2w.*

85.  $P = 2l + 2w$, $P = 32$, $w = 7$

$$32 = 2l + 2(7)$$
$$32 = 2l + 14$$
$$32 - 14 = 2l + 14 - 14$$
$$18 = 2l$$
$$\frac{18}{2} = \frac{2l}{2}$$
$$9 = l$$

The length of the rectangle is 9 inches.

87.  Since each hors d'oeuvre has an area of 1 square inch, the length and width of each hors d'oeuvre is 1 inch. 16 hors d'oeuvres fit along the longest sides of the serving tray, so the length of the tray is 16 inches. The perimeter of the tray is 52 inches.

$P = 2l + 2w$, $P = 52$, $l = 16$

$$52 = 2(16) + 2w$$
$$52 = 32 + 2w$$
$$52 - 32 = 32 + 2w - 32$$
$$20 = 2w$$
$$\frac{20}{2} = \frac{2w}{2}$$
$$10 = w$$

The width of the tray is 10 inches, so 10 hors d'oeuvres fit along each of the tray's shortest sides.

## Cumulative Skills Review

1.  $$\frac{3 \text{ course credits}}{\$750} = \frac{12 \text{ course credits}}{\$3000}$$

3.  $110.4z = 552$          Check

$$\frac{110.4z}{110.4} = \frac{552}{110.4} \qquad 110.4(5) \overset{?}{=} 552$$
$$z = 5 \qquad\qquad 552 = 552^{\text{a}}$$

5.  $y = 2, v = 3$

$$8v - 17y - 3y + 21v = 29v - 20y$$
$$29(3) - 20(2) = 87 - 40 = 47$$

7.  $80.67 - 74.89 = 5.78$

Females are expected to live 5.78 years longer than males.

$$5.78 \text{ years} \times \frac{365 \text{ days}}{1 \text{ year}} = 2109.7 \approx 2110$$

Females are expected to live about 2110 days Longer than males.

9.  $23 + s = \ 55$          Check

$$\underline{-23 \qquad\ -23}$$
$$s = \ \ 32 \qquad 23 + 32 \overset{?}{=} 55$$
$$55 = 55^{\text{a}}$$

## Section 10.5 Solving Application Problems

### Concept Check

1. To solve an application problem, first  read  and understand the problem.

3. Next, translate the problem into an  equation .

5. Always  check  the solution.

### Guide Problems

7. A number decreased by 15 is 23.

$$x \quad - \quad 15 \quad = \quad 23$$

*For problems 9–19, use the key words and phrases found on page 738 in your text to translate each sentence into an equation. We let x = the unknown number in each problem; however, other variables may be used.*

9. A number increased by 7 is 15.
$$x + 7 = 15$$

11. Three times a number is 15.
$$3x = 15.$$

13. The product of −8 and a number is $\frac{1}{2}$.

$$-8x = \frac{1}{2}$$

15. Twice a number added to 8 is 38.
$$8 + 2x = 38$$

17. Eighteen decreased by two-thirds of a number is 10.

$$18 - \frac{2}{3}x = 10$$

19. The product of 3 and the sum of a number and 10 is 36.

$$3(x + 10) = 36$$

*For problems 21–47, use the six-step process to solve each problem.*
> *1: Read and understand the problem.*
> *2: Assign a variable to the unknown quantity.*
> *3: Translate the problem into an equation.*
> *4: Solve the equation.*
> *5: Check the solution.*
> *6: Clearly state the result using units, if necessary.*

21. A number increased by 8 is 15. Find the number.
Let $x$ = the number.
$$x + 8 = 15 \qquad \text{Check}$$
$$x + 8 - 8 = 15 - 8$$
$$x = 7$$
$$7 + 8 \overset{?}{=} 15$$
$$15 = 15^{\text{a}}$$

The number is 7.

23. Three times a number is $\frac{1}{4}$. Find the number.

Let $x$ = the number.
$$3x = \frac{1}{4} \qquad \text{Check}$$
$$\frac{1}{3} \cdot 3x = \frac{1}{4} \cdot \frac{1}{3} \qquad 3\left(\frac{1}{12}\right) \overset{?}{=} \frac{1}{4}$$
$$x = \frac{1}{12} \qquad \frac{1}{4} = \frac{1}{4}^{\text{a}}$$

The number is $\frac{1}{12}$.

25. Five subtracted from twice a number is 3. Find the number.
    Let $x$ = the number.

$$2x - 5 = 3 \qquad \text{Check}$$
$$2x - 5 + 5 = 3 + 5$$
$$2x = 8 \qquad 2(4) - 5 \overset{?}{=} 3$$
$$\frac{2x}{2} = \frac{8}{2} \qquad 8 - 5 \overset{?}{=} 3$$
$$x = 4 \qquad 3 = 3^{\text{a}}$$

The number is 4.

27. The sum of a number and 4 is equal to twice the number. Find the number.
    Let $x$ = the number.

$$x + 4 = 2x \qquad \text{Check}$$
$$x + 4 - x = 2x - x \qquad 4 + 4 \overset{?}{=} 2(4)$$
$$4 = x \qquad 8 = 8^{\text{a}}$$

The number is 4.

29. Six less than a number is three times the sum of 4 and the number. Find the number.
    Let $x$ = the number.

$$x - 6 = 3(4 + x) \qquad \text{Check}$$
$$x - 6 = 12 + 3x$$
$$-9 - 6 \overset{?}{=} 3(4 + (-9))$$
$$x - 6 - x = 12 + 3x - x$$
$$-6 = 12 + 2x \qquad -15 \overset{?}{=} 3(-5)$$
$$-6 - 12 = 12 + 2x - 12 \qquad -15 = -15^{\text{a}}$$
$$-18 = 2x$$
$$\frac{-18}{2} = \frac{2x}{2}$$
$$-9 = x$$

The number is $-9$.

31. An airplane's speed with the wind at its back is 505 mph. If the wind speed is 55 mph, how fast does the plane fly without the wind?
    Let $x$ = the plane's speed without the wind. Then $x + 55$ = the plane's speed with the wind.

$$x + 55 = 505 \qquad \text{Check}$$
$$x + 55 - 55 = 505 - 55 \qquad 450 + 55 \overset{?}{=} 505$$
$$x = 450 \qquad 505 = 505^{\text{a}}$$

The plane's speed without the wind is 450 mph.

33. The length of a house is 21.5 feet more than its width. If the length of the house is 78 feet, what is the width?
    Let $w$ = the width of the house. Then $x + 21.5$ = the length of the house.

$$w + 21.5 = 78 \qquad \text{Check}$$
$$w + 21.5 - 21.5 = 78 - 21.5 \qquad 56.5 + 21.5 \overset{?}{=} 78$$
$$w = 56.5 \qquad 78 = 78^{\text{a}}$$

The width of the house is 56.5 feet.

35. Taipei 101, a building in Taipei, Taiwan, is 979 feet shorter than the Burj Dubai, a building in Dubai, UAE. If Taipei 101 is 1671 feet tall, find the height of the Burj Dubai.
    Let $x$ = the height of the Burj Dubai.
    Then $x - 979$ = the height of Taipei 101.

$$x - 979 = 1671 \qquad \text{Check}$$
$$x - 979 + 979 = 1671 + 979 \qquad 2650 - 1671 \overset{?}{=} 979$$
$$x = 2650 \qquad 979 = 979^{\text{a}}$$

The Burj Dubai is 2650 feet tall.

37. After conducting research on ocean kelp, a scientist found that it grows at a steady rate of 0.45 meters per day. If the kelp grew 12.6 meters, determine the number of days that the experiment lasted.
    Let $d$ = the number of days the experiment lasted.

$$0.45d = 12.6 \qquad \text{Check}$$
$$\frac{0.45d}{0.45} = \frac{12.6}{0.45} \qquad 0.45(28) \overset{?}{=} 12.6$$
$$d = 28 \qquad 12.6 = 12.6^{\text{a}}$$

The experiment lasted 28 days.

39. If a large 8-slice margherita pizza from Skeeter's Pizza costs $14.00, how much does each slice cost?
    Let $s$ = the cost of one slice of pizza.

$$8s = 14.00 \qquad \text{Check}$$
$$\frac{8s}{8} = \frac{14.00}{8} \qquad 8(1.75) \overset{?}{=} 14.00$$
$$s = 1.75 \qquad 14.00 = 14.00^{\text{a}}$$

One slice of pizza costs $1.75.

41. A parking garage charges $3.00 for the first hour or part thereof. Each additional hour or part thereof costs $1.50. If Eric has exactly $15.00 to spend on parking, how many hours can he park in this lot?

Let $h$ = the number of hours he can park in the lot.

$$3.00 + 1.50(h - 1) = 15.00$$
$$3.00 + 1.50h - 1.50 = 15.00$$
$$1.50 + 1.50h = 15.00$$
$$1.50 + 1.50h - 1.50 = 15.00 - 1.50$$
$$1.50h = 13.50$$
$$\frac{1.50h}{1.50} = \frac{13.50}{1.50}$$
$$h = 9$$

Check

$$3.00 + 1.50(9 - 1) \overset{?}{=} 15.00$$
$$3.00 + 1.50(8) \overset{?}{=} 15.00$$
$$3.00 + 12.00 \overset{?}{=} 15.00$$
$$15.00 = 15.00^{a}$$

Eric can park for 9 hours.

43. Carmen is selling tickets to a school function. The tickets are $7.50 for adults and $4.00 for students. She sells three times as many adult tickets as student tickets. If the ticket sales totaled $795.00, how many of each type of ticket did Carmen sell?

Let $s$ = the number of student tickets.
Then $3s$ = the number of adult tickets.

$$7.50(3s) + 4.00s = 795.00$$
$$22.50s + 4.00s = 795.00$$
$$26.50s = 795.00$$
$$\frac{26.50s}{26.50} = \frac{795.00}{26.50}$$
$$s = 30;$$
$$3s = 3 \cdot 30 = 90$$

Check

$$7.5(90) + 4.00(30) \overset{?}{=} 795.00$$
$$675.00 + 120.00 \overset{?}{=} 795.00$$
$$795.00 = 795.00^{a}$$

Carmen sold 30 student tickets and 90 adult tickets.

45. The lengths of the sides of a triangle are such that side $a$ if four times as long as side $b$, and side $c$ is 5 inches shorter than side $a$. The perimeter of the triangle is 40 inches. Determine the lengths of sides $a$, $b$, and $c$.

Let $b$ = the length of one side. Then $4b$ = the length of side $a$, and $a - 5 = 4b - 5$ = the length of side $c$.

$$4b + b + (4b - 5) = 40$$
$$9b - 5 = 40$$
$$9b - 5 + 5 = 40 + 5$$
$$9b = 45$$
$$\frac{9b}{9} = \frac{45}{9}$$
$$b = 5$$
$$a = 4b = 4 \cdot 5 = 20$$
$$c = a - 5 = 20 - 5 = 15$$

Check

$$5 + 20 + 15 \overset{?}{=} 40$$
$$40 = 40^{a}$$

The sides of the triangle are 20 inches, 5 inches, and 15 inches. (Note that the figure in the text is not drawn to scale.)

47. A 250-foot piece of rope is cut into three pieces. The first piece is twice as long as the third piece and the second piece is 4 feet longer than three times the third piece. What is the length of each piece?

Let $c$ = the length of the third piece. Then $2c$ = the length of the first piece, and $3c + 4$ = the length of the second piece.

$$2c + (3c + 4) + c = 250$$
$$6c + 4 = 250$$
$$6c + 4 - 4 = 250 - 4$$
$$6c = 246$$
$$\frac{6c}{6} = \frac{246}{6}$$
$$c = 41$$
$$2c = 82$$
$$3c + 4 = 127$$

Check

$$41 + 82 + 127 \overset{?}{=} 250$$
$$250 = 250^{a}$$

The three pieces of rope are 82 feet, 127 feet, and 41 feet.

## Cumulative Skills Review

1. $16 \cdot 3 - (5+1)^2 = 16 \cdot 3 - (6)^2$
$$= 16 \cdot 3 - 36$$
$$= 48 - 36$$
$$= 12$$

3. $\dfrac{23.12}{4} = 5.78$

5. 15% of $600 = 0.15 \times 600 = 90$

7. $\dfrac{7+4+6+4+2+9+5+3}{8} = \dfrac{40}{8} = 5$
The mean is 5.

9. $(7l - 6)8 = 56l - 48$

## Chapter 10 Numerical Facts of Life

1. The total population after $t$ seconds is the sum of the initial population and the number of births, decreased by the number of deaths, and then increased by the number of international migrants. This translates into

$$T = P + \frac{1}{7}t - \frac{1}{14}t + \frac{1}{26}t = P + \frac{26}{182}t - \frac{13}{182}t + \frac{7}{26}t = P + \frac{20}{182}t = P + \frac{10}{91}t$$

3. Since it takes 9.1 seconds to add 1 person, it takes 9,100,000 seconds to add 1,000,000 people. Consequently, it took approximately 105 days to add an additional 1,000,000 people to the population.

$$9,100,000 \text{ sec} \cdot \frac{1 \text{ min}}{60 \text{ sec}} \cdot \frac{1 \text{ hr}}{60 \text{ min}} \cdot \frac{1 \text{ day}}{24 \text{ hr}} = 105.32431 \text{ days} \approx 105 \text{ days}$$

Alternatively, we can solve the equation with $T = 301,000,000$ and $P = 300,000,000$ to find $t$.

$$301,000,000 = 300,000,000 + \frac{10}{91}t$$

$$301,000,000 - 300,000,000 = 300,000,000 + \frac{10}{91}t - 300,000,000$$

$$1,000,000 = \frac{10}{91}t$$

$$\frac{91}{10} \cdot 1,000,000 = \frac{10}{91}t \cdot \frac{91}{10}$$

$$9,100,000 = t$$

## Chapter 10 Review Exercises

*In Problems 1 through 4, evaluate each algebraic expression when x = 2 and y = 3.*

1. $2x - 3y$
$2(2) - 3(3) = 4 - 9 = -5$

2. $x^2 - y$
$2^2 - 3 = 4 - 3 = 1$

3. $xy + 6y - 3x - 18$
$2 \cdot 3 + 6 \cdot 3 - 3 \cdot 2 - 18 = 6 + 18 - 6 - 18 = 0$

4. $8x + 12y - 5xy$
$8 \cdot 2 + 12 \cdot 3 - 5 \cdot 2 \cdot 3 = 16 + 36 - 30 = 22$

*In exercises 5 through 12, combine like terms. Remember that like terms have the same variables and same exponents.*

5. $9a - 4b + 2a + 8b = (9a + 2a) + (-4b + 8b)$
$$= 11a + 4b$$

6. $4x + 7y - 5y + 12x = (4x + 12x) + (7y - 5y)$
$$= 16x + 2y$$

7. $4z + 5v + 6z - 8v = (5v - 8v) + (4z + 6z)$
$$= -3v + 10z$$

8. $2rt + 4r - 7rt + 5t - 6r$
$$= (4r - 6r) + (2rt - 7rt) + 5t$$
$$= -2r - 5rt + 5t$$

9. $a^2 + 5a - 4a - 8a^2 = (a^2 - 8a^2) + (5a - 4a)$
$$= -7a^2 + a$$

10. $m^3 + m^2n - 4m^3 + 8mn^2$
$$= (m^3 - 4m^3) + m^2n + 8mn^2$$
$$= -3m^3 + m^2n + 8mn^2$$

11. $gh + g^2 - gh + h^2 = g^2 + (gh - gh) + h^2$
$$= g^2 + h^2$$

12. $7x^3y + 3x^2y - 36x^3y + 5x^2y$
$$= (7x^3y - 36x^3y) + (3x^2y + 5x^2y)$$
$$= -29x^3y + 8x^2y$$

*For problems 13–16, apply the associative property of multiplication and the commutative property of multiplication as necessary to multiply the expression.*

13. $2(3x) = 6x$

14. $-5(5y) = -25y$

15. $2.1(3a) = 6.3a$

16. $\frac{1}{2}(8d) = 4d$

*For problems 17–24, use the distributive property $a(b + c) = ab + ac$ or $(a + b)c = ac + bc$.*

17. $3(x - 15) = 3x - 45$

18. $8(2k^2 + 1) = 16k^2 + 8$

19. $-(b - 7) = -b + 7$

20. $-(4z + 8 - y) = -4z - 8 + y$

21. $-3(8p + 7) = -24p - 21$

22. $-4(5t^2 - 2) = -20t^2 + 8$

23. $-2(-3m^2 + 2mn - 9n^2) = 6m^2 - 4mn + 18n^2$

24. $-5(7y^2 - 3yz + 2z^2) = -35y^2 + 15yz - 10z^2$

*In problems 25 through 32, determine whether the given value is a solution to the given equation by substituting the value for the unknown in the equation.*

25. $x = 2; \ 7x + 15 = 29$
$$7(2) + 15 \overset{?}{=} 29$$
$$14 + 15 \overset{?}{=} 29$$
$$29 = 29^a$$
2 is a solution.

26. $y = 2; \ 4 - (2y + 6) = -8$
$$4 - (2 \cdot 2 + 6) \overset{?}{=} -8$$
$$4 - (4 + 6) \overset{?}{=} -8$$
$$4 - 10 \overset{?}{=} -8$$
$$-6 \neq -8$$
2 is not a solution.

27. $b = 6;\ 5b - 9 = 7b + 3$

$$5(6) - 9 \overset{?}{=} 7(6) + 3$$

$$30 - 9 \overset{?}{=} 42 + 3$$

$$21 \neq 45$$

6 is not a solution.

28. $k = -20;\ k + 20 = 0$

$$-20 + 20 \overset{?}{=} 0$$

$$0 = 0^{a}$$

$-20$ is a solution.

29. $p = 3;\ p^2 - 3 = 5$

$$3^2 - 3 \overset{?}{=} 5$$

$$9 - 3 \overset{?}{=} 5$$

$$6 \neq 5$$

3 is not a solution.

30. $m = 4;\ m^2 + 9 = 25$

$$4^2 + 9 \overset{?}{=} 25$$

$$16 + 9 \overset{?}{=} 25$$

$$25 = 25^{a}$$

4 is a solution.

31. $a = 8;\ 63 - a^2 = 1$

$$63 - 8^2 \overset{?}{=} 1$$

$$63 - 64 \overset{?}{=} 1$$

$$-1 \neq 1$$

5 is not a solution.

32. $t = 10;\ 109 - t^2 = 9$

$$109 - 10^2 \overset{?}{=} 9$$

$$109 - 100 \overset{?}{=} 9$$

$$9 = 9^{a}$$

10 is a solution.

*Be sure to check each solution for problems 33–56.*

33. $v + 15 = 34$     Check

$v + 15 - 15 = 34 - 15$

$v = 19$

$$19 + 15 \overset{?}{=} 34$$

$$34 = 34^{a}$$

34. $x + 21 = 8$     Check

$x + 21 - 21 = 8 - 21$

$x = -13$

$$-13 + 21 \overset{?}{=} 8$$

$$8 = 8^{a}$$

35. $m - 2.3 = 5.4$     Check

$m - 2.3 + 2.3 = 5.4 + 2.3$

$m = 7.7$

$$7.7 - 2.3 \overset{?}{=} 5.4$$

$$5.4 = 5.4^{a}$$

36. $b - 5.7 = 8.4$     Check

$b - 5.7 + 5.7 = 8.4 + 5.7$

$b = 14.1$

$$14.1 - 5.7 \overset{?}{=} 8.4$$

$$8.4 = 8.4^{a}$$

37. $12 + p = 4$     Check

$12 + p - 12 = 4 - 12$

$p = -8$

$$12 + (-8) \overset{?}{=} 4$$

$$4 = 4^{a}$$

38. $20 + r = 7$     Check

$20 + r - 20 = 7 - 20$

$r = -13$

$$20 + (-13) \overset{?}{=} 7$$

$$7 = 7^{a}$$

39. $a + \dfrac{1}{3} = \dfrac{1}{2}$     Check

$a + \dfrac{1}{3} - \dfrac{1}{3} = \dfrac{1}{2} - \dfrac{1}{3}$

$a = \dfrac{3}{6} - \dfrac{2}{6} = \dfrac{1}{6}$

$$\dfrac{1}{6} + \dfrac{1}{3} \overset{?}{=} \dfrac{1}{2}$$

$$\dfrac{1}{6} + \dfrac{2}{6} \overset{?}{=} \dfrac{1}{2}$$

$$\dfrac{3}{6} \overset{?}{=} \dfrac{1}{2}$$

$$\dfrac{1}{2} = \dfrac{1}{2}^{a}$$

40. $n + \dfrac{2}{5} = \dfrac{1}{2}$     Check

$n + \dfrac{2}{5} - \dfrac{2}{5} = \dfrac{1}{2} - \dfrac{2}{5}$

$n = \dfrac{5}{10} - \dfrac{4}{10} = \dfrac{1}{10}$

$$\dfrac{1}{10} + \dfrac{2}{5} \overset{?}{=} \dfrac{1}{2}$$

$$\dfrac{1}{10} + \dfrac{4}{10} \overset{?}{=} \dfrac{1}{2}$$

$$\dfrac{5}{10} \overset{?}{=} \dfrac{1}{2}$$

$$\dfrac{1}{2} = \dfrac{1}{2}^{a}$$

41. $4k = 48$    Check

$$\frac{4k}{4} = \frac{48}{4} \qquad 4 \cdot 12 \overset{?}{=} 48$$

$$k = 12 \qquad\quad 48 = 48^{\text{a}}$$

42. $7q = 56$    Check

$$\frac{7q}{7} = \frac{56}{7} \qquad 7 \cdot 8 \overset{?}{=} 56$$

$$q = 8 \qquad\quad 56 = 56^{\text{a}}$$

43. $2z = 25$       Check

$$\frac{2z}{2} = \frac{25}{2} \qquad 2 \cdot \frac{25}{2} \overset{?}{=} 25$$

$$z = \frac{25}{2} \text{ or } 12\frac{1}{2} \qquad 25 = 25^{\text{a}}$$

44. $3h = 47$       Check

$$\frac{3h}{3} = \frac{47}{3} \qquad 3 \cdot \frac{47}{3} \overset{?}{=} 47$$

$$z = \frac{47}{3} \text{ or } 15\frac{2}{3} \qquad 47 = 47^{\text{a}}$$

45.    $-\dfrac{2}{5}t = 12$      Check

$$\left(-\frac{5}{2}\right)\left(-\frac{2}{5}t\right) = 12\left(-\frac{5}{2}\right) \qquad -\frac{2}{5}(-30) \overset{?}{=} 12$$

$$t = -30 \qquad\qquad\qquad\qquad\quad 12 = 12^{\text{a}}$$

46.    $-\dfrac{3}{7}t = 24$      Check

$$\left(-\frac{7}{3}\right)\left(-\frac{3}{7}t\right) = 24\left(-\frac{7}{3}\right) \qquad -\frac{3}{7}(-56) \overset{?}{=} 24$$

$$t = -56 \qquad\qquad\qquad\qquad\quad 24 = 24^{\text{a}}$$

47. $0.3p = 3.9$    Check

$$\frac{0.3p}{0.3} = \frac{3.9}{0.3} \qquad 0.3 \cdot 13 \overset{?}{=} 3.9$$

$$p = 13 \qquad\quad 3.9 = 3.9^{\text{a}}$$

48. $0.5x = 4.5$    Check

$$\frac{0.5x}{0.5} = \frac{4.5}{0.5} \qquad 0.5 \cdot 9 \overset{?}{=} 4.5$$

$$p = 9 \qquad\quad 4.5 = 4.5^{\text{a}}$$

49. $12t + 3 = 39$      Check

$$12t + 3 - 3 = 39 - 3$$
$$12t = 36 \qquad 12 \cdot 3 + 3 \overset{?}{=} 39$$
$$\frac{12t}{12} = \frac{36}{12} \qquad 36 + 3 \overset{?}{=} 39$$
$$t = 3 \qquad\qquad 39 = 39^{\text{a}}$$

50. $8b - 9 = 31$      Check

$$8b - 9 + 9 = 31 + 9$$
$$8b = 40 \qquad 8 \cdot 5 - 9 \overset{?}{=} 31$$
$$\frac{8b}{8} = \frac{40}{8} \qquad 40 - 9 \overset{?}{=} 31$$
$$b = 5 \qquad\qquad 31 = 31^{\text{a}}$$

51. $9x = 5x + 40$      Check

$$9x - 5x = 5x + 40 - 5x$$
$$4x = 40 \qquad 9 \cdot 10 \overset{?}{=} 5 \cdot 10 + 40$$
$$\frac{4x}{4} = \frac{40}{4} \qquad 90 \overset{?}{=} 50 + 40$$
$$t = 10 \qquad\quad 90 = 90^{\text{a}}$$

52. $7s = 3s + 52$      Check

$$7s - 3s = 3s + 52 - 3s$$
$$4s = 52 \qquad 7 \cdot 13 \overset{?}{=} 3 \cdot 13 + 52$$
$$\frac{4s}{4} = \frac{52}{4} \qquad 91 \overset{?}{=} 39 + 52$$
$$s = 13 \qquad\quad 91 = 91^{\text{a}}$$

53.    $2(m + 6) = 4(2m + 3)$      Check

$$2m + 12 = 8m + 12$$
$$2m + 12 - 2m = 8m + 12 - 2m \qquad 2(0 + 6) \overset{?}{=} 4(2 \cdot 0 + 3)$$
$$12 = 6m + 12 \qquad\qquad\qquad 2(6) \overset{?}{=} 4(0 + 3)$$
$$12 - 12 = 6m + 12 - 12 \qquad\qquad 12 \overset{?}{=} 4(3)$$
$$0 = 6m \qquad\qquad\qquad\qquad 12 = 12^{\text{a}}$$
$$\frac{0}{6} = \frac{6m}{6}$$
$$0 = m$$

**54.**

$$6(z-5) = 3(4z+8)$$
$$6z-30 = 12z+24$$
$$6z-30-12z = 12x+24-12z$$
$$-6z-30 = 24$$
$$-6z-30+30 = 24+30$$
$$-6z = 54$$
$$\frac{-6z}{-6} = \frac{54}{-6}$$
$$x = -9$$

Check

$$6(-9-5) \overset{?}{=} 3(4(-9)+8)$$
$$6(-14) \overset{?}{=} 3(-36+8)$$
$$-84 \overset{?}{=} 3(-28)$$
$$-84 = -84^{a}$$

**55.**

$$5(2n-11)-7n = 5$$
$$10n-55-7n = 5$$
$$3n-55 = 5$$
$$3n-55+55 = 5+55$$
$$3n = 60$$
$$\frac{3n}{3} = \frac{60}{3}$$
$$n = 20$$

Check

$$5(2 \cdot 20-11)-7 \cdot 20 \overset{?}{=} 5$$
$$5(40-11)-7 \cdot 20 \overset{?}{=} 5$$
$$5(29)-140 \overset{?}{=} 5$$
$$145-140 \overset{?}{=} 5$$
$$5 = 5^{a}$$

**56.**

$$7(3c+10)+2c = 1$$
$$21c+70+2c = 1$$
$$23c+70 = 1$$
$$23c+70-70 = 1-70$$
$$23c = -69$$
$$\frac{23c}{23} = \frac{-69}{23}$$
$$c = -3$$

Check

$$7(3(-3)+10)+2(-3) \overset{?}{=} 1$$
$$7(-9+10)+2(-3) \overset{?}{=} 1$$
$$7(1)+2(-3) \overset{?}{=} 1$$
$$7+(-6) \overset{?}{=} 1$$
$$1 = 1^{a}$$

*For problems 57–60, use the six-step process to solve each problem.*
  *1: Read and understand the problem.*
  *2: Assign a variable to the unknown quantity.*
  *3: Translate the problem into an equation.*
  *4: Solve the equation.*
  *5: Check the solution.*
  *6: Clearly state the result using units, if necessary.*

**57.** Write an algebraic expression to represent twelve subtracted from twice the difference between ten and a number.
Let $x$ = the number.
$$2(10-x)-12$$

**58.** An airplane flying at an altitude of 32,000 feet suddenly has to change altitude to 29,500 feet. What is the net change in altitude.
Let $x$ = the net change in altitude..
$$32,000-29,500 = x \quad \text{Check}$$
$$2500 = x$$
$$32,000-29,500 \overset{?}{=} 2500$$
$$2500 = 2500^{a}$$

The net change in altitude is 2500 feet.

59. A new radio sells for $58.99. If this is $17.68 above the wholesale price, find the wholesale price.

Let $x$ = the wholesale price.

$$x + \$17.68 = \$58.99$$

$$x + \$17.68 - \$17.68 = \$58.99 - \$17.68$$

$$x = \$41.31$$

Check

$$\$41.31 + \$17.68 \overset{?}{=} \$58.99$$

$$\$58.99 = \$58.99^a$$

The wholesale price is $41.31

60. Martha was paid $348.75 for 45 hours of work. Find her rate of pay.

Let $x$ = Martha's hourly rate of pay.

$$45x = \$348.75 \quad \text{Check}$$

$$\frac{45x}{45} = \frac{\$348.75}{45} \qquad 45 \cdot \$7.75 \overset{?}{=} \$348.75$$

$$x = \$7.75 \qquad \$348.75 = \$348.75^a$$

Martha earned $7.75 per hour.

61. There are 32 students in a beginning algebra class. The number of males is seven less than two times the number of females. Find the number of each.

Let $f$ = the number of females in the class. Then $2f - 7$ = the number of males.

$$f + (2f - 7) = 32 \qquad\qquad \text{Check}$$

$$f + 2f - 7 = 32 \qquad\qquad 13 + 19 \overset{?}{=} 32$$

$$3f - 7 = 32 \qquad\qquad\qquad 32 = 32^a$$

$$3f - 7 + 7 = 32 + 7$$

$$3f = 39$$

$$\frac{3f}{3} = \frac{39}{3}$$

$$f = 13$$

$$2f - 7 = 2(13) - 7 = 26 - 7 = 19$$

There are 13 females and 19 males in the class.

## Chapter 10 Assessment Test

1. $a^2 + 7(b - 4); a = 3, b = 8$

   $3^2 + 7(8 - 4) = 9 + 7(4)$

   $\qquad\qquad = 9 + 28 = 37$

2. $4b - a^2 + (a - b)^2; a = 3, b = 8$

   $4 \cdot 8 - 3^2 + (3 - 8)^2 = 4 \cdot 8 - 9 + (-5)^2$

   $\qquad\qquad\qquad = 32 - 9 + 25$

   $\qquad\qquad\qquad = 48$

3. $6m^4n^3 + 8m^4n - m^4n^3 + 12m^4m + 3m^4n^3 = \left(6m^4n^3 - m^4n^3 + 3m^4n^3\right) + \left(8m^4n + 12m^4n\right) = 8m^4n^3 + 20m^4n$

4. $a^2 + 2ab - b^2 + 5ab - 8a^2 + 3b^2 = \left(a^2 - 8a^2\right) + (2ab + 5ab) + \left(-b^2 + 3b^2\right) = -7a^2 + 7ab + 2b^2$

5. $-3(9d - 8) = -27d + 24$

6. $5(8m - 7) = 40m - 35$

7. $x = 12,\ 2x + 3(2x - 4) = 9x$

$2 \cdot 12 + 3(2 \cdot 12 - 4) \overset{?}{=} 9 \cdot 12$

$2 \cdot 12 + 3(24 - 4) \overset{?}{=} 9 \cdot 12$

$2 \cdot 12 + 3(20) \overset{?}{=} 9 \cdot 12$

$24 + 60 \overset{?}{=} 108$

$84 \neq 108$

12 is not a solution.

8. $y = -7,\ 6y + 14(y + 5) = 10y$

$6(-7) + 14(-7 + 5) \overset{?}{=} 10(-7)$

$6(-7) + 14(-2) \overset{?}{=} 10(-7)$

$-42 + (-28) \overset{?}{=} -70$

$-70 = -70^{\text{a}}$

$-7$ is a solution.

9. $x + 45 = 38$     Check

$x + 45 - 45 = 38 - 45$     $-7 + 45 \overset{?}{=} 38$

$x = -7$     $38 = 38^{\text{a}}$

10. $p + 12.5 = 9.5$     Check

$p + 12.5 - 12.5 = 9.5 - 12.5$     $-3 + 12.5 \overset{?}{=} 9.5$

$p = -3$     $9.5 = 9.5^{\text{a}}$

11. $9q = 108$     Check

$\dfrac{9q}{9} = \dfrac{108}{9}$     $9 \cdot 12 \overset{?}{=} 108$

$q = 12$     $108 = 108^{\text{a}}$

12. $\dfrac{1}{5} w = \dfrac{2}{3}$     Check

$5 \cdot \dfrac{1}{5} w = \dfrac{2}{3} \cdot 5$     $\dfrac{1}{5} \cdot \dfrac{10}{3} \overset{?}{=} \dfrac{2}{3}$

$w = \dfrac{10}{3}$ or $3\dfrac{1}{3}$     $\dfrac{2}{3} = \dfrac{2}{3}^{\text{a}}$

13. $7(x - 8) = 3(2x - 5)$

$7x - 56 = 6x - 15$

$7x - 56 - 6x = 6x - 15 - 6x$

$x - 56 = -15$

$x - 56 + 56 = -15 + 56$

$x = 41$

Check

$7(41 - 8) \overset{?}{=} 3(2 \cdot 41 - 5)$

$7(33) \overset{?}{=} 3(82 - 5)$

$231 \overset{?}{=} 3(77)$

$231 = 231^{\text{a}}$

14. $7(2t + 3) = 3(t - 4)$

$14t + 21 = 3t - 12$

$14t + 21 - 3t = 3t - 12 - 3t$

$11t + 21 = -12$

$11t + 21 - 21 = -12 - 21$

$11t = -33$

$\dfrac{11t}{11} = \dfrac{-33}{11}$

$t = -3$

Check

$7(2(-3) + 3) \overset{?}{=} 3(-3 - 4)$

$7(-6 + 3) \overset{?}{=} 3(-7)$

$7(-3) \overset{?}{=} -21$

$-21 = -21^{\text{a}}$

15.
$$8y + 3 = 12y$$

Check

$$8y + 3 - 8y = 12y - 8y$$

$$8 \cdot \frac{3}{4} + 3 \overset{?}{=} 12 \cdot \frac{3}{4}$$

$$3 = 4y$$

$$\frac{3}{4} = \frac{4y}{4}$$

$$6 + 3 \overset{?}{=} 9$$

$$9 = 9^a$$

$$\frac{3}{4} = y$$

16.
$$24(p + 1) = 3p$$

Check

$$24p + 24 = 3p$$

$$24\left(-\frac{8}{7} + 1\right) \overset{?}{=} 3\left(-\frac{8}{7}\right)$$

$$24p + 24 - 24p = 3p - 24p$$

$$24 = -21p$$

$$24\left(-\frac{8}{7} + \frac{7}{7}\right) \overset{?}{=} 3\left(-\frac{8}{7}\right)$$

$$\frac{24}{-21} = \frac{-21p}{-21}$$

$$24\left(-\frac{1}{7}\right) \overset{?}{=} 3\left(-\frac{8}{7}\right)$$

$$-\frac{8}{7} = p$$

$$-\frac{24}{7} = -\frac{24}{7}{}^a$$

*Write an algebraic expression to represent the given expression in problems 17 and 18.*

17. Three times a number decreased by 24.
Let $x$ = the number.
$3x - 24$

18. The product of three and the sum of the square of a number and 15.
Let $x$ = the number.
$$3\left(x^2 + 15\right)$$

19. The temperature this morning was 34 degrees. By noon, the temperature had risen 7 degrees. Find the new temperature.
Let $t$ = the new temperature.
$34 + 7 = t$    Check

$$41 = t$$

$$34 + 7 \overset{?}{=} 41$$

$$41 = 41^a$$

The new temperature was 41 degrees.

20. A jet flew a distance of 3300 miles in 6 hours. What was the average speed of the jet in terms of miles per hour?
Let $s$ = the average speed of the jet in miles per hour.

$$s = \frac{3300 \text{ miles}}{6 \text{ hours}} = 550 \text{ miles per hour}$$

The average speed of the jet was 550 miles per hour.

21. You are planning to advertise your car for sale on the Internet. *Car Showroom* charges $1.80 for a photo plus $0.09 per word. *Car Bazaar* charges $1.00 for the photo plus $0.11 per word. For what number of words will the charges be the same?
Let $w$ = the number of words.

$$1.80 + 0.09w = 1.00 + 0.11w$$

Check

$$1.80 + 0.09w - 0.09w = 1.00 + 0.11w - 0.09w$$

$$1.80 + 0.09(40) \overset{?}{=} 1.00 + 0.11(40)$$

$$1.80 = 1.00 + 0.02w$$

$$1.80 + 3.60 \overset{?}{=} 1.00 + 4.40$$

$$1.80 - 1.00 = 1.00 + 0.02w - 1.00$$

$$5.40 = 5.40^a$$

$$0.80 = 0.02w$$

$$\frac{0.80}{0.02} = \frac{0.02w}{0.02}$$

$$40 = w$$

The charges will be the same for 40 words.